ATMOSPHERE, OCEAN, AND CLIMATE DYNAMICS: AN INTRODUCTORY TEXT

This is Volume 93 in the
INTERNATIONAL GEOPHYSICS SERIES
A series of monographs and textbooks
Edited by RENATA DMOWSKA, DENNIS HARTMANN, and H. THOMAS ROSSBY
A complete list of books in this series appears at the end of this volume.

ATMOSPHERE, OCEAN, AND CLIMATE DYNAMICS: AN INTRODUCTORY TEXT

JOHN MARSHALL AND R. ALAN PLUMB
Massachusetts Institute of Technology
Cambridge, Massachusetts

AMSTERDAM • BOSTON • HEIDELBERG • LONDON
NEW YORK • OXFORD • PARIS • SAN DIEGO
SAN FRANCISCO • SINGAPORE • SYDNEY • TOKYO

Academic Press is an imprint of Elsevier

ELSEVIER

Elsevier Academic Press
30 Corporate Drive, Suite 400, Burlington, MA 01803, USA
525 B Street, Suite 1900, San Diego, California 92101-4495, USA
84 Theobald's Road, London WC1X 8RR, UK

This book is printed on acid-free paper. ∞

Library of Congress Cataloging-in-Publication Data
Atmosphere, ocean, and climate dynamics: an introductory text/editors
John Marshall and R. Alan Plumb.
    p. cm. – (International geophysics series; v. 93)
  ISBN 978-0-12-558691-7 (hardcover)
1. Atmospheric circulation. 2. Ocean-atmosphere interaction. 3. Ocean circulation.
4. Fluid dynamics. 5. Atmospheric thermodynamics. I. Marshall, John, 1954-
II. Plumb, R. Alan, 1948-
  QC880.4.A8A877 2007
  551.5'246–dc22

                                                                        2007034798

British Library Cataloguing in Publication Data
A catalogue record for this book is available from the British Library

ISBN 13: 978-0-12-558691-7

For all information on all Elsevier Academic Press publications
visit our Web site at *www.books.elsevier.com*

Printed and bound by CPI Group (UK) Ltd, Croydon, CR0 4YY

Transferred to Digital Print 2012

The material that makes up this book evolved from notes prepared for an undergraduate class that has been taught by the authors at MIT over a period of ten years or so. During this time, many people, especially the students taking the class and those assisting in its teaching, have contributed to the evolution of the material and to the correction of errors in both the text and the problem sets. We have also benefited from the advice of our colleagues at MIT and Harvard; we especially thank Ed Boyle, Kerry Emanuel, Mick Follows, Peter Huybers, Lodovica Illari, Julian Sachs, Eli Tziperman and Carl Wunsch for generously giving their time to provide comments on early drafts of the text.

Responsibility for the accuracy of the final text rests, of course, with the authors alone.

We would also like to thank Benno Blumenthal for his advice in using the IRI/LDEO Climate Data Library, Gordon (Bud) Brown for help with the laboratory equipment and Russell Windman for his advice on the book design and final preparation of the photographs and figures.

John Marshall and R. Alan Plumb

Contents

0.1 **Outline, scope, and rationale of the book**   xiii
0.2 **Preface**   xiv
   0.2.1   Natural fluid dynamics   xv
   0.2.2   Rotating fluid dynamics: GFD Lab 0   xvii
   0.2.3   Holicism   xix

1. **Characteristics of the atmosphere**   1

   1.1   Geometry   1
   1.2   Chemical composition of the atmosphere   2
   1.3   Physical properties of air   4
      1.3.1   Dry air   4
      1.3.2   Moist air   5
      1.3.3   GFD Lab I: Cloud formation on adiabatic expansion   7
   1.4   Problems   7

2. **The global energy balance**   9

   2.1   Planetary emission temperature   9
   2.2   The atmospheric absorption spectrum   13
   2.3   The greenhouse effect   14
      2.3.1   A simple greenhouse model   14
      2.3.2   A leaky greenhouse   16
      2.3.3   A more opaque greenhouse   16
      2.3.4   Climate feedbacks   19
   2.4   Further reading   20
   2.5   Problems   20

3. **The vertical structure of the atmosphere**   23

   3.1   Vertical distribution of temperature and greenhouse gases   23
      3.1.1   Typical temperature profile   23
      3.1.2   Atmospheric layers   24

3.2   The relationship between pressure and density: hydrostatic balance   26
3.3   Vertical structure of pressure and density   28
      3.3.1   Isothermal atmosphere   28
      3.3.2   Non-isothermal atmosphere   28
      3.3.3   Density   29
3.4   Further reading   29
3.5   Problems   29

4. Convection   31

4.1   The nature of convection   32
      4.1.1   Convection in a shallow fluid   32
      4.1.2   Instability   33
4.2   Convection in water   34
      4.2.1   Buoyancy   34
      4.2.2   Stability   35
      4.2.3   Energetics   36
      4.2.4   GFD Lab II: Convection   36
4.3   Dry convection in a compressible atmosphere   39
      4.3.1   The adiabatic lapse rate (in unsaturated air)   39
      4.3.2   Potential temperature   41
4.4   The atmosphere under stable conditions   42
      4.4.1   Gravity waves   42
      4.4.2   Temperature inversions   44
4.5   Moist convection   46
      4.5.1   Humidity   47
      4.5.2   Saturated adiabatic lapse rate   49
      4.5.3   Equivalent potential temperature   50
4.6   Convection in the atmosphere   50
      4.6.1   Types of convection   51
      4.6.2   Where does convection occur?   55
4.7   Radiative-convective equilibrium   56
4.8   Further reading   57
4.9   Problems   57

5. The meridional structure of the atmosphere   61

5.1   Radiative forcing and temperature   62
      5.1.1   Incoming radiation   62
      5.1.2   Outgoing radiation   63
      5.1.3   The energy balance of the atmosphere   64
      5.1.4   Meridional structure of temperature   64
5.2   Pressure and geopotential height   67
5.3   Moisture   70
5.4   Winds   73
      5.4.1   Distribution of winds   74
5.5   Further reading   78
5.6   Problems   78

6. The equations of fluid motion   81

    6.1   Differentiation following the motion   82

    6.2   Equation of motion for a nonrotating fluid   84

        6.2.1   Forces on a fluid parcel   84

        6.2.2   The equations of motion   86

        6.2.3   Hydrostatic balance   87

    6.3   Conservation of mass   87

        6.3.1   Incompressible flow   88

        6.3.2   Compressible flow   88

    6.4   Thermodynamic equation   89

    6.5   Integration, boundary conditions, and restrictions in application   89

    6.6   Equations of motion for a rotating fluid   90

        6.6.1   GFD Lab III: Radial inflow   90

        6.6.2   Transformation into rotating coordinates   93

        6.6.3   The rotating equations of motion   94

        6.6.4   GFD Labs IV and V: Experiments with Coriolis forces on a parabolic rotating table   96

        6.6.5   Putting things on the sphere   100

        6.6.6   GFD Lab VI: An experiment on the Earth's rotation   103

    6.7   Further reading   104

    6.8   Problems   104

7. Balanced flow   109

    7.1   Geostrophic motion   110

        7.1.1   The geostrophic wind in pressure coordinates   112

        7.1.2   Highs and lows; synoptic charts   114

        7.1.3   Balanced flow in the radial-inflow experiment   116

    7.2   The Taylor-Proudman theorem   117

        7.2.1   GFD Lab VII: Taylor columns   118

    7.3   The thermal wind equation   119

        7.3.1   GFD Lab VIII: The thermal wind relation   120

        7.3.2   The thermal wind equation and the Taylor-Proudman theorem   122

        7.3.3   GFD Lab IX: cylinder "collapse" under gravity and rotation   123

        7.3.4   Mutual adjustment of velocity and pressure   125

        7.3.5   Thermal wind in pressure coordinates   126

    7.4   Subgeostrophic flow: the Ekman layer   129

        7.4.1   GFD Lab X: Ekman layers: frictionally-induced cross-isobaric flow   130

        7.4.2   Ageostrophic flow in atmospheric highs and lows   130

        7.4.3   Planetary-scale ageostrophic flow   133

    7.5   Problems   135

8. The general circulation of the atmosphere   139

    8.1   Understanding the observed circulation   140

    8.2   A mechanistic view of the circulation   141

        8.2.1   The tropical Hadley circulation   142

        8.2.2   The extratropical circulation and GFD Lab XI: Baroclinic instability   145

8.3   Energetics of the thermal wind equation   149
    8.3.1   Potential energy for a fluid system   149
    8.3.2   Available potential energy   150
    8.3.3   Release of available potential energy in baroclinic instability   152
    8.3.4   Energetics in a compressible atmosphere   153
8.4   Large-scale atmospheric energy and momentum budget   154
    8.4.1   Energy transport   154
    8.4.2   Momentum transport   156
8.5   Latitudinal variations of climate   157
8.6   Further reading   158
8.7   Problems   159

9. The ocean and its circulation   163

9.1   Physical characteristics of the ocean   164
    9.1.1   The ocean basins   164
    9.1.2   The cryosphere   165
    9.1.3   Properties of seawater; equation of state   165
    9.1.4   Temperature, salinity, and temperature structure   168
    9.1.5   The mixed layer and thermocline   171
9.2   The observed mean circulation   176
9.3   Inferences from geostrophic and hydrostatic balance   182
    9.3.1   Ocean surface structure and geostrophic flow   183
    9.3.2   Geostrophic flow at depth   184
    9.3.3   Steric effects   186
    9.3.4   The dynamic method   187
9.4   Ocean eddies   188
    9.4.1   Observations of ocean eddies   188
9.5   Further reading   189
9.6   Problems   190

10. The wind-driven circulation   197

10.1   The wind stress and Ekman layers   198
    10.1.1   Balance of forces and transport in the Ekman layer   199
    10.1.2   Ekman pumping and suction and GFD Lab XII   201
    10.1.3   Ekman pumping and suction induced by large-scale wind patterns   203
10.2   Response of the interior ocean to Ekman pumping   206
    10.2.1   Interior balances   206
    10.2.2   Wind-driven gyres and western boundary currents   206
    10.2.3   Taylor-Proudman on the sphere   207
    10.2.4   GFD Lab XIII: Wind-driven ocean gyres   211
10.3   The depth-integrated circulation: Sverdrup theory   213
    10.3.1   Rationalization of position, sense of circulation, and volume transport of ocean gyres   214
10.4   Effects of stratification and topography   216
    10.4.1   Taylor-Proudman in a layered ocean   217
10.5   Baroclinic instability in the ocean   218
10.6   Further reading   220
10.7   Problems   220

11. **The thermohaline circulation of the ocean   223**

   11.1 Air-sea fluxes and surface property distributions   224
        11.1.1 Heat, freshwater, and buoyancy fluxes   224
        11.1.2 Interpretation of surface temperature distributions   231
        11.1.3 Sites of deep convection   232
   11.2 The observed thermohaline circulation   234
        11.2.1 Inferences from interior tracer distributions   234
        11.2.2 Time scales and intensity of thermohaline circulation   239
   11.3 Dynamical models of the thermohaline circulation   239
        11.3.1 Abyssal circulation schematic deduced from Taylor-Proudman
               on the sphere   239
        11.3.2 GFD Lab XIV: The abyssal circulation   241
        11.3.3 Why western boundary currents?   243
        11.3.4 GFD Lab XV: Source sink flow in a rotating basin   245
   11.4 Observations of abyssal ocean circulation   245
   11.5 The ocean heat budget and transport   247
        11.5.1 Meridional heat transport   248
        11.5.2 Mechanisms of ocean heat transport and the partition of heat transport
               between the atmosphere and ocean   251
   11.6 Freshwater transport by the ocean   255
   11.7 Further reading   256
   11.8 Problems   256

12. **Climate and climate variability   259**

   12.1 The ocean as a buffer of temperature change   261
        12.1.1 Nonseasonal changes in SST   262
   12.2 El Niño and the Southern Oscillation   264
        12.2.1 Interannual variability   264
        12.2.2 "Normal" conditions—equatorial upwelling and the Walker
               circulation   266
        12.2.3 ENSO   269
        12.2.4 Other modes of variability   273
   12.3 Paleoclimate   273
        12.3.1 Climate over Earth history   275
        12.3.2 Paleotemperatures over the past 70 million years: the $\delta^{18}O$ record   277
        12.3.3 Greenhouse climates   280
        12.3.4 Cold climates   280
        12.3.5 Glacial-interglacial cycles   282
        12.3.6 Global warming   291
   12.4 Further reading   292
   12.5 Problems   292

   Appendices   295

   A.1 Derivations   295
        A.1.1 The Planck function   295
        A.1.2 Computation of available potential energy   296
        A.1.3 Internal energy for a compressible atmosphere   296

A.2 Mathematical definitions and notation   296
    A.2.1   Taylor expansion   296
    A.2.2   Vector identities   297
    A.2.3   Polar and spherical coordinates   298
A.3 Use of foraminifera shells in paleoclimate   298
A.4 Laboratory experiments   299
    A.4.1   Rotating tables   299
    A.4.2   List of laboratory experiments   300
A.5 Figures and access to data over the web   302

References   303

Textbooks and reviews   303
Other references   303
References to paleo-data sources   304

Index   307

Online companion site:
http://books.elsevier.com/companions/9780125586917

0.1. OUTLINE, SCOPE, AND RATIONALE OF THE BOOK

This is an introductory text on the circulation of the atmosphere and ocean, with an emphasis on global scales. It has been written for undergraduate students who have no prior knowledge of meteorology and oceanography or training in fluid mechanics. We believe that the text will also be of use to beginning graduate students in the field of atmospheric, oceanic, and climate science. By the end of the book we hope that readers will have a good grasp of what the atmosphere and ocean look like on the large scale, and, through application of the laws of mechanics and thermodynamics, why they look like they do. We will also place our observations and understanding of the present climate in the context of how climate has evolved and changed over Earth's history.

The book is roughly divided in to three equal parts. The first third deals exclusively with the atmosphere (Chapters 1 to 5), the last third with the ocean and its role in climate (Chapters 9 to 12). Sandwiched in between we develop the necessary fluid dynamical background (Chapter 6 and 7). Our discussion of the general circulation of the atmosphere (Chapter 8), follows the dynamical chapters. The text can be used in a number of ways. It has been written so that those interested primarily in the atmosphere might focus on Chapters 1 to 8. Those interested in the ocean can begin at Chapter 9, referring back as necessary to the dynamical Chapters 6 and 7. It is our hope, however, that many will be interested in learning about both fluids. Indeed, one of the joys of working on this text—and using it as background material for undergraduate courses taught at the Massachusetts Institute of Technology (MIT)—has been our attempt to discuss the circulation of the atmosphere and ocean in a single framework and in the same spirit.

In our writing we have been led by observations rather than theory. We have not written a book about fluid dynamics illustrated by atmospheric and oceanic phenomena. Rather we hope that the observations take the lead, and theory is introduced when it is needed. Advanced dynamical ideas are only used if we deem it essential to bring order to the observations. We have also chosen not to unnecessarily formalize our discussion. Yet, as far as is possible, we have offered rigorous physical discussions expressed in mathematical form: we build (nearly) everything up from first principles, our explanations of the observations are guided by theory, and these guiding principles are, we hope, clearly espoused.

The majority of the observations described and interpreted here are available electronically via the companion Web site, http://books.elsevier.com/companions/9780125586917. We make much use of the remarkable database and web-browsing facilities developed at the Lamont Doherty Earth Observatory of Columbia University. Thus the raw data presented by figures on the pages of the book can be accessed and manipulated over the web, as described in Section A.5.

One particularly enjoyable aspect of the courses from which this book sprang has been the numerous laboratory experiments carried out in lectures as demonstrations, or studied in more detail in undergraduate laboratory courses. We hope that some of this flavor comes through on the written page. We have attempted to weave the experiments into the body of the text so that, in the spirit of the best musicals, the 'song and dance routines' seem natural rather than forced. The experiments we chose to describe are simple and informative, and for the most part do not require sophisticated apparatus. Video loops of the experiments can be viewed over the Web, but there is no real substitute for carrying them out oneself. We encourage you to try. Details of the equipment required to carry out the experiments, including the necessary rotating turntables, can be found in Section A.4.

Before getting on to the meat of our account, we now make some introductory remarks about the nature of the problems we are concerned with.

0.2. PREFACE

The circulation of the atmosphere and oceans is inherently complicated, involving the transfer of radiation through a semi-transparent medium of variable composition, phase changes between liquid water, ice and vapor, interactions between phenomena on scales from centimeters to the globe, and timescales from seconds to millennia. But one only has to look at a picture of the Earth from space, such as that shown in Fig. 1, to appreciate that organizing principles must be at work to bring such order and beauty.

This book is about the large-scale circulation of the atmosphere and ocean and the organizing fluid mechanical principles that shape it. We will learn how the unusual properties of rotating fluids manifest themselves in and profoundly influence the circulation of the atmosphere and ocean and the climate of the planet. The necessary fluid dynamics will be developed and explored in the context of phenomena that play important roles in climate, such as convection, weather systems, the Gulf Stream, and the thermohaline circulation of the ocean. Extensive use is made of laboratory experiments to isolate and illustrate key ideas. Any study of climate dynamics would be incomplete without a discussion of radiative transfer theory, and so we will also cover fundamental ideas on energy balance. In the final chapters we discuss the interaction of the atmosphere, ocean, and ice and how they collude together to control the climate of the Earth. The paleoclimate record suggests that the climate of the past has been very different from that of today. Thus we use the

FIGURE 1.   A view of Earth from space over the North Pole. The Arctic ice cap can be seen in the center. The white swirls are clouds associated with atmospheric weather patterns. Courtesy of NASA/JPL.

understanding gleaned from our study of the present climate to speculate on mechanisms that might drive climate change.

In these introductory remarks we draw out those distinctive features of the fluid mechanics of the atmosphere and ocean that endow its study with a unique flavor. We are dealing with what can be called "natural" fluids, which are energized thermally on a differentially-heated sphere whose rotation tightly constrains the motion.

0.2.1. Natural fluid dynamics

Fluid dynamics is commonly studied in engineering and applied mathematics departments. A typical context might be the following. In a fluid of constant density, shearing eddies develop whenever circumstances force a strong shear (velocity contrast over a short distance). For example, flow past a solid obstacle leads to a turbulent wake (see Fig. 2), as in the flow of water down a stream or in air blowing over an airfoil. The kinetic energy of the eddying motion comes from the kinetic energy of steady flow impinging on the obstacle.

The problem can be studied experimentally (by constructing a laboratory analog of the motion) or mathematically (by solving rather complicated differential equations). However, shear-induced turbulence, and indeed much of classical hydrodynamics,[1] is not *directly* applicable to the fluid dynamics of the atmosphere and ocean, because it assumes that the density, ρ, is constant, or more precisely, that the density only depends on the pressure,[2] p such that $\rho = \rho(p)$. The energy source for the eddies that form the turbulent wake in Fig. 2 is the kinetic energy of the incoming steady stream. There is a superficial resemblance to the ubiquitous large-scale eddies and swirls observed in the atmosphere, beautifully revealed by the water vapor images shown in Fig. 3. However, the energy for the eddies seen in Fig. 3 comes directly or indirectly from thermal rather than mechanical sources.

Let us consider for a moment what the atmosphere or ocean would be like if it were made up of a fluid in which the density is independent of temperature and so can be written $\rho = \rho(p)$. Because of the overwhelming influence of gravity, pressure increases downward in the atmosphere and ocean. If the arrangement is to be stable, light fluid

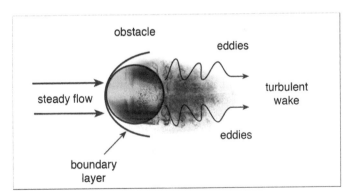

FIGURE 2.   Schematic diagram showing a fluid of constant density flowing past a solid obstacle, as might happen in the flow of water down a stream. Shearing eddies develop in a thin layer around the obstacle and result in a turbulent wake in the lee of the obstacle. The kinetic energy of the eddying motion comes from the kinetic energy of the steady flow impinging on the obstacle.

---

[1]See, for example, the treatise on classical hydrodynamics by Horace Lamb: Hydrodynamics, Cambridge, 1932.

[2]A fluid in which the density depends only on the pressure, $\rho = \rho(p)$, is called a barotropic fluid. A fluid in which the density depends on both temperature and pressure, $\rho = \rho(p, T)$, is called a baroclinic fluid.

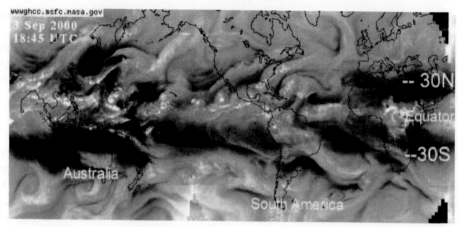

FIGURE 3.   A mosaic of satellite images showing the water vapor distribution over the globe at a height of 6–10 km above the surface. We see the organization of H_2O by the circulation; dry (sinking) areas in the subtropics ($\pm 30°$) are dark, moist (upwelling) regions of the equatorial band are bright. Jet streams of the middle latitudes appear as elongated dark regions with adjacent clouds and bright regions. From NASA.

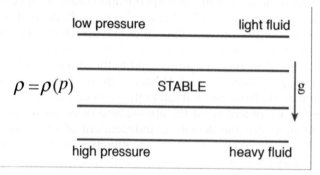

FIGURE 4.   Because of the overwhelming importance of gravity, pressure increases downward in the atmosphere and ocean. For gravitational stability, density must also increase downward—as sketched in the diagram—with heavy fluid below and light fluid above. If $\rho = \rho(p)$ only, then the density is independent of T, and the fluid cannot be brought into motion by heating and/or cooling.

must be on top of heavy fluid, and so the density must also increase downward, as sketched in Fig. 4. Now if ρ does not depend on T, the fluid cannot be brought into motion by heating/cooling. We conclude that if we make the assumption $\rho = \rho(p)$, then *life is abstracted out of the fluid*, because it cannot convert thermal energy into kinetic energy. But everywhere around us in the atmosphere and ocean we find fluid doing just that: acting as a natural heat engine, generating and maintaining its own motion by converting thermal energy into kinetic energy. The fluid can only do this because ρ is not just a function of p. If ρ depends on both pressure and temperature,[3] $\rho = \rho(p, T)$, as sketched in Fig. 5, then fluid heated by the Sun, for example, can become buoyant and rise in convection. Such a fluid can convect by converting thermal energy into kinetic energy—it is *full of life* because it can be energized thermally.

---

[3]The density of the ocean depends on salinity as well as temperature and pressure: $\rho = \rho(p, S, T)$. Then, for example, convection can be triggered from the surface layers of the ocean by the formation of ice; fresh water is locked up in the ice, leaving brackish and hence heavy water behind at the surface.

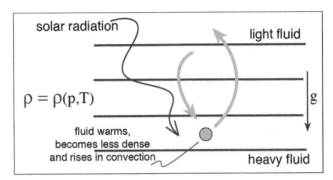

FIGURE 5. In contrast to Fig. 4, if ρ depends on both pressure and temperature, $\rho = \rho(p, T)$, then fluid heated by the Sun, for example, can become buoyant and rise in convection. Such a fluid can be energized thermally.

In meteorology and oceanography we are always dealing with fluids in which $\rho = \rho(p, T, \ldots)$ and turbulence is, in the main, thermally rather than mechanically maintained. Rather than classical hydrodynamics we are concerned with *natural aerodynamics* or *geophysical fluid dynamics*, the fluid dynamics of real fluids. The latter phrase, often going by the shorthand *GFD*, is now widely used to describe the kind of fluid dynamics we are dealing with.

0.2.2. Rotating fluid dynamics: GFD Lab 0

In climate dynamics we are not only dealing with the natural fluids just described, but also, because Earth is a rapidly rotating planet, with rotating fluids. As we shall see during the course of our study, rotation endows fluids with remarkable properties.

If one looks up the definition of *fluid* in the dictionary, one finds:

> something that can "change rapidly or easily" or
> "fill any container into which it is poured."

But fluid on a rotating planet cannot move in arbitrary paths because, if the scales of motion are sufficiently large and sluggish, they are profoundly aware of and affected by rotation. In a very real sense the atmosphere and ocean are not "fluid" at all; they are tightly constrained by the rotation of the Earth. This constraint makes the two fluids more similar than one might expect—the atmosphere and ocean can, and we would argue should, be studied together. This is what we set out to do in this text.

The unusual properties of rotating fluids can be demonstrated in the following very simple laboratory experiment.

GFD Lab 0: Rigidity imparted to rotating fluids

We take two tanks and place one on a rotating table and the other on a desk in the laboratory. We fill them with water to a depth of ~ 20 cm, set the rotating table turning at a speed of 15–20 revolutions per minute (see Section A.4 for discussion of rotating table) and leave them to settle down for 15 minutes or so. We gently agitate the two bodies of water—one's hand is best, using an up-down beating motion (try not to introduce a systematic swirl)—to generate motion, and wait ($\lesssim 1$ minute) for things to settle down a little, but not so long that the currents die away. We observe the motion by introducing dye (food coloring).

In the nonrotating tank the dye disperses much as we might intuitively expect. But in the rotating body of water something glorious happens. We see beautiful streaks of dye falling vertically; the vertical streaks become drawn out by horizontal fluid motion into vertical 'curtains' which wrap around one another. Try two different colors and watch the interleaving of fluid columns (see Fig. 6 here and Fig. 7.7 in Chapter 7)

The vertical columns, which are known as *Taylor columns* after G. I. Taylor who discovered them (see Chapter 7), are a result of the rigidity imparted to the fluid by the rotation of the tank. The water moves around in vertical columns which are aligned parallel to the rotation vector. It is in this sense that rotating fluids are rigid. As the horizontal spatial scales and timescales lengthen, rotation becomes an increasingly strong constraint on the motion of both the atmosphere and ocean.

On what scales might the atmosphere, ocean, or our laboratory experiment, "feel" the effect of rotation? Suppose that typical horizontal currents (atmospheric or oceanic, measured, as they are, in the rotating frame) are given by U, and the typical distance over which the current varies is L. Then the timescale of the motion is L/U. Let's compare this with τ_{rot}, the period of rotation, by defining a nondimensional number (known as the *Rossby number*; see Section 7.1):

$$R_o = \frac{U \times \tau_{rot}}{L}$$

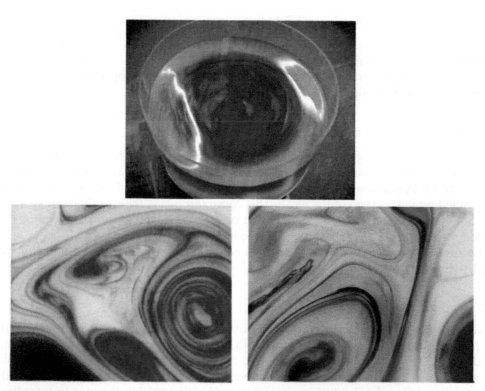

FIGURE 6. Taylor columns revealed by food coloring in the rotating tank. The water is allowed to come into solid body rotation and then gently stirred by hand. Dyes are used to visualize the flow. At the top we show the rotating cylinder of water with curtains of dye falling down from the surface. Below we show the beautiful patterns of dyes of different colors being stirred around one another by the rotationally constrained motion.

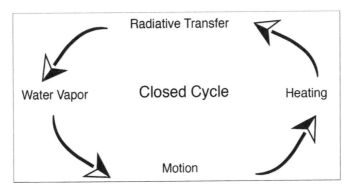

FIGURE 7. A schematic diagram showing the interplay between radiative transfer and circulation. The absorption of radiation by the atmosphere is very sensitive to the distribution of water vapor. But the water vapor distribution depends on the motion, which in turn depends on the heating, completing a closed cycle.

If R_o is much greater than one, then the timescale of the motion is short relative to a rotation period, and rotation will not significantly influence the motion. If R_o is much less than one, then the motion will be aware of rotation.

In our laboratory tank we observe horizontal swirling of perhaps $U \sim 1$ cm s^{-1} over the scale of the tank, $L \sim 30$ cm, which is rotating with a period $\tau_{\text{tank}} = 3$ s. This yields a Rossby number for the tank flow of $R_{o_{\text{tank}}} = 0.1$. Thus rotation will be an important constraint on the fluid motion, as we have witnessed by the presence of Taylor columns in Fig. 6.

Let us estimate R_o for large-scale flow in the atmosphere and ocean.

- ATMOSPHERE (e.g., for a weather system), discussed in Chapter 5 and Chapter 7: $L \sim 5000$ km, $U \sim 10$ m s^{-1}, and $\tau_{rot} = 1$ d $\approx 10^5$ s, giving $R_{o_{atmos}} = 0.2$, which suggests that rotation will be important.

- OCEAN (e.g., for the great gyres of the Atlantic or Pacific Oceans, described in Chapter 9): $L \sim 1000$ km, $U \sim 0.1$ m s^{-1}, giving $R_{o_{ocean}} = 0.01$, and rotation will be a controlling factor.

It is clear then that rotation will be of paramount importance in shaping the pattern of air and ocean currents on sufficiently large scales. Indeed, much of the structure and organization seen in Fig. 1 is shaped by rotation.

0.2.3. Holicism

There is another aspect that gives our study of climate dynamics its distinctive flavor. *The climate is a unity.* Only if great care is taken can it be broken up and the parts studied separately, since every aspect affects every other aspect. To illustrate this point, let us consider the interplay between the transfer of radiation through the atmosphere (known as radiative transfer) and the fluid motion. As we shall see in Chapter 2 and Chapter 3, the radiative temperature profile depends on, among other things, the distribution of water vapor, because water vapor strongly absorbs radiation in the same wavelengths that the Earth principally radiates. But the water vapor distribution is not given because it depends on the motion, as can be clearly seen in Fig. 3. The motion in turn depends on the heating, which depends on the radiative transfer. The closed cycle sketched in Fig. 7 must be studied as a whole.

So the background may be ordinary physics—classical mechanics and thermodynamics applied to a fluid on a rotating, differentially-heated sphere—but the study of the whole process has its own unique flavor. The approach is HOLISTIC rather than reductionist, because there is never a single cause. If one asks the question, "Why do swallows migrate south in the autumn?" string theory will never give us the answer; this kind of science may.

The companion Web site containing links to laboratory streaming videos and notes, solutions to end of chapter exercises, and an image gallery to complement the text can be found at http://books.elsevier.com/companions/9780125586917.

1

Characteristics of the atmosphere

1.1. Geometry
1.2. Chemical composition of the atmosphere
1.3. Physical properties of air
    1.3.1. Dry air
    1.3.2. Moist air
    1.3.3. GFD Lab I: Cloud formation on adiabatic expansion
1.4. Problems

The atmosphere (and ocean) are thin films of fluid on the spherical Earth under the influence of gravity, Earth's rotation, and differential heating by solar radiation. In this chapter we describe the chemical composition of the atmosphere and key physical properties of air. We discuss the equation of state of air (the connection between pressure, density, and temperature) and key properties of moist air. In particular, we will learn that warm air is generally more moist than cold air, a fact that has enormous implications for the climate of the planet.

1.1. GEOMETRY

The Earth is an almost perfect sphere with mean radius $a = 6370$ km, a surface gravity field $g = 9.81$ ms^{-2}, and a rotation period of $\tau_{Earth} = 24$ h, equivalent to an angular velocity $\Omega = 2\pi/\tau_{Earth} = 7.27 \times 10^{-5}$ s^{-1} (see Table 1.1).

The atmosphere which envelops the Earth is very thin; it fades rapidly away

TABLE 1.1. Some parameters of Earth.

Earth's rotation rate	Ω	7.27×10^{-5} s^{-1}
Surface gravity	g	9.81 ms^{-2}
Earth's mean radius	a	6.37×10^6 m
Surface area of Earth	$4\pi a^2$	5.09×10^{14} m^2
Area of Earth's disc	πa^2	1.27×10^{14} m^2

with altitude and does not have a definite top. As we shall see in Chapter 3, its density decreases approximately exponentially away from the surface, falling by a factor of e about every 7 km. About 80% of the mass of the atmosphere is contained below 10 km altitude. Fig. 1.1 shows, to scale, a shell of thickness 10 km on a sphere of radius 6370 km.

The thinness of the atmosphere allows us to make some simplifications. For one thing, we can take g to be constant (the fractional decrease in gravity from the Earth's surface to 10 km altitude is about 10^{-4} and so is negligible in most applications). We will see that we can often neglect the Earth's

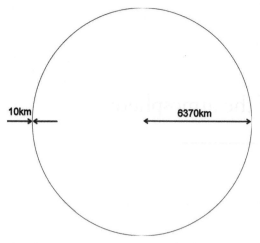

FIGURE 1.1. The thinness (to scale) of a shell of 10 km thickness on the Earth of radius 6370 km.

curvature and assume planar geometry. But there are of course (as we will also see) some aspects of spherical geometry that cannot be neglected.

Land covers about 30% of the surface of the Earth and, at the present time in Earth's history, about 70% of Earth's land is in the northern hemisphere (see Fig. 9.1). As Fig. 1.2 illustrates, the height of mountains rarely exceeds 2 km and so is a relatively small fraction of the vertical decay scale of the atmosphere. Thus, unlike the ocean, the atmosphere is not confined to basins. As it flows around the globe, air is deflected by topography but never completely blocked.

1.2. CHEMICAL COMPOSITION OF THE ATMOSPHERE

Air is a mixture of "permanent" gases (N_2, O_2) in constant ratio together with minor constituents (see Table 1.2). The molecular weight of the mixture that makes up air is 28.97, so that 22.4 liters of air at standard temperature and pressure (STP; $T = 273$ K and $p = 1013$ h Pa) weighs 28.97 g.

The composition of air is a direct consequence of the supply of elements from the Earth's interior and the presence of life on the surface. Photosynthesis by plants makes O_2; nitrogenous compounds from living organisms are returned to the atmosphere as N_2 from metabolism. Lightning converts N_2 into usable molecules for life. Two of the most important minor constituents are H_2O and CO_2; they play a central role in controlling the temperature of the Earth's surface (see Chapter 2) and sustaining life (living material is primarily composed of C, H and O).

Atmospheric water vapor is present in variable amounts (typically 0.5% by volume). It is primarily the result of evaporation from the ocean's surface. Unlike N_2 and O_2, water vapor—and to a lesser degree CO_2—is of great importance in radiative transfer (the passage of radiation through the atmosphere), because it strongly absorbs and emits in the infrared, the region of the spectrum (wavelengths about 10 μm)

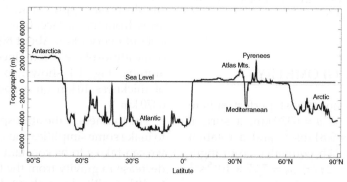

FIGURE 1.2. A north-south section of topography relative to sea level (in meters) along the Greenwich meridian (0° longitude) cutting through Fig. 9.1. Antarctica is over 2 km high, whereas the Arctic Ocean and the south Atlantic basin are about 5 km deep. Note how smooth the relief of the land is compared to that of the ocean floor.

TABLE 1.2. The most important atmospheric constituents. The chlorofluorocarbons (CFCs) CCl_2F_2 and CCl_3F are also known as CFC-12 and CFC-11, respectively. [N.B. (ppm, ppb, ppt) = parts per (million, billion, trillion)] The concentrations of some constituents are increasing systematically because of human activity. For example, the CO_2 concentration of 380 ppm was measured in 2004 (see Fig. 1.3); CFCs are now decreasing in concentration following restrictions on their production.

Chemical species	Molecular weight (g mol^{-1})	Proportion by volume	Chemical species	Molecular weight	Proportion by volume
N_2	28.01	78%	O_3	48.00	~500 ppb
O_2	32.00	21%	N_2O	44.01	310 ppb
Ar	39.95	0.93%	CO	28.01	120 ppb
H_2O (vapor)	18.02	~0.5%	NH_3	17.03	~100 ppb
CO_2	44.01	380 ppm	NO_2	46.00	~1 ppb
Ne	20.18	19 ppm	CCl_2F_2	120.91	480 ppt
He	4.00	5.2 ppm	CCl_3F	137.37	280 ppt
CH_4	16.04	1.7 ppm	SO_2	64.06	~200 ppt
Kr	83.8	1.1 ppm	H_2S	34.08	~200 ppt
H_2	2.02	~500 ppb	AIR	28.97	

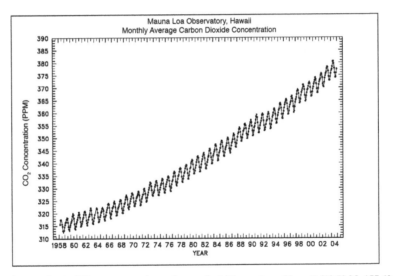

FIGURE 1.3. Atmospheric CO_2 concentrations observed at Mauna Loa, Hawaii (19.5° N, 155.6° W). Note the seasonal cycle superimposed on the long-term trend. The trend is due to anthropogenic emissions. The seasonal cycle is thought to be driven by the terrestrial biosphere: net consumption of CO_2 by biomass in the summertime (due to abundance of light and heat) and net respiration in wintertime.

at which Earth radiates energy back out to space (see Chapter 2). The CO_2 concentration in the atmosphere is controlled by such processes as photosynthesis and respiration, exchange between the ocean and the atmosphere, and, in the modern world, anthropogenic activities.

It is important to note that the proportion of some constituents (especially chemically or physically active species, such as H_2O) is variable in space and time. Moreover, several crucially important constituents (e.g., H_2O, CO_2, O_3) are present in very small concentrations, and so are sensitive to anthropogenic activity. For example, Fig. 1.3 shows the CO_2 concentration measured at the Hawaiian island of Mauna Loa. Atmospheric CO_2 concentration has risen from 315 ppm to 380 ppm over the past 50 years. Preindustrial levels of CO_2 were around 280 ppm; it is thought that over the course of Earth's history, CO_2 levels have greatly fluctuated. Atmospheric CO_2 concentrations were probably markedly different in warm as opposed to cold periods of Earth's climate. For example, at the last glacial maximum 20,000 years ago, CO_2 concentrations are thought to have been around 180 ppm. Reconstructions of atmospheric CO_2 levels over geologic time suggest that CO_2 concentrations were perhaps five times the present level 220 million years ago, and perhaps 20 times today's concentration between 450 and 550 million years ago, as we shall see in Section 12.3 and Fig. 12.14. If the curve shown in Fig. 1.3 continues its exponential rise, then by the end of the century, CO_2 concentrations will have reached levels—perhaps 600 ppm—not seen since 30 million years ago, a period of great warmth in Earth history.

1.3. PHYSICAL PROPERTIES OF AIR

Some important numbers for Earth's atmosphere are given in Table 1.3. Global mean surface pressure is 1.013×10^5 Pa = 1013 h Pa. (The hecto Pascal is now the official unit of atmospheric pressure [1 h Pa = 10^2 Pa], although the terminology "millibar" [1 mbar = 1 h Pa] is still in common use and will also be used here.)

The global mean density of air at the surface is $1.235\,\mathrm{kg\,m^{-3}}$. At this average density we require a column of air of about 7–8 km high to exert pressure equivalent to 1 atmosphere.

Throughout the region of our focus (the lowest 50 km of the atmosphere), the mean free path of atmospheric molecules is so short and molecular collisions so frequent that the atmosphere can be regarded as a *continuum fluid* in *local thermodynamic equilibrium* (LTE), and so the "blackbody" ideas to be developed in Chapter 2 are applicable. (These statements break down at sufficiently high altitude, $\gtrsim 80$ km, where the density becomes very low.)

1.3.1. Dry air

If in LTE, the atmosphere accurately obeys the perfect gas law,[1] then

$$p = \rho \frac{R_g}{m_a} T = \rho R T, \qquad (1\text{-}1)$$

where p is pressure, ρ, density, T, absolute temperature (measured in Kelvin), R_g, the universal gas constant

$$R_g = 8.3143\,\mathrm{J\,K^{-1}mol^{-1}},$$

the gas constant for dry air

$$R = \frac{R_g}{m_a} = 287\,\mathrm{J\,kg^{-1}K^{-1}},$$

and the mean molecular weight of dry air (see Table 1.2, last entry), $m_a = 28.97$ ($\times 10^{-3}$ kg mol^{-1}).

 [1] Robert Boyle (1627–1691) made important contributions to physics and chemistry and is best known for Boyle's law, describing an ideal gas. With the help of Robert Hooke, he showed among other things that sound did not travel in a vacuum, proved that flame required air, and investigated the elastic properties of air.

TABLE 1.3. Some atmospheric numbers.

Atmospheric mass	M_a	5.26×10^{18} kg
Global mean surface pressure	p_s	1.013×10^5 Pa
Global mean surface temperature	T_s	288 K
Global mean surface density	ρ_s	1.235 kg m^{-3}

TABLE 1.4. Properties of dry air at STP.

Specific heat at constant pressure	c_p	1005 J kg^{-1} K^{-1}
Specific heat at constant volume	c_v	718 J kg^{-1} K^{-1}
Ratio of specific heats	γ	1.40
Density at 273K, 1013mbar	ρ_0	1.293 kg m^{-3}
Viscosity at STP	μ	1.73×10^{-5} kg m^{-1} s^{-1}
Kinematic viscosity at STP	$v = \dfrac{\mu}{\rho_0}$	1.34×10^{-5} m^2 s^{-1}
Thermal conductivity at STP	K	2.40×10^{-2} W m^{-2} K^{-1}
Gas constant for dry air	R	287.05 J kg^{-1} K^{-1}

From Eq. 1-1 we see that it is only necessary to know any two of p, T, and ρ to specify the thermodynamic state of dry air completely. Thus at STP, Eq. 1-1 yields a density $\rho_0 = 1.293$ kg m^{-3}, as entered in Table 1.4, where some of the important physical parameters for dry air are listed.

Note that air, as distinct from liquids, is compressible (if p increases at constant T, ρ increases) and has a relatively large coefficient of thermal expansion (if T increases at constant p, ρ decreases). As we shall see, these properties have important consequences.

1.3.2. Moist air

Air is a mixture of gases, and the ideal gas law can be applied to the individual components. Thus if ρ_v and ρ_d are, respectively, the masses of water vapor and of dry air per unit volume (i.e., the partial densities) then the equations for the partial pressures (that is the pressure each component would exert at the same temperature as the mixture, if it alone occupied the volume that the mixture occupies) are:

$$e = \rho_v R_v T; \tag{1-2}$$

$$p_d = \rho_d R_d T, \tag{1-3}$$

where e is the partial pressure of water vapor, p_d is the partial pressure of dry air, R_v is the gas constant for water vapor, and R_d is the gas constant for dry air. By Dalton's law of partial pressures, the pressure of the mixture, p, is given by:

$$p = p_d + e.$$

In practice, because the amount of water vapor in the air is so small (see Table 1.2), we can assume that $p_d \gg e$, and so $p \simeq p_d$.

Now imagine that the air is in a box at temperature T, and suppose that the floor of the box is covered with water, as shown in Fig. 1.4. At equilibrium, the rate of evaporation will equal the rate of condensation, and the air is said to be *saturated* with water vapor. If we looked into the box, we would see a mist.[2] At this point, e has reached

---

[2] This is true provided there are plenty of *condensation nuclei*—tiny particles—around to ensure condensation takes place (see GFD Lab I).

a)

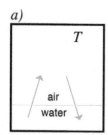

b)

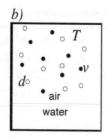

FIGURE 1.4.   Air over water in a box at temperature T. At equilibrium the rate of evaporation equals the rate of condensation. The air is saturated with water vapor, and the pressure exerted by the vapor is e_s, the saturated vapor pressure. On the right we show the mixture comprising dry 'd' and vapor 'v' components.

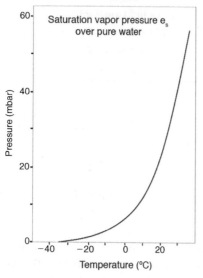

FIGURE 1.5.   Saturation vapor pressure e_s (in mbar) as a function of T in °C (solid curve). From Wallace & Hobbs, (2006).

the saturated vapor pressure, e_s. In fact, saturation occurs whenever the partial pressure of water exceeds the saturation vapor pressure e_s. As shown in Fig. 1.5, e_s is a function only of temperature and increases very rapidly with T. To a good approximation at typical atmospheric temperatures, $e_s(T)$ is given by:

$$e_s = Ae^{\beta T} \qquad (1\text{-}4)$$

where $A = 6.11$ hPa and $\beta = 0.067°\text{C}^{-1}$ are constants and T is in °C, a simplified statement of the Clausius-Clapeyron relationship. The saturated vapor pressure

increases *exponentially* with temperature, a property which is enormously important for the climate of the planet.

From Eq. 1-4 (see also Fig. 1.5) we note that $e_s = 16.7\,\text{hPa}$ at $T = 15°\text{C}$. From Table 1.2 we deduce that $R_v = R_g/m_v = 461.39\,\text{J kg}^{-1}\,\text{K}^{-1}$ and so, using Eq. 1-2, at saturation $\rho_v = 0.0126\,\text{kg m}^{-3}$. This is the maximum amount of water vapor per unit volume that can be held by the atmosphere at this temperature.

The $e_s(T)$ curve shown in Fig. 1.5 has the following very important climatic consequences:

- The moisture content of the atmosphere decays rapidly with height, because T decreases with height, from the Earth's surface up to 10 km or so. In Chapter 3 we will see that at the surface the mean temperature is about 15°C, but falls to about −50°C at a height of 10 km (see Fig. 3.1). We see from Fig. 1.5 that $e_s \longrightarrow 0$ at this temperature. Thus most of the atmosphere's water vapor is located in the lowest few km. Moreover, its horizontal distribution is very inhomogeneous, with much more vapor in the warm tropics than in cooler higher latitudes. As will be discussed in Chapter 2, this is crucially important in the transfer of radiation through the atmosphere.

- Air in the tropics tends to be much more moist than air over the poles, simply because it is warmer in the tropics than in polar latitudes; see Section 5.3.

- Precipitation occurs when moist air is cooled by convection, and causes H_2O concentrations to be driven back to their value at saturation at a given T; see Section 4.5.

- In cold periods of Earth's history, such as the last glacial maximum 20,000 years ago, the atmosphere was probably much more arid than in warmer periods. Conversely, warm climates tend to be much more moist; see Section 12.3.

1.3.3. GFD Lab I: Cloud formation on adiabatic expansion

The sensitive dependence of saturation vapor pressure on temperature can be readily demonstrated by taking a carboy and pouring warm water into it to a depth of a few cm, as shown in Fig. 1.6. We leave it for a few minutes to allow the air above the warm water to become saturated with water vapor. We rapidly reduce the pressure in the bottle by sucking at the top of the carboy. You can use your lungs to suck the air out, or a vacuum cleaner. One might expect that the rapid adiabatic expansion of the air would reduce its temperature and hence lower the saturated vapor pressure sufficiently that the vapor would condense to form water droplets, a "cloud in the jar." To one's disappointment, this does not happen.

The process of condensation of vapor to form a water droplet requires condensation nuclei, which are small particles on which the vapor can condense. We can introduce such particles into the carboy by dropping in a lighted match and repeating the experiment. Now on decompression we do indeed observe a thick cloud forming which disappears again when the pressure returns to normal, as shown in Fig. 1.6 (right).

In the bottom kilometer or so of the atmosphere there are almost always abundant condensation nuclei, because of the presence of sulfate aerosols, dust, smoke from fires, and ocean salt. Clouds consist of liquid water droplets (or ice particles) that are formed by condensation of water vapor onto these particles when T falls below the dew point, which is the temperature to which air must be cooled (at constant pressure and constant water vapor content) to reach saturation.

A common atmospheric example of the phenomenon studied in our bottle is the formation of fog due to radiational cooling of a shallow, moist layer of air near the surface. On clear, calm nights, cooling due to radiation can drop the temperature to the dew point and cause fog formation, as shown in the photograph of early morning mist on a New England lake (Fig. 1.7).

The sonic boom pictured in Fig. 1.8 is a particularly spectacular consequence of the sensitive dependence of e_s on T: just as in our bottle, condensation of water is caused by the rapid expansion and subsequent adiabatic cooling of air parcels induced by the shock waves resulting from the jet going through the sound barrier.

FIGURE 1.6. Warm water is poured into a carboy to a depth of 10 cm or so, as shown on the left. We leave it for a few minutes and throw in a lighted match to provide condensation nuclei. We rapidly reduce the pressure in the bottle by sucking at the top. The adiabatic expansion of the air reduces its temperature and hence the saturated vapor pressure, causing the vapor to consense and form water droplets, as shown on the right.

1.4. PROBLEMS

1. Given that the acceleration due to gravity decays with height from the centre of the Earth following an inverse square law, what is the percentage change in g from the Earth's surface to an altitude of 100 km? (See also Problem 6 of Chapter 3.)

FIGURE 1.7.   Dawn mist rising from Basin Brook Reservoir, White Mountain National Forest, July 25, 2004. Photograph: Russell Windman.

FIGURE 1.8.   A photograph of the sound barrier being broken by a US Navy jet as it crosses the Pacific Ocean at the speed of sound just 75 feet above the ocean. Condensation of water is caused by the rapid expansion and subsequent adiabatic cooling of air parcels induced by the shock (expansion/compression) waves caused by the plane outrunning the sound waves in front of it. Photograph was taken by John Gay from the top of an aircraft carrier. The photo won First Prize in the science and technology division of the World Press Photo 2000 contest.

2. Compute the mean pressure at the Earth's surface given the total mass of the atmosphere, M_a (Table 1.3), the acceleration due to gravity, g, and the radius of the Earth, a (Table 1.1).

3. Express your answer to Problem 2 in terms of the number of apples per square meter required to exert the same pressure. You may assume that a typical apple weighs 0.2 kg. If the average density of air is 5 apples per m^3 (in apple units), calculate how high the apples would have to be stacked at this density to exert a surface pressure equal to 1000 h Pa. Compare your estimate to the scale height, H, given by Eq. 3.6 in Section 3.3.

4. Using (i) Eq. 1-4, which relates the saturation vapor pressure of H_2O to temperature T, and (ii) the equation of state of water vapor, $e = \rho_v R_v T$ (see discussion in Section 1.3.2), compute the maximum amount of water vapor per unit volume that air can hold at the surface, where $T_s = 288$ K, and at a height of 10 km where (from Fig. 3.1) $T_{10\ km} = 220$ K. Express your answer in $kg\ m^{-3}$. What are the implications of your results for the distribution of water vapor in the atmosphere?

CHAPTER

2

The global energy balance

2.1. Planetary emission temperature
2.2. The atmospheric absorption spectrum
2.3. The greenhouse effect
    2.3.1. A simple greenhouse model
    2.3.2. A leaky greenhouse
    2.3.3. A more opaque greenhouse
    2.3.4. Climate feedbacks
2.4. Further reading
2.5. Problems

We consider now the general problem of the radiative equilibrium temperature of the Earth. The Earth is bathed in solar radiation and absorbs much of that incident upon it. To maintain equilibrium it must warm up and radiate energy away at the same rate it is received, as depicted in Fig. 2.1. We will see that the emission temperature of the Earth is 255 K and that a body at this temperature radiates energy primarily in the infrared (IR). But the atmosphere is strongly absorbing at these wavelengths due to the presence of trace gases—principally the triatomic molecules H_2O and CO_2—which absorb and emit in the infrared, thus raising the surface temperature above that of the emission temperature, a mechanism that has become known as the *greenhouse effect*.

2.1. PLANETARY EMISSION TEMPERATURE

The Earth receives almost all of its energy from the Sun. At the present time in its evolution the Sun emits energy at a rate of $Q = 3.87 \times 10^{26}$ W. The flux of solar energy at the Earth, called the *solar constant*, depends on the distance of the Earth from the Sun, r, and is given by the inverse square law, $S_0 = Q/4\pi r^2$. Of course, because of variations in the Earth's orbit (see Sections 5.1.1 and 12.3.5) the solar constant is not really constant; the terrestrial value

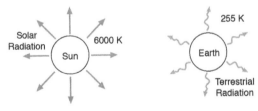

FIGURE 2.1. The Earth radiates energy at the same rate it is received from the Sun. The Earth's emission temperature is 255 K, and that of the Sun is 6000 K. The outgoing terrestrial radiation peaks in the infrared spectrum; the incoming solar radiation peaks at shorter wavelengths, in the visible spectrum.

TABLE 2.1. Properties of some of the planets. S_0 is the solar constant at a distance r from the Sun, α_p is the planetary albedo, T_e is the emission temperature computed from Eq. 2-4, T_m is the measured emission temperature, and T_s is the global mean surface temperature. The rotation period, τ, is given in Earth days.

	r 10^9 m	S_0 W m^{-2}	α_p	T_e K	T_m K	T_s K	τ Earth days
Venus	108	2632	0.77	227	230	760	243
Earth	150	1367	0.30	255	250	288	1.00
Mars	228	589	0.24	211	220	230	1.03
Jupiter	780	51	0.51	.103	130	134	0.41

$S_0 = 1367\,\text{Wm}^{-2}$, set out in Table 2.1, along with that for other planets, is an average corresponding to the average distance of Earth from the Sun, $r = 150 \times 10^9$ m.

The way radiation interacts with an atmosphere depends on the wavelength as well as the intensity of the radiative flux. The relation between the energy flux and wavelength, which is the spectrum, is plotted in Fig. 2.2. The Sun emits radiation that is primarily in the visible part of the spectrum, corresponding to the colors of the rainbow—red, orange, yellow, green, blue, indigo and violet—with the energy flux decreasing toward longer (infrared, IR) and shorter (ultraviolet, UV) wavelengths.

Why does the spectrum have this pattern? Such behavior is characteristic of the radiation emitted by incandescent material, as can be observed for example in a coal fire. The hottest parts of the fire are almost white and emit the most intense radiation, with a wavelength that is shorter than that coming from the warm parts of the fire, which glow red. The coldest parts of the fire do not seem to be radiating at all, but are, in fact, radiating in the infrared. Experiment

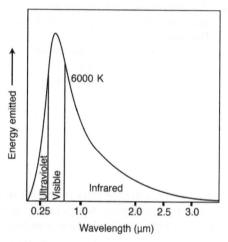

FIGURE 2.2. The energy emitted from the Sun plotted against wavelength based on a black body curve with $T = T_{Sun}$. Most of the energy is in the visible spectrum, and 95% of the total energy lies between 0.25 and 2.5μm (10^{-6}m).

and theory show that the wavelength at which the intensity of radiation is maximum, and the flux of emitted radiation, depend only on the temperature of the source. The theoretical spectrum, one of the jewels of physics, was worked out by Planck,[1] and is known as the *Planck* or

[1] In 1900 Max Planck (1858–1947) combined the formulae of Wien and Rayleigh, describing the distribution of energy as a function of wavelength of the radiation in a cavity at temperature T, to arrive at what is now known as Planck's radiation curve. He went on to a complete theoretical deduction, introduced quanta of energy, and set the scene for the development of quantum mechanics.

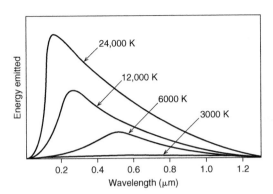

FIGURE 2.3. The energy emitted at different wavelengths for blackbodies at several temperatures. The function $B_\lambda(T)$, Eq. A-1, is plotted.

TABLE 2.2. Albedos for different surfaces. Note that the albedo of clouds is highly variable and depends on the type and form. See also the horizontal map of albedo shown in Fig. 2.5.

Type of surface	Albedo (%)
Ocean	2–10
Forest	6–18
Cities	14–18
Grass	7–25
Soil	10–20
Grassland	16–20
Desert (sand)	35–45
Ice	20–70
Cloud (thin, thick stratus)	30, 60–70
Snow (old)	40–60
Snow (fresh)	75–95

blackbody spectrum. (A brief theoretical background to the Planck spectrum is given in Appendix A.1.1.) It is plotted as a function of temperature in Fig. 2.3. Note that the hotter the radiating body, the more energy it emits at shorter wavelengths. If the observed radiation spectrum of the Sun is fitted to the blackbody curve by using T as a free parameter, we deduce that the blackbody temperature of the Sun is about 6000 K.

Let us consider the energy balance of the Earth as in Fig. 2.4, which shows the Earth intercepting the solar energy flux and radiating terrestrial energy. If at the location of the (mean) Earth orbit, the incoming solar energy flux is $S_0 = 1367$ W m^{-2}, then, given that the cross-sectional area of the Earth intercepting the solar energy flux is πa^2, where a is the radius of the Earth (Fig. 2.4),

$$\text{solar power incident on the Earth} =$$
$$S_0 \pi a^2 = 1.74 \times 10^{17} \text{ W,}$$

using the data in Table 1.1. Not all of this radiation is absorbed by the Earth; a significant fraction is reflected. The ratio of reflected to incident solar energy is called the *albedo*, α. As set out in Table 2.2 and the map of surface albedo shown in Fig. 2.5, α depends on the nature of the reflecting surface and is large for clouds, light surfaces such as deserts, and (especially) snow and ice. Under the present terrestrial conditions of cloudiness and snow and ice cover, on

average a fraction $\alpha_p \simeq 0.30$ of the incoming solar radiation at the Earth is reflected back to space; α_p is known as the *planetary albedo* (see Table 2.1). Thus

$$\text{Solar radiation absorbed by the Earth} =$$
$$(1 - \alpha_p)S_0 \pi a^2 = 1.22 \times 10^{17} \text{ W. (2-1)}$$

In equilibrium, the total terrestrial flux radiated to space must balance the solar radiation absorbed by the Earth. If in total the spinning Earth radiates in all directions like a blackbody of uniform temperature T_e (known as the *effective planetary temperature*, or *emission temperature* of the Earth) the Stefan-Boltzmann law gives:

$$\text{Emitted radiation per unit area} = \sigma T_e^4 \quad (2\text{-}2)$$

where $\sigma = 5.67 \times 10^{-8}$ W m^{-2} K^{-4} is the Stefan-Boltzmann constant. So

$$\text{Emitted terrestrial radiation} = 4\pi a^2 \sigma T_e^4. \quad (2\text{-}3)$$

Note that Eq. 2-3 is a **definition** of *emission temperature* T_e. It is the temperature one would infer by looking back at Earth if a blackbody curve was fitted to the measured spectrum of outgoing radiation.

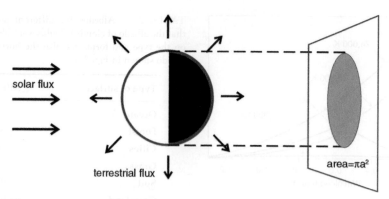

FIGURE 2.4. The spinning Earth is imagined to intercept solar energy over a disk of radius a and radiate terrestrial energy away isotropically from the sphere. Modified from Hartmann, 1994.

Surface Albedo

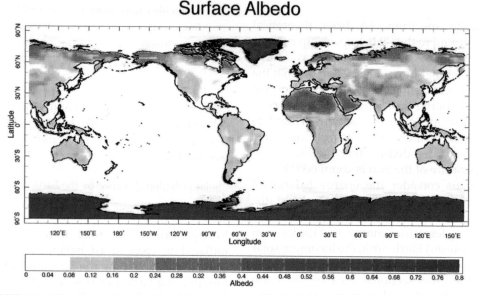

FIGURE 2.5. The albedo of the Earth's surface. Over the ocean the albedo is small (2–10%). It is larger over the land (typically 35–45% over desert regions) and is particularly high over snow and ice ($\sim$80%) (see Table 2.2).

Equating Eq. 2-1 with Eq. 2-3 gives

$$T_e = \left[\frac{S_0(1 - \alpha_p)}{4\sigma} \right]^{1/4}. \qquad (2\text{-}4)$$

Note that the radius of the Earth has cancelled out: T_e depends only on the planetary albedo and the distance of the Earth from the Sun. Putting in numbers, we find that the Earth has an emission temperature of 255 K. Table 2.1 lists the various parameters for some of the planets and compares approximate measured values, T_m, with T_e

computed from Eq. 2-4. The agreement is very good, except for Jupiter where it is thought that about one half of the energy input comes from the gravitational collapse of the planet (see Problem 3 at end of this chapter).

However, as can be seen from Table 2.1, the emission temperature of Earth is nearly 40 K cooler than the globally averaged observed surface temperature, which is $T_s = 288$ K. As we shall discuss in Section 2.3, $T_s \neq T_e$ because: (1) radiation is absorbed within the atmosphere, principally by its water vapor blanket, and (2) fluid motions—air currents—carry heat both vertically and horizontally.

2.2. THE ATMOSPHERIC ABSORPTION SPECTRUM

A property of the blackbody radiation curve is that the wavelength of maximum energy emission, λ_m, satisfies

$$\lambda_m T = \text{constant.} \qquad (2\text{-}5)$$

This is known as *Wien's displacement law*. Since the solar emission temperature is about 6000 K, the maximum of the solar spectrum is (see Fig. 2.2) at about $0.6\,\mu$m (in the visible spectrum), and we have determined $T_e = 255$ K for the Earth, it follows that the peak of the terrestrial spectrum is at

$$\lambda_m^{Earth} = 0.6\,\mu\text{m} \times \frac{6000}{255} \simeq 14\,\mu\text{m}.$$

Thus the Earth's radiation to space is primarily in the infrared spectrum. Normalized (see Appendix A.1.1) blackbody spectra for the Sun and Earth are shown in Fig. 2.6. The two spectra hardly overlap, which greatly simplifies thinking about radiative transfer.

Also shown in Fig. 2.6 is the atmospheric absorption spectrum; this is the fraction of radiation at each wavelength that is

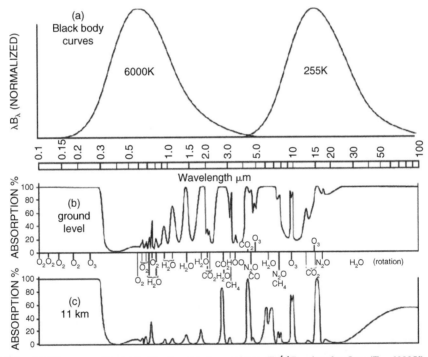

FIGURE 2.6. (a) The normalized blackbody emission spectra, $T^{-4}\lambda B_\lambda$, for the Sun ($T = 6000$ K) and Earth ($T = 255$ K) as a function of $\ln \lambda$ (top), where B_λ is the blackbody function (see Eq. A-2) and λ is the wavelength (see Appendix A.1.1 for further discussion). (b) The fraction of radiation absorbed while passing from the ground to the top of the atmosphere as a function of wavelength. (c) The fraction of radiation absorbed from the tropopause (typically at a height of 11 km) to the top of the atmosphere as a function of wavelength. The atmospheric molecules contributing the important absorption features at each frequency are also indicated. After Goody and Yung (1989).

absorbed on a single vertical path through the atmosphere. From it we see that:

- The atmosphere is almost completely transparent in the visible spectrum, at the peak of the solar spectrum.

- The atmosphere is very opaque in the UV spectrum.

- The atmosphere has variable opacity across the IR spectrum. It is almost completely opaque at some wavelengths, and transparent at others.

- N_2 does not figure at all in absorption, and O_2 absorbs only in the far UV (where there is little solar energy flux) and, a little, in the near IR. The dominant constituents of the atmosphere are incredibly transparent across almost the whole spectral range of importance.

- The absorption of terrestrial radiation is dominated by triatomic molecules—O_3 in the UV; H_2O, CO_2, and others in the IR—because triatomic molecules have rotational and vibrational modes that can easily be excited by IR radiation. These molecules are present in tiny concentrations (see Table 1.2) but play a key role in the absorption of terrestrial radiation (see Fig. 2.6). They are known as *greenhouse gases*. This is the fundamental reason why atmospheric radiation may be so vulnerable to the human-induced changes in composition shown in Fig. 1.3.

2.3. THE GREENHOUSE EFFECT

The global average mean surface temperature of the Earth is 288 K (Table 2.1). Previously we deduced that the emission temperature of the Earth is 255 K, which is considerably lower. Why? We saw from Fig. 2.6 that the atmosphere is rather opaque to IR radiation, so we cannot think of terrestrial radiation as being radiated into space directly from the surface. Much of the radiation emanating from the surface will be absorbed, primarily by H_2O, before passing through the atmosphere. On average, the emission to space will emanate from some level in the atmosphere (typically about 5 km) such that the region above that level is mostly transparent to IR radiation. It is this region of the atmosphere, rather than the surface, that must be at the emission temperature. Thus radiation from the atmosphere will be directed downward as well as upward, and hence the surface will receive not only the net solar radiation, but IR from the atmosphere as well. Because the surface feels more incoming radiation than if the atmosphere were not present (or were completely transparent to IR) it becomes warmer than T_e. This has become known as the *greenhouse effect*.[2]

2.3.1. A simple greenhouse model

Consider Fig. 2.7. Since the atmosphere is thin, let us simplify things by considering a planar geometry, in which the incoming radiation per unit area is equal to the average flux per unit area striking the Earth. This average incoming solar energy *per unit area of the Earth's surface* is

$$\text{average solar energy flux} = \frac{\text{intercepted incoming radiation}}{\text{Earth's surface area}}$$
$$= \frac{S_0 \pi a^2}{4\pi a^2} = \frac{S_0}{4}. \tag{2-6}$$

We will represent the atmosphere by a single layer of temperature T_a, and, in this first calculation, assume: (1) that it is completely transparent to shortwave solar radiation, and (2) that it is completely opaque to IR radiation (i.e., it absorbs all the IR radiating

---

[2]It is interesting to note that the domestic greenhouse does not work in this manner! A greenhouse made of plastic window panes, rather than conventional glass, is effective even though plastic (unlike glass) does not have significant absorption bands in the IR. The greenhouse works because its windows allow energy in and its walls prevent the warm air from rising or blowing away.

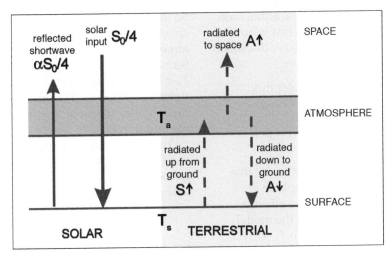

FIGURE 2.7. The simplest greenhouse model, comprising a surface at temperature T_s, and an atmospheric layer at temperature T_a, subject to incoming solar radiation $S_0/4$. The terrestrial radiation upwelling from the ground is assumed to be completely absorbed by the atmospheric layer.

from the ground) so that the layer emitting to space is also "seen" by the ground. Now, since the whole Earth-atmosphere system must be in equilibrium (on average), the net flux into the system must vanish. The average net solar flux per unit area is, from Eq. 2-6, and allowing for reflection, $1/4\left(1 - \alpha_p\right) S_0$, whereas the terrestrial radiation emitted to space per unit area is, using Eq. 2-2:

$$A \uparrow = \sigma T_a^4.$$

Equating them, we find:

$$\sigma T_a^4 = \frac{1}{4}\left(1 - \alpha_p\right) S_0 = \sigma T_e^4, \quad (2\text{-}7)$$

using the definition of T_e, Eq. 2-4. We see that the atmosphere is at the emission temperature (naturally, because it is this region that is emitting to space).

At the surface, the average incoming shortwave flux is also $1/4\left(1 - \alpha_p\right) S_0$, but there is also a downwelling flux emitted by the atmosphere,

$$A \downarrow = \sigma T_a^4 = \sigma T_e^4.$$

The flux radiating upward from the ground is

$$S \uparrow = \sigma T_s^4,$$

where T_s is the surface temperature. Since, in equilibrium, the net flux at the ground must be zero,

$$S \uparrow = \frac{1}{4}\left(1 - \alpha_p\right) S_0 + A \downarrow,$$

whence

$$\sigma T_s^4 = \frac{1}{4}\left(1 - \alpha_p\right) S_0 + \sigma T_e^4 = 2\sigma T_e^4, \quad (2\text{-}8)$$

where we have used Eq. 2-7. Therefore

$$T_s = 2^{1/4} T_e. \quad (2\text{-}9)$$

So the presence of an absorbing atmosphere, as depicted here, increases the surface temperature by a factor $2^{1/4} = 1.19$. This arises as a direct consequence of absorption of terrestrial radiation by the atmosphere, which in turn re-radiates IR back down to the surface, thus increasing the net downward radiative flux at the surface. Note that $A \downarrow$ is of the same order—in fact in this simple model, equal to—the solar radiation that strikes the ground. This is true of more complex models, and indeed observations show that the downwelled radiation from the atmosphere can exceed that due to the direct solar flux.

Applying this factor to our calculated value $T_e = 255\,\text{K}$, we predict $T_s = 2^{1/4} \times$

255 = 303 K. This is closer to the actual mean surface temperature of 288 K but is now an overestimate! The model we have discussed is clearly an oversimplification:

- For one thing, not all the solar flux incident on the top of the atmosphere reaches the surface; typically, 20%–25% is absorbed within the atmosphere (including by clouds).

- For another, we saw in Section 2.2 that IR absorption by the atmosphere is incomplete. The greenhouse effect is actually less strong than in the model assumed previously, and so T_s will be less than the value implied by Eq. 2-9. We shall analyze this by modifying Fig. 2.7 to permit partial transmission of IR through the atmosphere—a leaky greenhouse model.

2.3.2. A leaky greenhouse

Consider Fig. 2.8. We suppose the atmosphere has absorptivity ϵ, that is, a fraction ϵ of the IR upwelling from the surface is absorbed within the atmosphere (so the case of Fig. 2.7 corresponds to $\epsilon = 1$). Now again if we insist that in equilibrium the net flux at the top of the atmosphere vanishes, we get

$$\frac{1}{4}(1 - \alpha_p)S_0 = A\uparrow + (1 - \epsilon)S\uparrow. \qquad (2\text{-}10)$$

Zero net flux at the surface gives

$$\frac{1}{4}(1 - \alpha_p)S_0 + A\downarrow = S\uparrow. \qquad (2\text{-}11)$$

Since at equilibrium, $A\uparrow = A\downarrow$, we have

$$S\uparrow = \sigma T_s^4 = \frac{1}{2(2 - \epsilon)}(1 - \alpha_p)S_0 = \frac{2}{(2 - \epsilon)}\sigma T_e^4.$$
$$(2\text{-}12)$$

Therefore,

$$T_s = \left(\frac{2}{2 - \epsilon}\right)^{1/4} T_e. \qquad (2\text{-}13)$$

So in the limit $\epsilon \to 0$ (transparent atmosphere), $T_s = T_e$, and for $\epsilon \to 1$ (opaque atmosphere), $T_s = 2^{1/4}T_e$, as found in Section 2.3.1. In general, when $0 < \epsilon < 1$, $T_e < T_s < 2^{1/4}T_e$. Thus partial transparency of the atmosphere to IR radiation—a "leaky" greenhouse—reduces the warming effect we found in Eq. 2-9.

To find the atmospheric temperature, we need to invoke *Kirchhoff's law*,[3] such that the emissivity of the atmosphere is equal to its absorptivity. Thus

$$A\uparrow = A\downarrow = \epsilon\sigma T_a^4. \qquad (2\text{-}14)$$

We can now use Eqs. 2-14, 2-10, 2-11, and 2-12 to find

$$T_a = \left(\frac{1}{2 - \epsilon}\right)^{1/4} T_e = \left(\frac{1}{2}\right)^{1/4} T_s.$$

So the atmosphere is, for $\epsilon < 1$, cooler than T_e (since the emission is then only partly from the atmosphere). Note, however, that $T_a < T_s$; the atmosphere is *always* cooler than the ground.

2.3.3. A more opaque greenhouse

Previously we considered a leaky greenhouse. To take the other extreme, suppose that the atmosphere is so opaque that even a shallow layer will absorb *all* the IR radiation passing through it. Now the assumption implicit in Fig. 2.7—that space and the surface both "see" the same atmospheric layer—is wrong. We can elaborate our model to include a second totally absorbing layer in the atmosphere, as illustrated in Fig. 2.9. Of course, to do the calculation correctly (rather than just to illustrate the principles), we would divide the atmosphere into an infinite number of thin layers, allow for the presence of clouds, treat each wavelength in Fig. 2.6 separately, allow for atmospheric absorption

---

[3]Kirchhoff's law states that the emittance of a body, which is the ratio of the actual emitted flux to the flux that would be emitted by a blackbody at the same temperature, equals its absorptance.

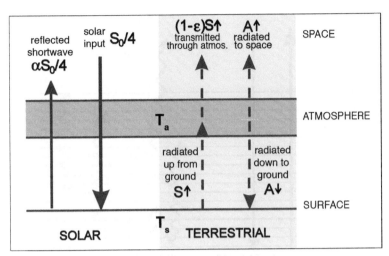

FIGURE 2.8. A leaky greenhouse. In contrast to Fig. 2.7, the atmosphere now absorbs only a fraction, ε, of the terrestrial radiation upwelling from the ground.

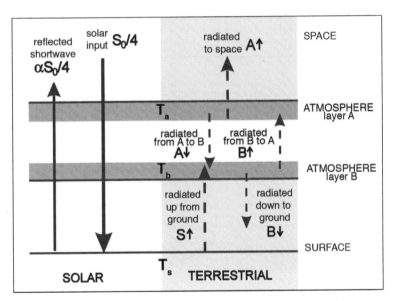

FIGURE 2.9. An "opaque" greenhouse made up of two layers of atmosphere. Each layer completely absorbs the IR radiation impinging on it.

layer-by-layer—which depends on the vertical distribution of absorbers, particularly H_2O, CO_2, and O_3 (see section 3.1.2)—and do the required budgets for each layer and at the surface (we are not going to do this). An incomplete schematic of how this might look for a rather opaque atmosphere is shown in Fig. 2.10.

The resulting profile, which would be the actual mean atmospheric temperature profile *if heat transport in the atmosphere occurred only through radiative transfer*, is known as **the radiative equilibrium temperature profile**. It is shown in Fig. 2.11. In particular, note the presence of a large temperature discontinuity at the surface in the radiative

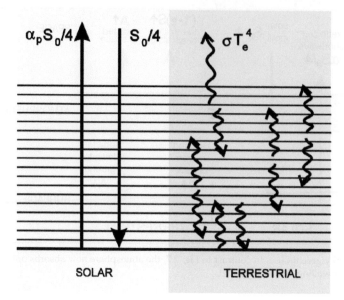

FIGURE 2.10.   Schematic of a radiative transfer model with many layers.

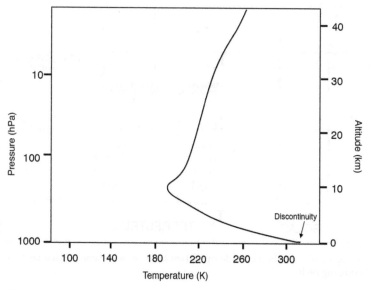

FIGURE 2.11.   The radiative equilibrium profile of the atmosphere obtained by carrying out the calculation schematized in Fig. 2.10. The absorbers are H_2O, O_3, and CO_2. The effects of both terrestrial radiation and solar radiation are included. Note the discontinuity at the surface. Modified from Wells (1997).

equilibrium profile, which is not observed in practice. (Recall from our analysis of Fig. 2.8 that we found that the atmosphere in our slab model is always colder than the surface.) The reason this discontinuity is produced in radiative equilibrium is that, although there is some absorption within the troposphere, both of solar and terrestrial radiation, most solar radiation is absorbed at the surface. The reason such a discontinuity is not observed in nature is that it would (and does) lead to *convection* in the atmosphere, which introduces an additional mode of *dynamical* heat transport. Because of the presence of convection in the lower atmosphere, the observed profile differs substantially from that obtained by the radiative calculation described above. This is discussed at some length in Chapter 4.

Before going on in Chapter 3 to a discussion of the observed vertical profile of temperature in the atmosphere, we briefly discuss what our simple greenhouse models tell us about climate feedbacks and sensitivity to changes in radiative forcing.

2.3.4. Climate feedbacks

The greenhouse models described previously illustrate several important radiative feedbacks that play a central role in regulating the climate of the planet. Following Hartmann (1994) we suppose that a perturbation to the climate system can be represented as an additional energy input dQ (units $W\,m^{-2}$) and study the resultant change in global-mean surface temperature, dT_s. Thus we define $\partial T_s/\partial Q$ to be a measure of climate sensitivity.

The most important negative feedback regulating the temperature of the planet is the dependence of the outgoing longwave radiation on temperature. If the planet warms up, then it radiates more heat back out to space. Thus using Eq. 2-2 and setting $\delta Q = \delta\left(\sigma T_e^4\right) = 4T_e^3\delta T_s$, where it has been assumed that T_e and T_s differ by a constant, implies a climate sensitivity associated with blackbody radiation of

$$\frac{\partial T_s}{\partial Q}_{BB} = \left(4\sigma T_e^3\right)^{-1} = 0.26\frac{K}{W\,m^{-2}}\,, \quad (2\text{-}15)$$

where we have inserted numbers setting $T_e = 255\,K$. Thus for every $1\,W\,m^{-2}$ increase in the forcing of energy balance at the surface, T_s will increase by about a quarter of a degree. This is rather small when one notes that a $1\,W\,m^{-2}$ change in surface forcing

demands a change in solar forcing of about $6\,W\,m^{-2}$, on taking into account geometrical and albedo effects (see Problem 7 at the end of this chapter).

A powerful positive climate feedback results from the temperature dependence of saturated water vapor pressure, e_s, on T; see Eq. 1-4. If the temperature increases, the amount of water that can be held at saturation increases. Since H_2O is the main greenhouse gas, this further raises surface temperature. From Eq. 1-4 we find that

$$\frac{de_s}{e_e} = \beta dT,$$

and so, given that $\beta = 0.067°C^{-1}$, a $1°C$ change in temperature leads to a full 7% change in saturated specific humidity. The observed relative humidity of the atmosphere (that is the ratio of actual to the saturated specific humidity; see Section 5.3) does not vary significantly, even during the seasonal cycle when air temperatures vary markedly. One consequence of the presence of this blanket of H_2O is that the emission of terrestrial radiation from the surface depends less on T_s than suggested by the Stefan-Boltzmann law. When Stefan-Boltzmann and water vapor feedbacks are combined, calculations show that the climate sensitivity is

$$\frac{\partial T_s}{\partial Q}_{BB\text{ and }H_2O} = 0.5\frac{K}{W\,m^{-2}}\,,$$

which is twice that of Eq. 2-15.

The albedos of ice and clouds also play a very important role in climate sensitivity. The primary effect of ice cover is its high albedo relative to typical land surfaces or the ocean (see Table 2.2 and Fig. 2.5). If sea ice, for example, were to expand into low albedo regions, the amount of solar energy absorbed at the surface would be reduced, causing further cooling and enhancing the expansion of ice. Clouds, because of their high reflectivity, typically double the albedo of the Earth from 15% to 30%, and so have a major impact on the radiative balance of the planet. However it is not known to what extent the amount or type of cloud (both of which are important for climate, as we will see in Chapter 4) is sensitive to the state of the climate or how they might change as the climate evolves over time. Unfortunately our understanding of cloud/radiative feedbacks is one of the greatest uncertainties in climate science.

2.4. FURTHER READING

More advanced treatments of radiative transfer theory can be found in the texts of Houghton (1986) and Andrews (2000). Hartmann (1994) has a thorough discussion of greenhouse models, radiative/convective processes, and their role in climate and climate feedbacks.

2.5. PROBLEMS

1. At present the emission temperature of the Earth is 255 K, and its albedo is 30%. How would the emission temperature change if:

   (a) the albedo was reduced to 10% (and all else were held fixed)?

   (b) the infrared absorptivity of the atmosphere (ϵ in Fig. 2.8) was doubled, but albedo remained fixed at 30%?

2. Suppose that the Earth is, after all, flat. Specifically, consider it to be a thin circular disk (of radius 6370 km), orbiting the Sun at the same distance as the Earth; the planetary albedo is 30%. The vector normal to one face of this disk always points directly towards the Sun, and the disk is made of perfectly conducting material, so both faces of the disk are at the same temperature. Calculate the emission temperature of this disk, and compare with Eq. 2-4 for a spherical Earth.

3. Consider the thermal balance of Jupiter.

   (a) Assuming a balance between incoming and outgoing radiation, and given the data in Table 2.1, calculate the emission temperature for Jupiter.

   (b) In fact, Jupiter has an internal heat source resulting from its gravitational collapse. The measured emission temperature T_e defined by

$$\sigma T_e^4 = \begin{array}{l}\text{(outgoing flux of} \\ \text{planetary radiation per} \\ \text{unit surface area)}\end{array}$$

   is 130 K. Comment in view of your theoretical prediction in part (a). Modify your expression for emission temperature for the case where a planet has an internal heat source giving a surface heat flux Q per unit area. Calculate the magnitude of Jupiter's internal heat source.

   (c) It is believed that the source of Q on Jupiter is the release of gravitational potential energy by a slow contraction of the planet. On the simplest assumption that Jupiter is of uniform density and remains so as it contracts, calculate the annual change in its radius a_{jup} required to produce your value of Q. (Only one half of the released gravitational energy is convertible to heat, the remainder appearing as the additional kinetic energy required to preserve the angular momentum of the planet.)

   [A uniform sphere of mass M and radius a has a gravitational potential energy of $-\frac{3}{5} G \frac{M^2}{a}$, where G is the gravitational constant $= 6.7 \times 10^{-11}\,\text{kg}^{-1}\,\text{m}^3\,\text{s}^{-2}$. The mass of Jupiter is 2×10^{27} kg and its radius is $a_{jup} = 7.1 \times 10^7$ m.]

4. Consider the "two-slab" greenhouse model illustrated in Fig. 2.9, in which the atmosphere is represented by two perfectly absorbing layers of temperature T_a and T_b.

   Determine T_a, T_b, and the surface temperature T_s in terms of the emission temperature T_e.

5. Consider an atmosphere that is completely transparent to shortwave (solar) radiation, but very opaque to infrared (IR) terrestrial radiation. Specifically, assume that it can be represented by N slabs of atmosphere, each of which is completely absorbing of IR, as depicted in Fig. 2.12 (not all layers are shown).

   (a) By considering the radiative equilibrium of the surface, show that the surface must be warmer than the lowest atmospheric layer.

   (b) By considering the radiative equilibrium of the nth layer, show that, in equilibrium,

   $$2T_n^4 = T_{n+1}^4 + T_{n-1}^4, \qquad (2\text{-}16)$$

   where T_n is the temperature of the nth layer, for $n > 1$. Hence argue

that the equilibrium surface temperature is

$$T_s = (N+1)^{1/4} T_e,$$

where T_e is the planetary emission temperature. [Hint: Use your answer to part (a); determine T_1 and use Eq. 2-16 to get a relationship for temperature differences between adjacent layers.]

6. Determine the emission temperature of the planet Venus. You may assume the following: the mean radius of Venus' orbit is 0.72 times that of the Earth's orbit; the solar flux S_o decreases as the square of the distance from the Sun and has a value of $1367\,\mathrm{W\,m^{-2}}$ at the mean Earth orbit; Venus' planetary albedo = 0.77.

   The observed mean surface temperature of the planet Venus is about 750 K (see Table 2.1). How many layers of the N−layer model considered in Problem 5 would be required to achieve this degree of warming? Comment.

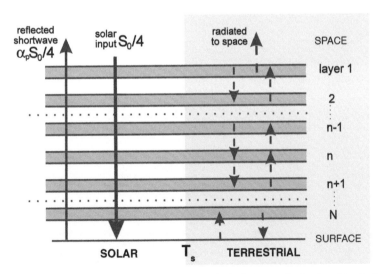

FIGURE 2.12.   An atmosphere made up of N slabs, each of which is completely absorbing in the IR spectrum.

7. Climate feedback due to Stefan-Boltzmann.

   (a) Show that the globally averaged incident solar flux at the ground is $\frac{1}{4}(1 - \alpha_p)S_0$.

   (b) If the outgoing longwave radiation from the Earth's surface were governed by the Stefan-Boltzmann law, then we showed in Eq. 2-15 that for every 1 W m^{-2} increase in the forcing of the surface energy balance, the surface temperature will increase by about a quarter of a degree. Use your answer to (a) to estimate by how much one would have to increase the solar constant to achieve a 1°C increase in surface temperature? You may assume that the albedo of Earth is 0.3 and does not change.

3

The vertical structure of the atmosphere

3.1. Vertical distribution of temperature and greenhouse gases
    3.1.1. Typical temperature profile
    3.1.2. Atmospheric layers
3.2. The relationship between pressure and density: Hydrostatic balance
3.3. Vertical structure of pressure and density
    3.3.1. Isothermal atmosphere
    3.3.2. Non-isothermal atmosphere
    3.3.3. Density
3.4. Further reading
3.5. Problems

In this chapter we discuss the observed vertical distribution of temperature, water vapor, and greenhouse gases in the atmosphere. The observed temperature distribution is compared to the radiative equilibrium profile discussed in Chapter 2. We go on to calculate the implied distribution of pressure and density, assuming the atmosphere to be in hydrostatic balance, and compare with observations. We discover that the atmosphere does not have a distinct top. Rather, the density and pressure decay with height by a factor of e every 7–8 km.

3.1. VERTICAL DISTRIBUTION OF TEMPERATURE AND GREENHOUSE GASES

3.1.1. Typical temperature profile

Temperature varies greatly both vertically and horizontally (as well as temporally) throughout the atmosphere. However,

despite horizontal variations, the vertical structure of temperature is qualitatively similar everywhere, and so it is meaningful to think of (and to attempt to explain) a "typical" temperature profile. (We look at horizontal variations in Chapter 5.) A typical temperature profile (characteristic of 40° N in December) up to about 100 km is shown in Fig. 3.1.

The profile is not governed by a simple law and is rather complicated. Note, however, that the (mass-weighted) mean temperature is close to 255 K, the emission temperature computed in the last chapter (remember almost all the mass of the atmosphere is in the bottom 10 km). The heating effect of solar radiation can be readily seen: there are three "hot spots" corresponding to regions where solar radiation is absorbed at different wavelengths in the thermosphere, the stratopause, and the troposphere. These maxima separate the atmosphere neatly into different layers.

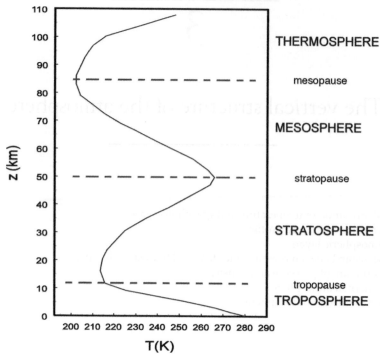

FIGURE 3.1.   Vertical temperature profile for the "US standard atmosphere" at 40° N in December.

3.1.2. Atmospheric layers

Coming down from the top of the atmosphere, the first hot spot evident in Fig. 3.1 is the **thermosphere,** where the temperature is very high and variable. It is here that very short wavelength UV is absorbed by oxygen (cf. Section 2.2), thus heating the region. Molecules (including O_2 as well as CO_2, the dominant IR emitter at this altitude) are dissociated (photolyzed) by high-energy UV ($\lambda < 0.1\ \mu$m). Therefore, because of the scarcity of polyatomic molecules, IR loss of energy is weak, so the temperature of the region gets very high (as high as 1000 K). The air is so tenuous that assumptions of local thermodynamic equilibrium, as in blackbody radiation, are not applicable. At and above these altitudes, the atmosphere becomes ionized (the **ionosphere**), causing reflection of radio waves, a property of the upper atmosphere that is of great practical importance.

Below the *mesopause* at about 80–90 km altitude, temperature increases, moving down through the **mesosphere** to reach a maximum at the *stratopause*, near 50 km, the second hot spot. This maximum is a direct result of absorption of medium wavelength UV (0.1 μm to 0.35 μm) by ozone. It is interesting to note that ozone concentration peaks much lower down in the atmosphere, at heights of 20–30 km, as illustrated in Fig. 3.2. This is because the ozone layer is very opaque to UV (cf. Fig. 2.6), so most of the UV flux is absorbed in the upper parts of the layer, and there is little left to be absorbed at lower altitudes.

The reason for the existence of ozone at these levels is that it is produced here, as a by-product of the *photodissociation (photolysis)* of molecular oxygen, producing atomic oxygen, which may then combine with molecular oxygen thus:

$$O_2 + h\nu \quad \rightarrow \quad O + O,$$
$$O + O_2 + M \quad \rightarrow \quad O_3 + M, \qquad (3\text{-}1)$$

where $h\nu$ is the energy of incoming photons (ν is their frequency and h is Planck's

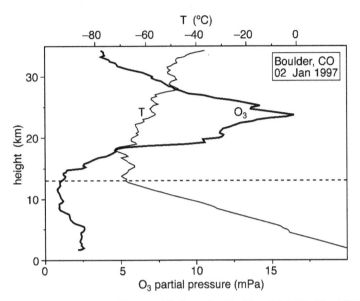

FIGURE 3.2.   A typical winter ozone profile in middle latitudes (Boulder, CO, USA, 2 Jan 1997). The heavy curve shows the profile of ozone partial pressure (mPa), the light curve temperature (°C) plotted against altitude up to about 33 km. The dashed horizontal line shows the approximate position of the tropopause. Balloon data courtesy of NOAA Climate Monitoring and Diagnostics Laboratory.

constant) and M is any third body needed to carry the excess energy.

The resulting ozone, through its radiative properties, is the reason for the existence of the **stratosphere**.[1] It is also one of the primary practical reasons to be interested in stratospheric behavior, since (as we saw in Chapter 2) ozone is the primary absorber of solar UV and thus shields life at the surface (including us) from the damaging effects of this radiation. The stratosphere, as its name suggests, is highly stratified and poorly mixed (*stratus*, meaning "layered"), with long residence times for particles ejected into it (for example by volcanos) from the troposphere below. It is close to radiative equilibrium.

Below the *tropopause*, which is located at altitudes of 8–16 km (depending on latitude and season), temperature increases strongly moving down through the **troposphere** (*tropos*, meaning 'turn') to the surface, the third hot spot. It contains about 85% of the atmosphere's mass and essentially all the water vapor, the primary greenhouse gas, as illustrated in Fig. 3.3. Note that the distribution of water vapor is in large part a consequence of the Clausius-Clapeyron relation, Eq. 1.4 and rapidly decays with height as T decreases.

From the vertical distribution of O_3 and H_2O, shown in Fig. 3.2 and Fig. 3.3, and of CO_2 (which is well mixed in the vertical) a radiative equilibrium profile can

[1] Léon Philippe Teisserenc de Bort (1855–1913). French meteorologist who pioneered the use of unmanned, high-flying, instrumented balloons and discovered the stratosphere. He was the first to identify the temperature inversion at the tropopause. In 1902 he suggested that the atmosphere was divided into two layers.

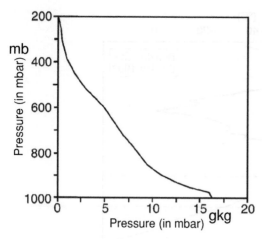

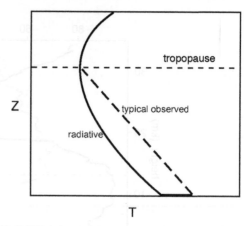

FIGURE 3.3. The global average vertical distribution of water vapor (in $g\,kg^{-1}$) plotted against pressure (in mbar).

FIGURE 3.4. A schematized radiative equilibrium profile in the troposphere (cf. Fig. 2.11) (solid) and a schematized observed profile (dashed). Below the tropopause, the troposphere is stirred by convection and weather systems and is not in radiative balance. Above the tropopause, dynamical heat transport is less important, and the observed T is close to the radiative profile.

be calculated, using the methods outlined in Chapter 2. This profile was shown in Fig. 2.11. The troposphere is warmed in part through absorption of radiation by H_2O and CO_2; the stratosphere is warmed, indeed created, through absorption of radiation by O_3.

It is within the troposphere that almost everything we classify as "weather" is located (and, of course, it is where we happen to live); it will be the primary focus of our attention. As we shall see in Chapter 4, its thermal structure cannot be satisfactorily explained solely by radiative balances. The troposphere is in large part warmed by convection from the lower surface. Temperature profiles observed in the troposphere and as calculated from radiative equilibrium are illustrated schematically in Fig. 3.4. The observed profile is rather different from the radiative equilibrium profile of Fig. 2.11. As noted at the end of Chapter 2, the temperature discontinuity at the surface in the radiative equilibrium profile is not observed in practice. This discontinuity in temperature triggers a convective mode of vertical heat transport, which is the subject of Chapter 4.

Having described the observed T profile, we now go on to discuss the associated p and ρ profiles.

3.2. THE RELATIONSHIP BETWEEN PRESSURE AND DENSITY: HYDROSTATIC BALANCE

Let us imagine that the atmospheric T profile is as observed, for example, in Fig. 3.1. What is the implied vertical distribution of pressure p and density ρ? If the atmosphere were at rest, or static, then pressure at any level would depend on the weight of the fluid above that level. This balance, which we now discuss in detail, is called *hydrostatic balance*.

Consider Fig. 3.5, which depicts a vertical column of air of horizontal cross-sectional area δA and height δz. Pressure $p(z)$ and density $\rho(z)$ of the air are both expected to be functions of height z (they may be functions of $x, y,$ and t also). If the pressure at the bottom of the cylinder is $p_B = p(z)$, then that at the top is

$$p_T = p(z + \delta z)$$
$$= p(z) + \delta p,$$

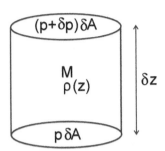

FIGURE 3.5. A vertical column of air of density ρ, horizontal cross-sectional area δA, height δz, and mass $M = \rho \delta A \delta z$. The pressure on the lower surface is p, the pressure on the upper surface is $p + \delta p$.

where δp is the change in pressure moving from z to $z + \delta z$. Assuming δz to be small,

$$\delta p = \frac{\partial p}{\partial z}\delta z. \qquad (3\text{-}2)$$

Now, the mass of the cylinder is

$$M = \rho \delta A \delta z.$$

If the cylinder of air is not accelerating, it must be subjected to zero net force. The vertical forces (upward being positive) are:

  i) gravitational force
     $F_g = -gM = -g\rho\delta A\delta z$,

 ii) pressure force acting at the top face,
     $F_T = -\left(p + \delta p\right)\delta A$, and

iii) pressure force acting at the bottom face, $F_B = p\delta A$.

Setting the net force $F_g + F_T + F_B$ to zero gives $\delta p + g\rho\delta z = 0$, and using Eq. 3-2 we obtain

$$\frac{\partial p}{\partial z} + g\rho = 0. \qquad (3\text{-}3)$$

Eq. 3-3 is the equation of **hydrostatic balance.** It describes how pressure decreases with height in proportion to the weight of the overlying atmosphere. Note that since p must vanish as $z \to \infty$—the atmosphere fades away[2]—we can integrate Eq. 3-3 from z to ∞ to give the pressure at any height

$$p(z) = g\int_{z}^{\infty}\rho\,dz. \qquad (3\text{-}4)$$

Here $\int_{z}^{\infty}\rho\,dz$ is just the mass per unit area of the atmospheric column above z. The surface pressure is then related to the total mass of the atmosphere above: Eq. 3-4 implies that $p_s = \frac{gM_a}{\text{surface area of Earth}}$. Thus, from measurements of surface pressure, one can deduce the mass of the atmosphere (see Table 1.3).

The only important assumption made in the derivation of Eq. 3-3 was the neglect of any vertical acceleration of the cylinder (in which case, the net force need not be zero). This is an excellent approximation under almost all circumstances in both the atmosphere and ocean. It can become suspect, however, in very vigorous small-scale systems in both the atmosphere and ocean (e.g., convection; tornados; violent thunderstorms; and deep, polar convection in the ocean; see Chapters 4 and 11). We discuss hydrostatic balance in the context of the equations of motion of a fluid in Section 6.2.

Note that Eq. 3-3 does not tell us what $p(z)$ is, since we do not know a priori what $\rho(z)$ is. In order to determine $p(z)$ we must invoke an equation of state to tell us the

---

[2] Blaise Pascal (1623–1662), a physicist and mathematician of prodigious talents and accomplishments, was also intensely interested in the variation of atmospheric pressure with height and its application to the measurement of mountain heights. In 1648 he observed that the pressure of the atmosphere decreases with height and, to his own satisfaction, deduced that a vacuum existed above the atmosphere. The unit of pressure is named after him.

connection between ρ and p, as described in the next section.

3.3. VERTICAL STRUCTURE OF PRESSURE AND DENSITY

Using the equation of state of air, Eq. 1-1, we may rewrite Eq. 3-3 as

$$\frac{\partial p}{\partial z} = -\frac{gp}{RT}. \qquad (3\text{-}5)$$

In general, this has not helped, since we have replaced the two unknowns, p and ρ, by p and T. However, unlike p and ρ, which vary by many orders of magnitude from the surface to, say, 100 km altitude, the variation of T is much less. In the profile in Fig. 3.1, for example, T lies in the range 200–280 K, thus varying by no more than 15% from a value of 240 K. So for the present purpose, we may replace T by a typical mean value to get a feel for how p and ρ vary.

3.3.1. Isothermal atmosphere

If $T = T_0$, a constant, we have

$$\frac{\partial p}{\partial z} = -\frac{gp}{RT_0} = -\frac{p}{H},$$

where H, the *scale height*, is a constant (neglecting, as noted in Chapter 1, the small dependence of g on z) with the value

$$H = \frac{RT_0}{g}. \qquad (3\text{-}6)$$

If H is constant, the solution for p is, noting that by definition, $p = p_s$ at the surface $z = 0$,

$$p(z) = p_s \exp\left(-\frac{z}{H}\right). \qquad (3\text{-}7)$$

Alternatively, by taking the logarithm of both sides we may write z in terms of p thus:

$$z = H \ln\left(\frac{p_s}{p}\right). \qquad (3\text{-}8)$$

Thus pressure decreases exponentially with height, with e–folding height H. For the troposphere, if we choose a representative value $T_0 = 250$ K, then $H = 7.31$ km. Therefore, for example, in such an atmosphere p is 100 hPa, or one tenth of surface pressure, at a height of $z = H \times (\ln 10) = 16.83$ km. This is quite close to the observed height of the 100 hPa surface. Note, very roughly, the 300 hPa surface is at a height of about 9 km and the 500 hPa surface at a height of about 5.5 km.

3.3.2. Non-isothermal atmosphere

What happens if T is not constant? In this case we can still define a local scale height

$$H(z) = \frac{RT(z)}{g}, \qquad (3\text{-}9)$$

such that

$$\frac{\partial p}{\partial z} = -\frac{p}{H(z)},$$

where $H(z)$ is the local scale height. Therefore

$$\frac{1}{p}\frac{\partial p}{\partial z} = \frac{\partial \ln p}{\partial z} = -\frac{1}{H(z)}, \qquad (3\text{-}10)$$

whence

$$\ln p = \int_0^z \frac{dz'}{H(z')} + \text{constant},$$

or

$$p(z) = p_s \exp\left(-\int_0^z \frac{dz'}{H(z')}\right). \qquad (3\text{-}11)$$

Note that if $H(z) = H$, the constant value considered in the previous section, Eq. 3-11 reduces to Eq. 3-7.

In fact, despite its simplicity, the isothermal result, Eq. 3-7, yields profiles that are a good approximation to reality. Fig. 3.6 shows the actual pressure profile for 40° N in December (corresponding to the temperature profile in Fig. 3.1) (solid), together with the profile given by Eq. 3-7, with $H = 6.80$ km (dashed). Agreement between the two is generally good (to some extent, the value of H was chosen to optimize this). The differences can easily be understood

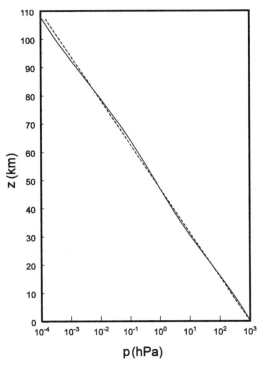

FIGURE 3.6. Observed profile of pressure (solid) plotted against a theoretical profile (dashed) based on Eq. 3-7 with $H = 6.8$ km.

from Eqs. 3-9 and 3-10. In regions where, for example, temperatures are warmer than the reference value ($T_0 = gH/R = 237.08$ K for $H = 6.80$ km), such as (cf. Fig. 3.1) in the lower troposphere, near the stratopause and in the thermosphere, the observed pressure decreases less rapidly with height than predicted by the isothermal profile.

3.3.3. Density

For the isothermal case, the density profile follows trivially from Eq. 3-7, by combining it with the gas law Eq. 1.1:

$$\rho(z) = \frac{p_s}{RT_0} \exp\left(-\frac{z}{H}\right) . \qquad (3\text{-}12)$$

Thus in this case, density also falls off exponentially at the same rate as p. One consequence of Eq. 3-12 is that, as noted at the start of Chapter 1, about 80% of the mass of the atmosphere lies below an altitude of 10 km.

For the nonisothermal atmosphere with temperature $T(z)$, it follows from Eq. 3-11 and the equation of state that

$$\rho(z) = \frac{p_s}{RT(z)} \exp\left(-\int_0^z \frac{dz'}{H(z')}\right) . \qquad (3\text{-}13)$$

3.4. FURTHER READING

A thorough discussion of the role of the various trace gases in the atmospheric radiative balance can be found in Chapter 3 of Andrews (2000).

3.5. PROBLEMS

1. Use the hydrostatic equation to show that the mass of a vertical column of air of unit cross section, extending from the ground to great height, is $\frac{p_s}{g}$, where p_s is the surface pressure. Insert numbers to estimate the mass on a column or air of area $1\,\text{m}^2$. Use your answer to estimate the total mass of the atmosphere.

2. Using the hydrostatic equation, derive an expression for the pressure at the center of a planet in terms of its surface gravity, radius a, and density ρ, assuming that the latter does not vary with depth. Insert values appropriate for the Earth and evaluate the central pressure. [Hint: the gravity at radius r is $g(r) = Gm(r)/r^2$, where $m(r)$ is the mass inside a radius r and $G = 6.67 \times 10^{-11}\,\text{kg}^{-1}\,\text{m}^3\,\text{s}^{-2}$ is the gravitational constant. You may assume the density of rock is $2000\,\text{kg}\,\text{m}^{-3}$.]

3. Consider a horizontally uniform atmosphere in hydrostatic balance. The atmosphere is isothermal, with temperature of $-10°C$. Surface pressure is 1000 mbar.

(a) Consider the level that divides the atmosphere into two equal parts by mass (i.e., one half of the atmospheric mass is above this level). What is the altitude, pressure, and density at this level?

(b) Repeat the calculation of part (a) for the level below which lies 90% of the atmospheric mass.

4. Derive an expression for the hydrostatic atmospheric pressure at height z above the surface in terms of the surface pressure p_s and the surface temperature T_s for an atmosphere with constant lapse rate of temperature $\Gamma = -dT/dz$. Express your results in terms of the dry adiabatic lapse rate $\Gamma_d = \frac{g}{c_p}$ (see Section 4.3.1). Calculate the height at which the pressure is 0.1 of its surface value (assume a surface temperature of 290 K and a uniform lapse rate of 10 K km^{-1}).

5. Spectroscopic measurements show that a mass of water vapor of more than 3 kg m^{-2} in a column of atmosphere is opaque to the "terrestrial" waveband. Given that water vapor typically has a density of 10^{-2} kg m^{-3} at sea level (see Fig. 3.3) and decays in the vertical as $e^{-\left(\frac{z}{b}\right)}$, where z is the height above the surface and $b \sim 3$ km, estimate at what height the atmosphere becomes transparent to terrestrial radiation.

By inspection of the observed vertical temperature profile shown in Fig. 3.1, deduce the temperature of the atmosphere at this height. How does it compare to the emission temperature of the Earth, $T_e = 255$ K, discussed in Chapter 2? Comment on your answer.

6. Make use of your answer to Problem 1 of Chapter 1 to estimate the error incurred in p at 100 km through use of Eq. 3-11 if a constant value of g is assumed.

4

Convection

4.1. The nature of convection
    4.1.1. Convection in a shallow fluid
    4.1.2. Instability
4.2. Convection in water
    4.2.1. Buoyancy
    4.2.2. Stability
    4.2.3. Energetics
    4.2.4. GFD Lab II: Convection
4.3. Dry convection in a compressible atmosphere
    4.3.1. The adiabatic lapse rate (in unsaturated air)
    4.3.2. Potential temperature
4.4. The atmosphere under stable conditions
    4.4.1. Gravity waves
    4.4.2. Temperature inversions
4.5. Moist convection
    4.5.1. Humidity
    4.5.2. Saturated adiabatic lapse rate
    4.5.3. Equivalent potential temperature
4.6. Convection in the atmosphere
    4.6.1. Types of convection
    4.6.2. Where does convection occur?
4.7. Radiative-convective equilibrium
4.8. Further reading
4.9. Problems

We learned in Chapters 2 and 3 that terrestrial radiation emanates to space primarily from the upper troposphere, rather than the ground; much of what radiates from the surface is absorbed within the atmosphere. The surface is thus warmed by both direct solar radiation and downwelling terrestrial radiation from the atmosphere. In consequence, in radiative equilibrium, the surface is warmer than the overlying atmosphere. However, this state is *unstable* to convective motions that develop, as sketched in Fig. 4.1, and transport heat upward from the surface. In the troposphere, therefore, equilibrium is not established solely by radiative processes. In this chapter, we discuss the nature of the convective process and its role in determining the *radiative-convective* balance of the troposphere.

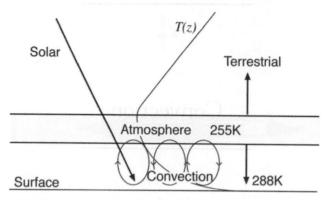

FIGURE 4.1. Solar radiation warms the Earth's surface, triggering convection, which carries heat vertically to the emission level from which, because the atmosphere above this level is transparent in the IR spectrum, energy can be radiated out to space. The surface temperature of about 288 K is significantly higher than the temperature at the emission level, 255 K, because the energy flux from the surface must balance not just the incoming solar radiation but also downwelling IR radiation from the atmosphere above. An idealized radiative equilibrium temperature profile, $T(z)$, is superimposed (cf. Fig. 2.11).

As we are about to explore, the conditions under which convection occurs depend on the characteristics of the fluid. The theory appropriate in a moist, compressible atmosphere is somewhat more complicated than in an incompressible medium like water. Accordingly, we will discuss convection in a sequence of cases of increasing complexity. After making some general remarks about the nature of convection, we will describe the incompressible case, using a theoretical approach and a laboratory experiment, in Section 4.2. Convection in an unsaturated compressible atmosphere is discussed in Section 4.3; the effects of latent heat consequent on condensation of moisture are addressed in Section 4.5 following a discussion of the atmosphere under stable conditions in Section 4.4.

4.1. THE NATURE OF CONVECTION

4.1.1. Convection in a shallow fluid

When a fluid such as water is heated from below (or cooled from above), it develops overturning motions. It may seem obvious that this must occur, because the tendency of the heating (or cooling) is to make the fluid top-heavy.[1] Consider the shallow, horizontally infinite fluid shown in Fig. 4.2. Let the heating be applied uniformly at the base; then we may expect the fluid to have a horizontally uniform temperature, so $T = T(z)$ only. This will be top-heavy (warmer and therefore lighter fluid below cold, dense fluid above). But as we have seen, gravitational forces can be balanced by a vertical pressure gradient in

---

[1] John William Strutt—Lord Rayleigh (1842–1919)—set the study of convection on a firm theoretical basis in his seminal studies in the 1900s. In one of his last articles, published in 1916, he attempted to explain what is now known as Rayleigh-Benard Convection. His work remains the starting point for most of the modern theories of convection.

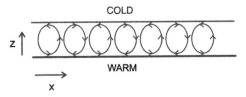

FIGURE 4.2. Schematic of shallow convection in a fluid, such as water, triggered by warming from below and/or cooling from above.

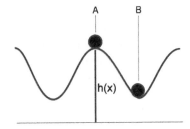

FIGURE 4.3. Stability and instability of a ball on a curved surface.

hydrostatic balance with the density field (Eq. 3-3), in which case any fluid parcel will experience zero net force, even if heavy fluid is above light fluid. Nevertheless, observations and experiments show that convection develops,[2] as sketched in the figure. Why? There are two parts to the question:

1. Why do motions develop when the equilibrium state just discussed has no net forces anywhere?

2. Why are the motions horizontally inhomogeneous when the external forcing (the heating) is horizontally uniform?

The answer is that the motions are not directly forced, but (like many types of motion in the atmosphere and ocean) arise from an *instability* of the fluid in the presence of heating. We therefore begin our discussion of convection by reminding ourselves of the nature of instability.

4.1.2. Instability

For any system that possesses some equilibrium state, instability will arise if, in response to a perturbation, the system tends to drive the perturbation further from the equilibrium state. A simple and familiar example is a ball on a curved surface, as in Fig. 4.3.

A ball that is stationary and exactly at a peak (point A, $x = x_A$) is in equilibrium, but of course this state is unstable. If the ball is displaced a small distance δx from A, it finds itself on a downward slope, and therefore is accelerated further. To be specific, the component of gravitational acceleration along the slope is, for small slope, $-g\,dh/dx$. The slope at $x_A + \delta x$ is, making a Taylor expansion about $x = x_A$ (see Appendix A.2.1),

$$\frac{dh}{dx}(x_A + \delta x) \simeq \frac{dh}{dx}(x_A) + \left(\frac{d^2h}{dx^2}\right)_A \delta x$$
$$= \left(\frac{d^2h}{dx^2}\right)_A \delta x$$

for small δx, since the slope is zero at x_A. Hence the equation of motion for the ball is

$$\frac{d^2}{dt^2}(\delta x) = -g\frac{dh}{dx} = -g\left(\frac{d^2h}{dx^2}\right)_A \delta x. \quad (4\text{-}1)$$

This has solutions

$$\delta x(t) = c_1 e^{\sigma_+ t} + c_2 e^{\sigma_- t} \quad (4\text{-}2)$$

where c_1 and c_2 are constants and

$$\sigma_\pm = \pm\sqrt{-g\left(\frac{d^2h}{dx^2}\right)_A}.$$

At a peak, where dh/dx changes from positive to negative with increasing x, $d^2h/dx^2 < 0$, both roots for σ are real, and the first term in Eq. 4-2, describes an exponentially growing perturbation; the state (of the ball

---

[2]We are considering here the stability of a fluid that has no (or rather very small) viscosity and diffusivity. Rayleigh noted that convection of a viscous fluid heated from below does not always occur; for example, oatmeal or polenta burns if it is not kept stirred because the high viscosity can prevent convection currents.

at A in Fig. 4.3) is *unstable*. (Exponential growth will occur only for as long as our assumption of a small perturbation is valid and will obviously break down before the ball reaches a valley.) On the other hand, for an equilibrium state in which $d^2h/dx^2 > 0$ (such as the valley at B), the roots for σ are imaginary, Eq. 4-2 yields oscillatory solutions, and the state is *stable*; any perturbation will simply produce an oscillation as the ball rolls back and forth across the valley.

A consideration of the energetics of our ball is also instructive. Rather than writing down and solving differential equations describing the motion of the ball, as in Eq. 4-1, the stability condition can be deduced by a consideration of the energetics. If we perturb the ball in a valley, it moves up the hill and we have to do work (add energy) to increase the potential energy of the ball. Hence, in the absence of any external energy source, the ball will return to its position at the bottom of the valley. But if the ball is perturbed on a crest it moves downslope, its potential energy *decreases* and its kinetic energy *increases*. Thus we may deduce that state A in Fig. 4.3 is unstable and state B is stable.

As we shall now see, the state of heavy fluid over light fluid is an unstable one; the fluid will overturn and return itself to a stable state of lower potential energy.

4.2. CONVECTION IN WATER

4.2.1. Buoyancy

Objects that are lighter than water bounce back to the surface when immersed, as has been understood since the time of Archimedes (287–212 BC). But what if the "object" is a parcel[3] of the fluid itself, as sketched in Fig. 4.4? Consider the stability of such a parcel in an incompressible liquid.

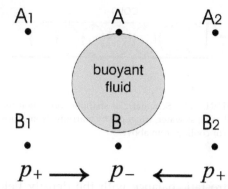

FIGURE 4.4.   A parcel of light, buoyant fluid surrounded by resting, homogeneous, heavier fluid in hydrostatic balance, Eq. 3.3. The fluid above points A_1, A, and A_2 has the same density, and hence, as can be deduced by consideration of hydrostatic balance, the pressures at the A points are all the same. But the pressure at B is lower than at B_1 or B_2 because the column of fluid above it is lighter. There is thus a pressure gradient force which drives fluid inwards toward B, forcing the light fluid upward.

We will suppose that density depends on temperature and not on pressure. Imagine that the parcel shaded in Fig. 4.4 is warmer, and hence less dense, than its surroundings.

If there is no motion, then the fluid will be in hydrostatic balance. Since ρ is uniform above, the pressure at A_1, A, and A_2 will be the same. But, because there is lighter fluid in the column above B than above either point B_1 or B_2, from Eq. 3-4 we see that the hydrostatic pressure at B will be less than at B_1 and B_2. Since fluid has a tendency to flow from regions of high pressure to low pressure, fluid will begin to move toward the low pressure region at B and tend to equalize the pressure along B_1BB_2; the pressure at B will tend to increase and apply an upward force to the buoyant fluid which will therefore begin to move upwards. Thus the light fluid will rise.

In fact (as we will see in Section 4.4) the acceleration of the parcel of fluid is not g but $g\,\Delta\rho/\rho_P$, where $\Delta\rho = (\rho_P - \rho_E)$, ρ_P is the density of the parcel, and ρ_E is the density

---

[3]A "parcel" of fluid is imagined to have a small but finite dimension, to be thermally isolated from the environment, and always to be at the same pressure as its immediate environment.

of the environment. It is common to speak of the buoyancy, b, of the parcel, defined as

$$b = -g \frac{(\rho_P - \rho_E)}{\rho_P} \qquad (4\text{-}3)$$

If $\rho_P < \rho_E$ then the parcel is *positively buoyant* and rises; if $\rho_P > \rho_E$ the parcel is *negatively buoyant* and sinks; if $\rho_P = \rho_E$ the parcel is *neutrally buoyant* and neither sinks or rises.

We will now consider this problem in terms of the stability of a perturbed fluid parcel.

4.2.2. Stability

Suppose we have a horizontally uniform state with temperature $T(z)$ and density $\rho(z)$. T and ρ are assumed here to be related by an equation of state

$$\rho = \rho_{ref}(1 - \alpha[T - T_{ref}]). \qquad (4\text{-}4)$$

Equation 4-4 is a good approximation for (fresh) water in typical circumstances, where ρ_{ref} is a constant reference value of the density and α is the coefficient of thermal expansion at $T = T_{ref}$. (A more detailed discussion of the equation of state for water will be given in Section 9.1.3.) Again we focus attention on a single fluid parcel, initially located at height z_1. It has temperature $T_1 = T(z_1)$ and density $\rho_1 = \rho(z_1)$, the same as its environment; it is therefore *neutrally buoyant* and thus in equilibrium. Now let us displace this fluid parcel a small vertical distance to $z_2 = z_1 + \delta z$, as shown in Fig. 4.5. We need to determine the buoyancy of the parcel when it arrives at height z_2.

Suppose the displacement is done sufficiently rapidly that the parcel does not lose or gain heat on the way, so the displacement is *adiabatic*. This is a reasonable assumption because the temperature of the parcel can only change by diffusion, which is a slow process compared to typical fluid movements and can be neglected here. Since the parcel is incompressible, it will not contract nor expand, and thus it will do no work on its surroundings; its internal energy and hence its temperature T will be conserved.

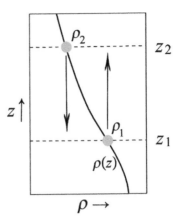

FIGURE 4.5. We consider a fluid parcel initially located at height z_1 in an environment whose density is $\rho(z)$. It has density $\rho_1 = \rho(z_1)$, the same as its environment at height z_1. It is now displaced adiabatically a small vertical distance to $z_2 = z_1 + \delta z$, where its density is compared to that of the environment.

Therefore the temperature of the perturbed parcel at z_2 will still be T_1, and its density will still be $\rho_P = \rho_1$. The environment, however, has density

$$\rho_E = \rho(z_2) \simeq \rho_1 + \left(\frac{d\rho}{dz} \right)_E \delta z,$$

where $(d\rho/dz)_E$ is the environmental density gradient. The buoyancy of the parcel just depends on the difference between its density and that of its environment; using Eq. 4-3, we find that

$$b = \frac{g}{\rho_1} \left(\frac{d\rho}{dz} \right)_E \delta z.$$

The parcel will therefore be

$$\left. \begin{array}{l} \text{positively} \\ \text{neutrally} \\ \text{negatively} \end{array} \right\} \text{buoyant if } \left(\frac{d\rho}{dz} \right)_E \left\{ \begin{array}{l} > 0 \\ = 0 \\ < 0 \end{array} \right. .$$
$$(4\text{-}5)$$

If the parcel is positively buoyant (the situation sketched in Fig. 4.4), it will keep on rising at an accelerating rate. Therefore an incompressible liquid is **unstable if density increases with height** (in the absence of viscous and diffusive effects). This is the familiar "top-heavy" condition. It is this instability that leads to the convective motions discussed above. Using Eq. 4-4, the stability condition can also be expressed in terms of temperature as

$$\left.\begin{array}{r} \text{unstable} \\ \text{neutral} \\ \text{stable} \end{array}\right\} \text{ if } \left(\frac{dT}{dz}\right)_E \left\{\begin{array}{l} < 0 \\ = 0 \\ > 0 \end{array}\right. \quad (4\text{-}6)$$

Note that Eq. 4-6 is appropriate for an incompressible fluid whose density depends only on temperature.

4.2.3. Energetics

Consider now our problem from yet another angle, in terms of energy conversion. We know that if the potential energy of a parcel can be reduced, just like the ball on the top of a hill in Fig. 4.3, the lost potential energy will be converted into kinetic energy of the parcel's motion. Unlike the case of the ball on the hill, however, when dealing with fluids we cannot discuss the potential energy of a single parcel in isolation, since any movement of the parcel requires rearrangement of the surrounding fluid; rather, we must consider the potential energy difference between two realizable states of the fluid. In the present case, the simplest way to do so is to consider the potential energy consequences when two parcels are interchanged.

Consider then two parcels of incompressible fluid of equal volume at differing heights, z_1 and z_2 as sketched in Fig. 4.5. They have the same density as their respective environments. Because the parcels are incompressible they do not expand or contract as p changes and so do not do work on, nor have work done on them by, the environment. This greatly simplifies consideration of energetics. The potential energy of the initial state is

$$PE_{initial} = g\left(\rho_1 z_1 + \rho_2 z_2\right).$$

Now interchange the parcels. The potential energy of the final state, after swapping, is

$$PE_{final} = g\left(\rho_1 z_2 + \rho_2 z_1\right).$$

The change in potential energy, ΔPE, is therefore given by

$$\begin{aligned} \Delta PE &= PE_{final} - PE_{initial} \qquad (4\text{-}7) \\ &= -g\left(\rho_2 - \rho_1\right)\left(z_2 - z_1\right) \\ &\simeq -g\left(\frac{d\rho}{dz}\right)_E \left(z_2 - z_1\right)^2 \end{aligned}$$

if $z_2 - z_1$ is small, where $\left(d\rho/dz\right)_E = \dfrac{\left(\rho_2 - \rho_1\right)}{\left(z_2 - z_1\right)}$ is the mean density gradient of the environmental state. Note that the factor $g\left(z_2 - z_1\right)^2$ is always positive and so the sign of ΔPE depends on that of $\left(d\rho/dz\right)_E$. Hence, if $\left(d\rho/dz\right)_E > 0$, rearrangement leads to a decrease in ΔPE and thus to the growth of the kinetic energy of the parcels; therefore a disturbance is able to grow, and the system will be unstable. But if $\left(d\rho/dz\right)_E < 0$, then $\Delta PE > 0$, and potential energy cannot be released by exchanging parcels. So we again arrive at the stability criterion, Eq. 4-6. This energetic approach is simple but very powerful. It should be emphasized, however, that we have only demonstrated the possibility of instability. To show that instability is a fact, one must carry out a stability analysis analogous to that carried out in Section 4.1.2 for the ball on the curved surface (a simple example is given in Section 4.4) in which the details of the perturbation are worked out. However, whenever energetic considerations point to the possibility of convective instability, exact solutions of the governing dynamical equations almost invariably show that instability is a fact, provided diffusion and viscosity are sufficiently weak.

4.2.4. GFD Lab II: Convection

We can study convection in the laboratory using the apparatus shown in Fig. 4.6.

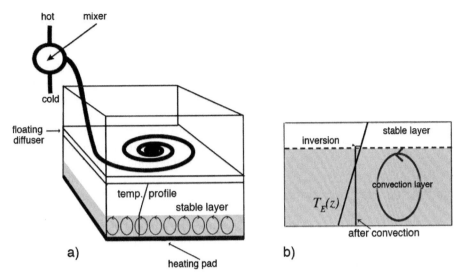

FIGURE 4.6.   (a) A sketch of the laboratory apparatus used to study convection. A stable stratification is set up in a 50 cm × 50 cm × 50 cm tank by slowly filling it up with water whose temperature is slowly increased with time. This is done using (1) a mixer, which mixes hot and cold water together, and (2) a diffuser, which floats on the top of the rising water and ensures that the warming water floats on the top without generating turbulence. Using the hot and cold water supply we can achieve a temperature difference of 20°C over the depth of the tank. The temperature profile is measured and recorded using thermometers attached to the side of the tank. Heating at the base is supplied by a heating pad. The motion of the fluid is made visible by sprinkling a very small amount of potassium permanganate evenly over the base of the tank (which turns the water pink) after the stable stratification has been set up and just before turning on the heating pad. (b) Schematic of evolving convective boundary layer heated from below. The initial linear temperature profile is T_E. The convection layer is mixed by convection to a uniform temperature. Fluid parcels overshoot into the stable stratification above, creating the inversion evident in Fig. 4.7. Both the temperature of the convection layer and its depth slowly increase with time.

A heating pad at the base of the tank triggers convection in a fluid that is initially stratified by temperature. Convection carries heat from the heating pad into the body of the fluid, distributing it over the convection layer much like convection carries heat away from the Earth's surface.

Thermals can be seen to rise from the heating pad, entraining fluid as they rise. Parcels overshoot the level at which they become neutrally buoyant and brush the stratified layer above, generating gravity waves on the inversion (see Fig. 4.7 and Section 4.4) before sinking back into the convecting layer beneath. Successive thermals rise higher as the layer deepens. The net effect of the convection is to erode the vertical stratification, returning the fluid to a state of neutral stability—in this case a state in which the temperature of the convecting

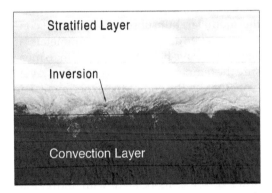

FIGURE 4.7.   A snapshot of the convecting boundary layer in the laboratory experiment. Note the undulations on the inversion caused by convection overshooting the well mixed layer below into the stratified layer above.

layer is close to uniform, as sketched in the schematic on the right side of Fig. 4.6.

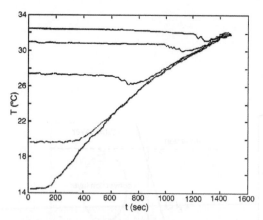

FIGURE 4.8.   Temperature time series measured by five thermometers spanning the depth of the fluid at equal intervals. The lowest thermometer is close to the heating pad. We see that the ambient fluid initially has a roughly constant stratification, somewhat higher near the top than in the body of the fluid. The heating pad was switched on at $t = 150$ sec. Note how all the readings converge onto one line as the well mixed convection layer deepens over time.

Figure 4.8 shows time series of T measured by thermometers at various heights above the heating pad (see legend for details). Initially, there is a temperature difference of 18°C from top to bottom. After the heating pad is switched on, T increases with time, first for the lowermost thermometer, but subsequently, as the convecting layer deepens, for thermometers at each successive height as they begin to measure the temperature of the convecting layer. Note that by the end of the experiment T is rising simultaneously at all heights within the convection layer. We see then that the convection layer is well mixed, or essentially of uniform temperature. Closer inspection of the $T(t)$ curves reveals fluctuations of order $\pm\,0.1°C$ associated with individual convective events within the fluid. Note also that T increases at a rate that is less than linear (this is the subject of Problem 3 at the end of this chapter).

Law of vertical heat transport

We can use the energetic considerations discussed in Section 4.2.3 to develop a sim-

ple "law of vertical heat transport" for the convection in our tank, which turns out to be a useful model of heat transport in cumulus convection. To quantify the transport of heat (or of any other fluid property) we need to define its *flux*. Since the quantity of interest here is the vertical flux, consider fluid moving across a horizontal plane with velocity w; the volume of fluid crossing unit area of the plane during a small time interval δt is just $w\,\delta t$. The heat content of the fluid per unit volume is $\rho c_p T$, where c_p is the specific heat of water; accordingly, the heat flux—the amount of heat transported across unit volume per unit time—is

$$\mathcal{H} = \rho c_p w T. \qquad (4\text{-}8)$$

In a convecting fluid, this quantity will fluctuate rapidly, and so it will be appropriate to average the flux over the horizontal plane and in time over many convective events. In our experiment we can think of half of the fluid at any level moving upward with typical velocity w_c and temperature $T + \Delta T$, and equal amounts of cool fluid moving downward with velocity $-w_c$ and temperature T. Then the net flux, averaged horizontally, is just $\frac{1}{2}\rho_{ref} c_p w_c \Delta T$.

Now we found that the change in potential energy resulting from the interchange of the two small parcels of (incompressible) fluid is given by Eq. 4.7. Let us now assume that the potential energy released in convection (as light fluid rises and dense fluid sinks) is acquired by the kinetic energy (KE) of the convective motion:

$$KE = 3 \times \frac{1}{2}\rho_{ref} w_c^2 = \Delta PE$$
$$= -g\,(\rho_2 - \rho_1)\,(z_2 - z_1)$$

where we have assumed that the convective motion is isotropic in the three directions of space with typical speed w_c. Using our equation of state for water, Eq. 4-4, we may simplify this to:

$$w_c^2 \simeq \frac{2}{3}\alpha g \Delta z \Delta T \qquad (4\text{-}9)$$

where ΔT is the difference in temperature between the upwelling and downwelling parcels that are exchanged over a height $\Delta z = z_2 - z_1$. Using Eq. 4-9 in Eq. 4-8 yields the following "law" of vertical heat transfer for the convection in our tank:

$$H = \frac{1}{2}\rho_{ref} c_p w \Delta T \simeq \frac{1}{2}\rho_{ref} c_p \left(\frac{2}{3}\alpha g \Delta z\right)^{1/2} \Delta T^{3/2}.$$

(4-10)

In the convection experiment shown in Fig. 4.7, the heating pad supplied energy at around $H = 4000\,\mathrm{W\,m^{-2}}$. If the convection penetrates over a vertical scale $\Delta z = 0.2\,\mathrm{m}$, then Eq. 4-10 implies $\Delta T \simeq 0.1\,\mathrm{K}$ if $\alpha = 2 \times 10^{-4}\,\mathrm{K^{-1}}$ and $c_p = 4000\,\mathrm{J\,kg^{-1}\,K^{-1}}$. Eq. 4-9 then implies a parcel speed of $\simeq 0.5\,\mathrm{cm\,s^{-1}}$. This is not atypical of what is observed in the laboratory experiment.

Thus convection transfers heat vertically away from the pad. Even though the convection layer is very well mixed, with small temperature fluctuations (as can be seen in the $T(t)$ observations in Fig. 4.8) the motions are sufficiently vigorous to accomplish the required transfer.

4.3. DRY CONVECTION IN A COMPRESSIBLE ATMOSPHERE

Before we can apply the foregoing ideas to atmospheric convection, we must take into account the fact that the atmosphere is a compressible fluid in which $\rho = \rho(p, T)$; specifically, since the atmosphere closely obeys the perfect gas law, $\rho = p/RT$. For now we will assume a dry atmosphere, deferring consideration of the effects of moisture until Section 4.5. The parcel and environmental pressure, temperature, and density at $z = z_1$ in Fig. 4.5 are $p_1 = $

$p(z_1)$, $T_1 = T(z_1)$, and $\rho_1 = p_1/RT_1$. The real difference from the incompressible case comes when we consider the adiabatic displacement of the parcel to z_2. As the parcel rises, it moves into an environment of lower pressure. The parcel will adjust to this pressure; in doing so it will expand, doing work on its surroundings, and thus cool. So the *parcel temperature is not conserved* during displacement, even if that displacement occurs adiabatically. To compute the buoyancy of the parcel in Fig. 4.5 when it arrives at z_2, we need to determine what happens to its temperature.

4.3.1. The adiabatic lapse rate (in unsaturated air)

Consider a parcel of ideal gas of unit mass with a volume V, so that $\rho V = 1$. If an amount of heat, δQ, is exchanged by the parcel with its surroundings then applying the first law of thermodynamics $\delta Q = dU + dW$, where dU is the change in energy and dW is the change in external work done,[4] gives us

$$\delta Q = c_v dT + p\, dV, \qquad (4\text{-}11)$$

where $c_v dT$ is the change in internal energy due to a change in parcel temperature of dT and $p\, dV$ is the work done by the parcel on its surroundings by expanding an amount dV. Here c_v is the specific heat at constant volume.

Our goal now is to rearrange Eq. 4-11 to express it in terms of dT and dp so that we can deduce how dT depends on dp. To that end we note that, because $\rho V = 1$,

$$dV = d\left(\frac{1}{\rho}\right) = -\frac{1}{\rho^2} d\rho.$$

[4] Rudolf Clausius (1822–1888) the Polish physicist, brought the science of thermodynamics into existence. He was the first to precisely formulate the laws of thermodynamics stating that the energy of the universe is constant and that its entropy tends to a maximum. The expression $\delta Q = dU + dW$ is due to Clausius.

Thus

$$pdV = -\frac{p}{\rho^2}d\rho.$$

Repeated use of $p = \rho RT$ yields

$$dp = RT\,d\rho + \rho R\,dT,$$

and

$$p\,dV = -\frac{p}{\rho^2}d\rho = -\frac{p}{\rho^2 RT}dp + \frac{p}{\rho T}dT$$

$$= -\frac{dp}{\rho} + R\,dT.$$

The first law, Eq. 4-11 can then be written:

$$\delta Q = (R + c_v)\,dT - \frac{dp}{\rho} \qquad (4\text{-}12)$$

$$= c_p\,dT - \frac{dp}{\rho},$$

where $c_p = R + c_v$ is the specific heat at constant pressure.

For adiabatic motions, $\delta Q = 0$, whence

$$c_p\,dT = \frac{dp}{\rho}. \qquad (4\text{-}13)$$

Now if the environment is in hydrostatic balance then, from Eq. 3-3, $dp = -g\rho_E\,dz$, where ρ_E is the density of the environment (since the parcel and environmental pressures must be locally equal). Before being perturbed, the parcel's density was equal to that of the environment. If the displacement of the parcel is sufficiently small, its density is still almost equal to that of the environment, $\rho \simeq \rho_E$, and so under adiabatic displacement the parcel's temperature will change according to

$$\frac{dT}{dz} = -\frac{g}{c_p} = -\Gamma_d, \qquad (4\text{-}14)$$

where Γ_d is known as the *dry adiabatic lapse rate*, the rate at which the parcel's temperature decreases with height under adiabatic displacement. Given $c_p = 1005$ J kg^{-1}K^{-1} (Table 1.4), we find $\Gamma_d \simeq 10$ K km^{-1}.

To determine whether the parcel experiences a restoring force on being displaced from z_1 to z_2 in Fig. 4.5, we must compare its density to that of the environment. At z_2, the environment has pressure p_2, temperature $T_2 \simeq T_1 + (dT/dz)_E\,\delta z$, where $(dT/dz)_E$ is the environmental lapse rate, and density $\rho_2 = p_2/RT_2$. The parcel, on the other hand, has pressure p_2, temperature $T_P = T_1 - \Gamma_d\delta z$, and density $\rho_P = p_2/RT_P$. Therefore the parcel will be positively, neutrally, or negatively buoyant according to whether T_P is greater than, equal to, or less than T_2. Thus our stability condition can be written

$$\left.\begin{array}{c} \text{unstable} \\ \text{neutral} \\ \text{stable} \end{array}\right\} \text{ if } \left(\frac{dT}{dz}\right)_E \left\{\begin{array}{c} < -\Gamma_d \\ = -\Gamma_d \\ > -\Gamma_d \end{array}\right. .$$
$$(4\text{-}15)$$

Therefore, a compressible atmosphere is **unstable if temperature decreases with height faster than the adiabatic lapse rate**. This is no longer a simple "top-heavy" criterion (as we saw in Section 3.3, atmospheric density must decrease with height under all circumstances). Because of the influence of adiabatic expansion, the temperature must decrease with height more rapidly than the finite rate Γ_d for instability to occur.

The lower tropospheric lapse rate in the tropics is, from Fig. 4.9,

$$\left(\frac{dT}{dz}\right)_E \simeq \frac{T(500\text{mbar}) - T(1000\text{mbar})}{Z(500\text{mbar}) - Z(1000\text{mbar})}$$

$$= \frac{(270 - 295)\text{ K}}{(5.546 - 0.127)\text{ km}}$$

$$\simeq -4.6\,\text{K km}^{-1},$$

or about 50% of the adiabatic value. On the basis of our stability results, we would expect no convection, and thus no convective heat transport. In fact the tropical atmosphere, and indeed the atmosphere as a whole, is almost always *stable* to dry convection; the situation is as sketched in the schematic, Fig. 4.10. We will see in Section 4.5 that it is the release of latent heat, when water vapor condenses on expansion and cooling, that leads to convective

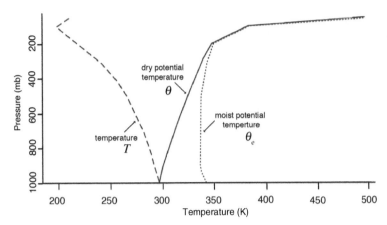

FIGURE 4.9.   Climatological atmospheric temperature T (dashed), potential temperature θ (solid), and moist potential temperature θ_e (dotted) as a function of pressure, averaged over the tropical belt ±30°

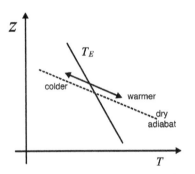

FIGURE 4.10.   The atmosphere is nearly always stable to dry processes. A parcel displaced upwards (downwards) in an adiabatic process moves along a dry adiabat (the dotted line) and cools down (warms up) at a rate that is faster than that of the environment, $\partial T_E/\partial z$. Since the parcel always has the same pressure as the environment, it is not only colder (warmer) but also denser (lighter). The parcel therefore experiences a force pulling it back toward its reference height.

instability in the troposphere and thus to its ability to transport heat vertically. But before going on, we will introduce the very important and useful concept of potential temperature, a temperature-like variable that *is* conserved in adiabatic motion. This will enable us to simplify the stability condition.

4.3.2. Potential temperature

The nonconservation of T under adiabatic displacement makes T a less-than-ideal

measure of atmospheric thermodynamics. However, we can identify a quantity called *potential temperature* that *is* conserved under adiabatic displacement.

Using the perfect gas law Eq. 1.1, our adiabatic statement, Eq. 4-13, can be rearranged thus

$$c_p dT = RT\frac{dp}{p},$$

$$\frac{dT}{T} = \frac{R}{c_p}\frac{dp}{p} = \kappa\frac{dp}{p},$$

where $\kappa = R/c_p = 2/7$ for a perfect diatomic gas like the atmosphere. Thus, noting that $d\ln x = dx/x$, the last equation can be written

$$d\ln T - \kappa d\ln p = 0 \quad \text{or} \quad \frac{T}{p^\kappa} = \text{const.} \quad (4\text{-}16)$$

Potential temperature, θ, is defined as

$$\theta = T\left(\frac{p_0}{p}\right)^\kappa; \qquad (4\text{-}17)$$

by convention, the constant reference pressure p_0 is taken to be 1000 mbar. It then follows from Eq. 4-17, using Eq. 4-16, that under adiabatic displacements

$$\frac{d\theta}{\theta} = \frac{dT}{T} - \kappa\frac{dp}{p} = 0.$$

Unlike T, θ is conserved in a compressible fluid under (dry) adiabatic conditions.

From its definition, Eq. 4-17, we can see that θ is the temperature a parcel of air would have if it were expanded or compressed adiabatically from its existing p and T to the standard pressure p_0. It allows one, for example, to directly determine how the temperature of an air parcel will change as it is moved around adiabatically; if we know its θ, all we need to know at any instant is its pressure, and then Eq. 4-17 allows us to determine its temperature at that instant. For example, from the climatological profile shown in Fig. 4.9, a parcel of air at 300 mbar has $T = 229$ K ($-44°$C) and $\theta = 323$K; if the parcel were brought down to the ground ($p = p_0$) adiabatically, thus conserving θ, its temperature would be $T = \theta = 323$ K ($50°$C).

We can express the stability of the column to dry adiabatic processes in terms of θ as follows. Let's return to our air parcel in Fig. 4.5. At the undisturbed position z_1, it has environmental temperature and pressure and therefore also environmental potential temperature $\theta_1 = \theta_E(z_1)$, where $\theta_E(z)$ is the environmental profile. Since the parcel preserves θ in adiabatic motion, it still has $\theta = \theta_1$ when displaced to z_2. The parcel pressure is the same as that of its environment and so, from Eq. 4-17, it is warmer (or cooler) than its environment according to whether θ_1 is greater (or lesser) than $\theta_E(z_2)$. Since $\theta_E(z_2) \simeq \theta_E(z_1) + (d\theta/dz)_E \delta z = \theta_1 + (d\theta/dz)_E \delta z$, the parcel is

$$\left. \begin{matrix} \text{unstable} \\ \text{neutral} \\ \text{stable} \end{matrix} \right\} \text{ if } \left(\frac{d\theta}{dz} \right)_E \left\{ \begin{matrix} < 0 \\ = 0 \\ > 0 \end{matrix} \right. . \quad (4\text{-}18)$$

Note that Eq. 4-18 has the same form as Eq. 4-6 for an incompressible fluid, but is now expressed in terms of θ rather than T. So another way of expressing the instability criterion is that a compressible atmosphere is unstable if **potential temperature decreases with height**.

Figure. 4.9 shows climatological T and θ as functions of pressure up to 100 mbar, averaged over the tropical belt. Note that $d\theta/dz > 0$, but that $dT/dz < 0$. As noted earlier, we see that the climatological state of the atmosphere is stable to *dry* convection. However, dry convection is often observed in hot arid regions, such as deserts (e.g., the Sahara desert or Arizona) where the surface can become very hot and dry. This state of affairs is sketched in Fig. 4.11. Air parcels rise from the surface and follow a dry adiabat (conserving potential temperature) until their temperature matches that of the environment, when they will become neutrally buoyant. (In reality, the rising parcels have nonzero momentum, so they may overshoot the level of neutral buoyancy, just as observed in our laboratory convection experiment. Conversely, they may also entrain cooler air from the environment, thus reducing their buoyancy and limiting their upward penetration.) So in Fig. 4.11 (left), if the surface temperature is T_1 (or T_2), convection will extend to an altitude z_1 (or z_2). This is the atmospheric and therefore compressible analogue of the convection of an incompressible fluid (water) studied in GFD Lab II, Section 4.2.4. The analogy becomes even clearer if one views the same process in terms of θ, as sketched in Fig. 4.11 (right). The convective layer is of uniform θ corresponding to neutral stability, just as observed in the laboratory experiment (cf. Fig. 4.6).

4.4. THE ATMOSPHERE UNDER STABLE CONDITIONS

4.4.1. Gravity waves

The ball perched on the peak in Fig. 4.3 is unstable, just like the atmosphere under convectively unstable conditions. We have seen, however, that the atmosphere is mostly stable to dry processes (i.e., $dT/dz > -\Gamma_d$) and so the analogy is with the ball in the valley: when disturbed, a dry air parcel will simply oscillate about its mean position.

To analyze this situation, we once again consider the buoyancy forces on a displaced

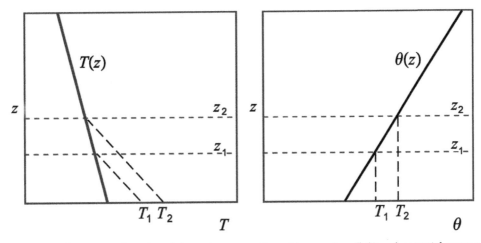

FIGURE 4.11. Dry convection viewed from the perspective of temperature (left) and potential temperature (right). Air parcels rise from the surface and follow a dry adiabat until their temperature matches that of the environment, when they will become neutrally buoyant. If the surface temperature is T_1 (or T_2), convection will extend to an altitude z_1 (or z_2). The same process viewed in terms of potential temperature is simpler. The stable layer above has $d\theta/dz > 0$; convection returns the overturning layer to a state of uniform θ corresponding to neutral stability, just as observed in the laboratory experiment (cf. the right frame of Fig. 4.6). Note that by definition, $\theta = T$ at $p = 1000\,\mathrm{hPa}$.

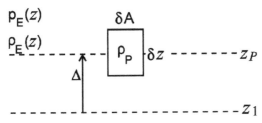

FIGURE 4.12. A parcel displaced a distance Δ from height z_1 to height z_P. The density of the parcel is ρ_P, and that of the environment, ρ_E.

air parcel. Consider Fig. 4.12. An air parcel has been displaced upward adiabatically a distance Δ from level z_1 to level $z_P = z_1 + \Delta$. The environment has density profile $\rho_E(z)$, and a corresponding pressure field $p_E(z)$ in hydrostatic balance with the density. The parcel's pressure must equal the environmental pressure; the parcel's density is $\rho_P = p_P/RT_P = p_E(z_P)/RT_P$. We suppose the parcel has height δz and cross-sectional area δA.

Now the forces acting on the parcel are (following the arguments and notation given in Section 3.2):

i) gravity, $F_g = -g\,\rho_P\,\delta A \delta z$ (downward), and

ii) net pressure force
$F_T + F_B = -\delta p_E\,\delta A = g\rho_E \delta A \delta z$ (upward)

where we have used hydrostatic balance. Hence the net force on the parcel is

$$F_g + F_T + F_B = g\,(\rho_E - \rho_P)\,\delta A\,\delta z.$$

The parcel's mass is $\rho_P \delta z \delta A$, and the equation of motion for the parcel is therefore

$$\rho_P\,\delta z\,\delta A\,\frac{d^2\Delta}{dt^2} = g\,(\rho_E - \rho_P)\,\delta A\,\delta z,$$

so that

$$\frac{d^2\Delta}{dt^2} = -g\left(\frac{\rho_P - \rho_E}{\rho_P}\right). \qquad (4\text{-}19)$$

The quantity $b = -g\,(\rho_P - \rho_E)/\rho_P$ is, of course, the buoyancy of the parcel as defined in Eq. 4-3: if $\rho_P > \rho_E$ the parcel will be negatively buoyant. Now, because the parcel always has the same pressure as the environment, we may write its buoyancy, using the ideal gas law, Eq. 1-1, and the definition of potential temperature, Eq. 4-17, as

$$b = g\frac{\rho_E - \rho_P}{\rho_P} = \frac{g}{T_E}(T_P - T_E) = \frac{g}{\theta_E}(\theta_P - \theta_E).$$
(4-20)

For small Δ, $\theta_E(z + \Delta) = \theta_E(z_1) + \Delta d\theta_E/dz$. Moreover, since the potential temperature of the parcel is conserved, $\theta_P = \theta_E(z_1)$, and so $\theta_P - \theta_E = -\Delta d\theta_E/dz$, enabling Eq. 4-19 to be written:

$$\frac{d^2\Delta}{dt^2} = -\frac{g}{\theta_E}\frac{d\theta_E}{dz}\Delta = -N^2\Delta,$$
(4-21)

where we define

$$N^2 = \frac{g}{\theta_E}\frac{d\theta_E}{dz},$$
(4-22)

which depends only on the vertical variation of θ_E. Note that, under the stable conditions assumed here, $N^2 > 0$, so N is real. Then Eq. 4-21 has *oscillatory* solutions of the form

$$\Delta = \Delta_1 \cos Nt + \Delta_2 \sin Nt,$$

where Δ_1 and Δ_2 are constants set by initial conditions, and N, defined by Eq. 4-22, is the frequency of the oscillation. It is for this reason that the quantity N (with units of s^{-1}) is known as the *buoyancy frequency*.

Thus in the stable case, the restoring force associated with stratification allows the existence of waves, which are known as *internal gravity waves*, and which are in fact analogous to those commonly seen on water surfaces. The latter, known as *surface gravity waves*, owe their existence to the stable, "bottom heavy," density difference at the water-air interface. Internal gravity waves owe their existence to a continuous, internal, stable stratification.

Under typical tropospheric conditions (see the θ profile in Fig. 4.9), we estimate

$$N^2 = \frac{9.81\,\mathrm{m\,s^{-2}}}{300\,\mathrm{K}} \times \frac{340\,\mathrm{K} - 300\,\mathrm{K}}{10^4\,\mathrm{m}}$$
$$= 1.3 \times 10^{-4}\,\mathrm{s^{-2}},$$

whence internal gravity waves in the atmosphere have a typical period of $2\pi/N \simeq$ 9 min.[5] Internal gravity waves are ubiquitous in the atmosphere and are continually excited by, for example, horizontal winds blowing over hills and mountains, and convective plumes buffeting a stable layer above, among other things. On occasion, when the air is nearly saturated, they are made visible by the presence of regular bands of clouds in the crest of each wave (note, not all visible bands of clouds are produced in this way). These clouds often have a characteristic "lens" shape, hence the name *lenticular*. (See Figs. 4.13 and 4.14.) One can (very roughly) estimate an expected horizontal wavelength for the waves as follows. If each parcel of air oscillates at frequency N but is carried along by the wind at speed u such that a stationary pattern results, as in Fig. 4.15, then the expected wavelength is $2\pi u/N$, around 5–10 km for the N estimated above and a wind of 10–20 m s^{-1}. Regular cloud features of this type are especially dramatic in the vicinity of mountain ranges (such as the Sierra Nevada and the Continental Divide of North America), as illustrated in Fig. 4.13 and shown in the photograph in Fig. 4.14. Another rather spectacular example of internal gravity waves in the atmosphere is shown in Fig. 4.15 and has an uncanny resemblance to the wake left by a ship (superimposed on the figure).

We shall see in Chapter 9 that the interior of the ocean is also stably stratified; internal gravity waves are a ubiquitous feature of the ocean too. Features indicative of such waves have also been observed in the Martian and Jovian atmospheres.

4.4.2. Temperature inversions

In abnormal situations in the troposphere in which T increases with height, the atmosphere is very stable; the restoring force

---

[5]The frequency given by Eq. 4-22 is an upper limit. We have considered the case in which the parcels of air oscillate exactly vertically. For parcel oscillations at an angle α to the vertical, the frequency is $N \cos\alpha$.

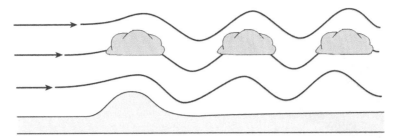

FIGURE 4.13. A schematic diagram illustrating the formation of mountain waves (also known as lee waves). The presence of the mountain disturbs the air flow and produces a train of downstream waves (cf. the analogous situation of water in a river flowing over a large submerged rock, producing a downstream surface wave train). Directly over the mountain, a distinct cloud type known as lenticular ("lens-like") cloud is frequently produced. Downstream and aloft, cloud bands may mark the parts of the wave train in which air has been uplifted (and thus cooled to saturation).

FIGURE 4.14. Looking downwind at a series of lenticular wave clouds in the lee of the Continental Divide of North America. Photo courtesy of Dale Durran, University of Washington.

on a lifted air parcel is large, and the atmosphere is thus particularly resistive to vertical motion. Such "inversions" can be produced in several ways. Low-level inversions (at altitudes of a few hundred meters) are commonly produced during calm winter nights from radiative cooling of the surface (see Fig. 4.16a).

Note that the inversion may be self-reinforcing; under conditions of slight wind, turbulence carries heat from aloft and limits the cooling of the surface. If an inversion forms, the resistance to vertical motion suppresses the turbulence and allows the surface layer to cool, thus strengthening the

inversion. (Fog may then form in the cold surface layer below the inversion.) Apart from its thermal impact, the inversion may trap surface air and thus allow pollutants to build up in the surface layer.

A second type of low-level inversion, common in many subtropical regions of the Earth, is known as the *trade-wind*, or just *trade*, inversion. As we shall see in Chapter 5, air in the subtropics is, on average, descending and thus *warms* adiabatically, according to Eq. 4-14, as it does so. As shown in Fig. 4.16b, this can produce a persistent inversion (at altitudes between 400m and 2 km, depending on location). Many

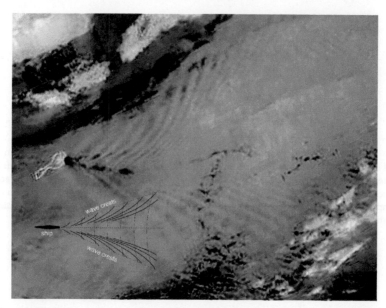

FIGURE 4.15.   Atmospheric gravity waves formed in the lee of Jan Mayen island (only 50 km long, situated 375 miles north-northeast of Iceland), observed in February 2000. The wind is blowing from the WSW. A volcano—called Beerenberg—forms the north end of the island and rises to a height of over 2 km. Note the similarity between the atmospheric wake and that formed on a water surface by a ship, superimposed on the figure.

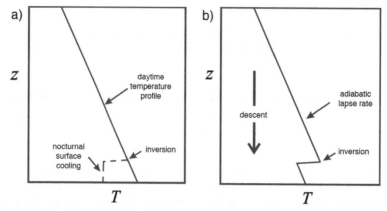

FIGURE 4.16.   (a) Low-level inversions are commonly produced during calm winter nights from radiative cooling of the surface. (b) A trade inversion created by descent and adiabatic warming typical of subtropical regions.

subtropical areas have pollution problems that are exacerbated by the trade inversion (see Fig. 4.17). In some cases, the vertical trapping of air by the inversion can be compounded by horizontal trapping by mountains. Los Angeles and Mexico City (and many other cities) suffer from this situation.

4.5. MOIST CONVECTION

We have seen that the atmosphere in most places and at most times is stable to dry convection. Nevertheless, convection is common in most locations over the globe (more so in some locations than in others, as we will discuss in Section 4.6.2). There is

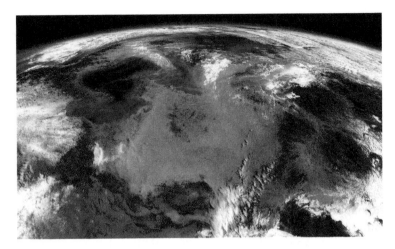

FIGURE 4.17.   (Top) A satellite image showing dense haze associated with pollution over eastern China. The view looks eastward across the Yellow Sea toward Korea. Provided by the SeaWiFS Project, NASA/Goddard Space Flight Center. (Bottom) Temperature inversions in Los Angeles often trap the pollutants from automobile exhaust and other pollution sources near the ground.

one very important property of air that we have not yet incorporated into our discussion of convection. Air is *moist*, and if a moist air parcel is lifted, it cools adiabatically; if this cooling is enough to saturate the parcel, some water vapor condenses to form a cloud. The corresponding latent heat release adds buoyancy to the parcel, thus favoring instability. This kind of convection is called *moist convection*. To derive a stability

condition for moist convection, we must first discuss how to describe the moisture content of air.

4.5.1. Humidity

The moisture content of air is conveniently expressed in terms of humidity. The **specific humidity**, q, is a measure of the mass of water vapor to the mass of air per unit volume defined thus:

$$q = \frac{\rho_v}{\rho},\qquad (4\text{-}23)$$

where $\rho = \rho_d + \rho_v$ (see Section 1.3.2) is the total mass of air (dry air plus water vapor) per unit volume. Note that in the absence of mixing or of condensation, specific humidity is conserved by a moving air parcel, since the masses of both water and air within the parcel must be separately conserved.

The **saturation-specific humidity**, q_*, is the specific humidity at which saturation occurs. Since both water vapor and dry air behave as perfect gases, using Eq. 1-2 to express Eq. 4-23 at saturation, we define q_* thus:

$$q_* = \frac{e_s/R_v T}{p/RT} = \left(\frac{R}{R_v}\right)\frac{e_s}{p},\qquad (4\text{-}24)$$

where $e_s(T)$ is the saturated partial pressure of water vapor plotted in Fig. 1.5. Note that q_* is a function of temperature *and* pressure. In particular, at fixed p it is a strongly increasing function of T.

Relative humidity, U, is the ratio of the specific humidity to the *saturation specific humidity*, q_*, often expressed as a percentage thus:

$$U = \frac{q}{q_*} \times 100\% \qquad (4\text{-}25)$$

Now near the Earth's surface the moisture content of air is usually fairly close to saturation (e.g., throughout the lower tropical atmosphere, the relative humidity of air is close to 80%, as will be seen in Chapter 5). If such an air parcel is lifted, the pressure will decrease and it will cool. From Eq. 4-24, decreasing pressure alone would make q_* increase with altitude. However, the exponential dependence of e_s on T discussed in Chapter 1 overwhelms the pressure dependence, and consequently q_* decreases rapidly with altitude. So as the air parcel is lifted, conserving its q, it does not usually have to rise very far before $q > q_*$.

The level at which this occurs is called the *condensation level* z_c (Fig. 4.18). At and above z_c, excess vapor will condense so that $q = q_*$. Moreover, since q_* will continue to decrease as the parcel is lifted further, q will decrease correspondingly. Such condensation is visible, for example, as convective clouds. As the vapor condenses, latent heat release partly offsets the cooling due to adiabatic expansion. Thus we expect the moist parcel to be more buoyant than if it were dry. As illustrated in Fig. 4.18, above z_c the parcel's temperature falls off more slowly (contrast with the dry convection case, Fig. 4.11) until neutral buoyancy is reached at z_t, the cloud top. Clearly, the warmer or moister the surface air, the higher the cloud top will be.

Below the condensation level we expect a parcel undergoing convection to follow a dry adiabat. But how does its temperature change in the saturated layer above? It follows a *saturated adiabat*, as we now describe.

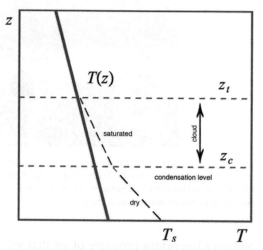

FIGURE 4.18. The temperature of a moist air parcel lifted in convection from the surface at temperature T_s will follow a dry adiabat until condensation occurs at the condensation level z_c. Above z_c, excess vapor will condense, releasing latent heat and warming the parcel, offsetting its cooling at the dry adiabatic rate due to expansion. Thus a moist parcel cools less rapidly (following a moist adiabat) than a dry one, until neutral buoyancy is reached at z_t, the cloud top. This should be compared to the case of dry convection shown in Fig. 4.11.

4.5.2. Saturated adiabatic lapse rate

Let us return to the case of a small vertical displacement of an air parcel. If the air is unsaturated, no condensation will occur, and so the results of Section 4.3.1 for dry air remain valid. However, if condensation does occur, there will be a release of latent heat in the amount $\delta Q = -L\,dq$ per unit mass of air, where L is the latent heat of condensation and dq is the change in specific humidity q. Thus we must modify Eq. 4-13 to

$$c_p\,dT = \frac{dp}{\rho} - L\,dq \qquad (4\text{-}26)$$

for an air parcel undergoing moist adiabatic displacement. Note that there is a minus sign here because if $dq < 0$, latent heat is released and the parcel warms. If the environment is in hydrostatic balance, $dp/\rho = -gdz$, and so

$$d\left(c_p T + gz + Lq\right) = 0. \qquad (4\text{-}27)$$

The term in parentheses is known as the *moist static energy* and comprises $c_p T + gz$, the dry static energy, and Lq, the latent heat content.

If the air parcel is always at saturation, we can replace q by q_* in Eq. 4-26. Now, since $q_* = q_*(p,T)$,

$$dq_* = \frac{\partial q_*}{\partial p}dp + \frac{\partial q_*}{\partial T}dT .$$

From Eq. 4-24,

$$\frac{\partial q_*}{\partial p} = -\left(\frac{R}{R_v}\right)\frac{e_s}{p^2} = -\frac{q_*}{p}$$

$$\frac{\partial q_*}{\partial T} = \left(\frac{R}{R_v}\right)\frac{1}{p}\frac{de_s}{dT} = \left(\frac{R}{R_v}\right)\frac{\beta e_s}{p} = \beta q_* ,$$

where we have used Eq. 1-4 to write $de_s/dT = \beta e_s$. Hence Eq. 4-26 gives setting $dq = dq_*$,

$$\left[c_p + L\beta q_*\right]dT = \frac{dp}{\rho}\left[1 + Lq_*\frac{\rho}{p}\right] .$$

Writing $dp/\rho = -g\,dz$ and rearranging,

$$-\frac{dT}{dz} = \Gamma_s = \Gamma_d\left[\frac{1 + Lq_*/RT}{1 + \beta Lq_*/c_p}\right] , \qquad (4\text{-}28)$$

where Γ_s is known as the *saturated adiabatic lapse rate.*[6] The factor in brackets on the right is always less than unity, so the saturated adiabatic lapse rate is less than the dry adiabatic lapse rate; at high altitudes, however, q_* is small and the difference becomes very small. Since q_* varies with p and T, one cannot ascribe a single number to Γ_s. It has typical tropospheric values ranging between $\Gamma_s \simeq 3\,\mathrm{K\,km^{-1}}$ in the moist, tropical lower troposphere and $\Gamma_s \simeq \Gamma_d = 10\,\mathrm{K\,km^{-1}}$ in the upper troposphere. A typical atmospheric temperature profile is sketched, along with dry and saturated adiabats, in Fig. 4.19.

The qualitative impact of condensation is straightforward; the release of latent heat makes the air parcel warmer and therefore more buoyant, and so the atmosphere is

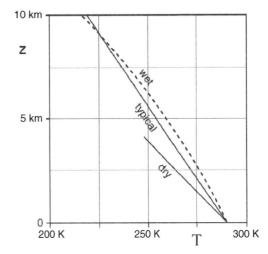

FIGURE 4.19. A schematic of tropospheric temperature profiles showing the dry adiabat, a typical wet adiabat, and a typical observed profile. Note that the dry adiabatic ascent of a parcel is typically cooler than the surroundings at all levels, whereas the wet adiabat is warmer up to about 10 km. The wet and dry lapse rates are close to one another in the upper troposphere, where the atmosphere is rather dry.

---

[6]Γ_s is also known as the *pseudo-adiabatic lapse rate.*

destabilized by the presence of moisture, i.e., a saturated atmosphere is unstable if

$$\frac{dT}{dz} < -\Gamma_s, \qquad (4\text{-}29)$$

where $\Gamma_s < \Gamma_d$. The resulting instability is known as *conditional instability*, since it is conditional on the air being saturated. The tropical troposphere is close to neutrality with respect to moist convection, meaning that it has $\partial T/\partial z \simeq -\Gamma_s$ (see below).

Lines that show the decrease in T of a parcel of air which is rising/sinking in the atmosphere under saturated adiabatic conditions are called saturated adiabats. As we now describe, we can define a temperature-like quantity that is conserved in moist processes and plays an analogous role to that of potential temperature in dry convection.

4.5.3. Equivalent potential temperature

Moist thermodynamics is complicated, but it is relatively straightforward to define a potential temperature that is conserved in moist processes. This quantity, known as equivalent potential temperature, θ_e, tends to be mixed by moist convection, just as dry potential temperature, θ, is mixed in dry convection.

We begin from the first law, Eq. 4-26. Making use of $p = \rho R T$ and $\kappa = R/c_p$, it can be rearranged thus:

$$d \ln T = \kappa d \ln p - \frac{L}{c_p T} dq .$$

From the definition of potential temperature, Eq. 4-17, $\ln \theta = \ln T - \kappa \ln p +$ constant, and so this can be written

$$d \ln \theta = -\frac{L}{c_p T} dq \simeq -d \left(\frac{Lq}{c_p T} \right),$$

where we have made the approximation of slipping the factor $L/c_p T$ inside the derivative, on the grounds that the fractional change in temperature is much less than that of specific humidity (this approximation is explored in Problem 6 at the end of

the chapter). Then we may conveniently define equivalent potential temperature to be

$$\theta_e = \theta \exp \left(\frac{Lq}{c_p T} \right), \qquad (4\text{-}30)$$

such that $d\theta_e = 0$ in adiabatic processes. The utility of θ_e is that:

1. It is conserved in both dry and wet adiabatic processes. (In the absence of condensation, q is conserved; hence both θ and θ_e are conserved. If condensation occurs, $q \to q_*(p,T)$; then θ_e is conserved but θ is not.)

2. If the air is dry, it reduces to dry potential temperature ($\theta_e \to \theta$ when $q \to 0$).

3. Vertical gradients of θ_e tend to be mixed away by moist convection, just like the T gradient in GFD Lab II.

This last point is vividly illustrated in Fig. 4.9, where climatological vertical profiles of T, θ and θ_e are plotted, averaged over the tropical belt $\pm 30°$ (see also Fig. 5.9). Note how θ increases with height, indicating that the tropical atmosphere is stable to dry convection. By contrast, the gradient of θ_e is weak, evidence that moist convection effectively returns the tropical atmosphere to a state that is close to neutral with respect to moist processes ($d\theta_e/dp \simeq 0$).

Having derived conditions for convective instability of a moist atmosphere, let us now review the kinds of convection we observe in the atmosphere and their geographical distribution.

4.6. CONVECTION IN THE ATMOSPHERE

We have seen that the atmosphere is normally stable in the absence of condensation. Hence most convection in the atmosphere is moist convection, accompanied by saturation and hence cloud formation. Downwelling air parcels do not become saturated because descending air

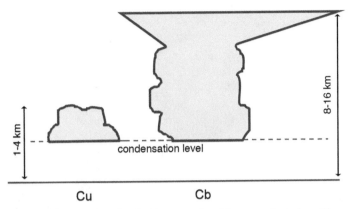

FIGURE 4.20. Schematic of convective clouds: Cu = cumulus; Cb = cumulonimbus. The condensation level is the level above which $q = q_*$. Cb clouds have a characteristic "anvil," where the cloud top spreads and is sheared out by strong upper level winds.

warms adiabatically. They would therefore become positively buoyant, if it were not for radiative cooling, a process that is much slower than latent heat release in the updrafts; so the descent must be slow. Thus moist convection comprises narrow, cloudy, vigorous updrafts, with larger areas of clear, dry, air slowly descending between.

4.6.1. Types of convection

Convective clouds have two main forms: *cumulus* (Cu) clouds (usually small, "fair-weather," and nonprecipitating) and *cumulonimbus* (Cb) clouds (usually associated with thunderstorms and heavy rain, and perhaps hail).[7]

Modest convection is common and usually shallow (up to just a few km) and capped by Cu clouds (see Figs. 4.20 and 4.21). They are typically 1–2 km tall and towers can be seen to grow this far in about 15 min, implying a vertical velocity of around $2\,\mathrm{m\,s^{-1}}$. Typical temperature

fluctuations are of order 0.1 K (see the application of parcel theory below). The sensible heat flux is $\rho c_p \overline{w'T'}$, where the primes denote differences from the time mean, represented by an overbar. Using the above estimates and Table 1.4, we obtain a heat flux of $200\,\mathrm{W\,m^{-2}}$, an impressive number and comparable to the radiative fluxes shown, for example, in Fig. 5.5.

Deep convection is common in the tropics (see Section 4.6.2) and occasionally elsewhere. It is manifested by huge Cb clouds, illustrated in Figs. 4.20, 4.22, and 4.23, the tops of which may reach the tropopause, and become so cold that the cloud top is sheared off by wind to form an *anvil* made of ice crystals. Vertical motions can reach tens of $\mathrm{m\,s^{-1}}$ with temperature fluctuations around 1 K. The vertical heat flux associated with individual cumulonimbus clouds is many $\mathrm{kW\,m^{-2}}$. However, they are intermittent both in space and time. They are the primary mechanism of vertical heat transport in the tropics.

---

[7]  Luke Howard (1772–1864). An English manufacturing chemist and pharmacist, Howard was also an amateur meteorologist. He wrote one of the first textbooks on weather and developed the basis for our cloud classification system; he is responsible for the cloud nomenclature now in standard use.

FIGURE 4.21. A well-distributed population of cumulus clouds over the midwestern United States. Photograph: Russell Windman, May 30, 2002, from an altitude of 35,000 ft.

FIGURE 4.22. A mature cumulonimbus (thunderstorm) cloud producing rain and hail on the Great Plains. The hail core is evident in the bright white streaks (center). As the updrafts rise through the cloud and into noticeably warmer air, the top of the cloud spreads out and flattens (top). From the University Corporation for Atmospheric Research.

Dynamics of cumulus convection

We have seen that at a certain height, rising moist air parcels become saturated and a cloud forms. What happens above the cloud base depends on the type of cloud. Cumulus clouds are observed to mix with their surroundings, entraining ambient air. The result is that they rapidly lose their buoyancy. This is illustrated in Fig. 4.24, where we sketch air parcels ascending a little along wet adiabats, followed by complete mixing (the short horizontal lines).

FIGURE 4.23. A supercell is a giant cumulonimbus storm with a deep rotating updraft. Supercells can produce large amounts of hail, torrential rainfall, strong winds, and sometimes tornadoes. The close-up view of the supercell thunderstorm in the picture shows a bulging dome of clouds extending above the flat, anvil top. This is caused by a very intense updraft that is strong enough to punch through the tropopause and into the stratosphere. At the time of this photograph, baseball-sized hail was falling and a tornado was causing havoc in southern Maryland. Photograph by Steven Maciejewski (April 28, 2002): reported by Kevin Ambrose.

One consequence evident from the figure is that the cumulus top is expected where the wet adiabat first runs parallel to the environmental profile, more or less observed in practice.

A simple model of heat transport in cumulus convection can be constructed as follows, in the spirit of the parcel theory developed in Section 4.2.4 to describe convection in our tank of water in GFD Lab II. From Eq. 4-20 we see that the buoyant acceleration of an air parcel is $g\frac{\Delta T}{T}$, where $\Delta T = T_P - T_E$. If the parcel eventually rises by a height Δz, the PE of the system will have decreased by an amount per unit mass of $g\frac{\Delta T}{T}\Delta z$. Equating this to acquired KE, equally distributed between horizontal and vertical components, we find that $\frac{3}{2}w^2 \simeq g\frac{\Delta z \Delta T}{T}$, which is Eq. 4-9 with α replaced by T^{-1}. The vertical heat transport by a population of cumulus clouds is then given by Eq. 4-10 with α replaced by

T^{-1}. This fairly crude model of convective vertical heat transport is a useful representation and yields realistic values. If convection carries heat at a rate of $200\,\mathrm{W\,m^{-2}}$ from the surface up to $1\,\mathrm{km}$, Eq. 4-10 predicts $\Delta T \simeq 0.1\,\mathrm{K}$ and $w \simeq 1.5\,\mathrm{m\,s^{-1}}$, roughly in accord with what is observed.

Cumulonimbus convection

Suppose we had imagined that the Cu cloud discussed in the previous section had a vertical scale of $10\,\mathrm{km}$ rather than the $1\,\mathrm{km}$ assumed. We will see in Chapter 5 that the wind at a height of $10\,\mathrm{km}$ is some 20–$30\,\mathrm{m\,s^{-1}}$ faster than at cloud base (see Fig. 5.20), and so the cumulus cloud would be ripped apart. However, Cb storms move at the same speed as some middle-level wind. Fig. 4.25 shows the flow relative to a cumulonimbus storm moving along with the wind at midlevels. It overtakes the potentially warm air near the surface, and so, relative to the storm, this air

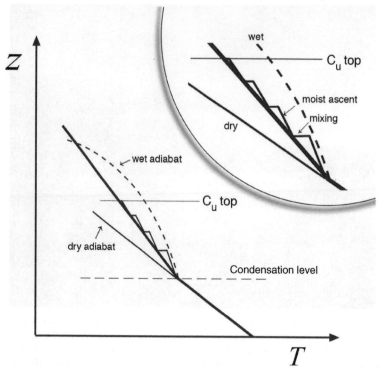

FIGURE 4.24. In cumulus convection, buoyant air parcels ascend along a moist adiabat but repeatedly mix with ambient fluid, reducing their buoyancy. Cumulus top is expected to occur at the level at which the wet adiabat first runs parallel to the environmental curve, as shown in the inset at top right of figure.

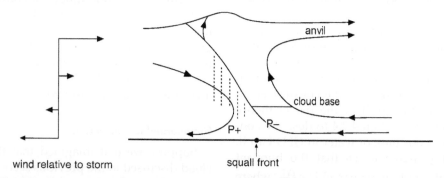

FIGURE 4.25. The pattern of flow relative to a cumulonimbus storm moving along with a middle-level wind. Ahead of the storm, air is sucked in ($p-$), ascends, and is expelled in the anvil. Upper-middle level air approaches the storm from behind and is brought down to the ground ($p+$). The "squall front" is a stagnation point relative to the storm which moves roughly at the speed of the storm. Heavy rain is represented by the vertical dotted lines in the updraught. Modified from Green (1999).

flows strongly toward the storm, ascends in the cloud and is eventually expelled as an anvil in a shallow, fast-moving sheet containing ice crystals. Release of latent heat in condensation (and associated heavy rain) in the updraft creates positive buoyancy and vertical acceleration, powering the motion. A region of low pressure is created at the surface just ahead of the storm, "sucking" low-level flow into it. At the same time, upper-middle level air approaches the storm from behind, is cooled by evaporation from the rain falling into it, and brought down to the ground where it creates a region of high pressure as it decelerates. The "squall front" is a stagnation point relative to the storm, which moves roughly at the speed of the storm.

In contrast to a cumulus cloud, there is hardly any mixing in a cumulonimbus cloud, because the flow is so streamlined. Consequently nearly all the potential energy released goes into kinetic energy of the updraft. Hence $\frac{1}{2}w^2 \sim g\frac{\Delta T}{T}\Delta z$, which yields $w \sim 25\,\mathrm{m\,s^{-1}}$ if $\Delta T \sim 1\,\mathrm{K}$ and $\Delta z \sim 10\,\mathrm{km}$, roughly in accord with observations. Updrafts of this magnitude are strong enough, for example, to suspend hailstones until they grow to a large size.

4.6.2. Where does convection occur?

The short answer to this question is, in fact, almost everywhere. However, deep convection is common in some places and rare in others. In general, tropical rainfall is associated with deep convection, which is most common in the three equatorial regions

where rainfall is most intense (Indonesia and the western equatorial Pacific Ocean, Amazonia, and equatorial Africa). Over the desert regions of the subtropics, it is uncommon. The contrast between these two areas is shown in the distribution of outgoing long-wave radiation (OLR) in Fig. 4.26.

OLR is the total radiative flux in the terrestrial wavelengths, measured by downward-looking satellite instruments. As discussed in Chapter 2, if we can think of this flux as emanating from a single layer in the atmosphere, then we can deduce the temperature of that layer (assuming blackbody radiation, Eq. 2.2). So OLR is a measure of the temperature of the emitting region. Note that the polar regions in Fig. 4.26 have low OLR: this is not very surprising, since these regions are cold. The OLR is also low, however, over the three equatorial regions mentioned previously. The radiation cannot be coming from the surface there, since the surface is warm; it must be (and is) coming from high altitudes (10–15 km), where the temperature is low, even in the tropics. As shown in Fig. 4.27, this happens because the radiation is coming from the tops of deep convective clouds; the low OLR is

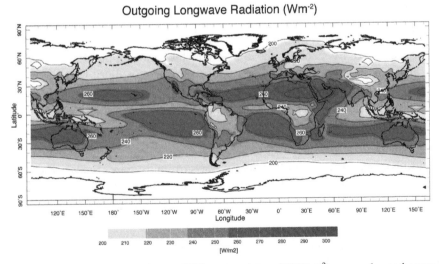

FIGURE 4.26.   Outgoing longwave radiation (OLR: contour interval 20 Wm^{-2}) averaged over the year. Note the high values over the subtropics and low values over the three wet regions on the equator: Indonesia, Amazonia, and equatorial Africa.

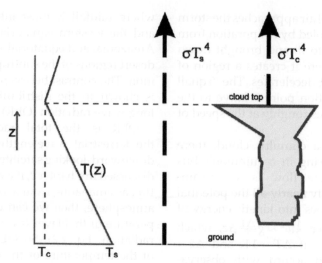

FIGURE 4.27. Schematic of IR radiation from the ground (at temperature $T_s \simeq 300\,\mathrm{K}$ in the tropics) and from the tops of deep convective clouds (at temperature $T_c \simeq 220\,\mathrm{K}$).

indicative of deep convection. Note that in the subtropical regions, especially over the deserts and cooler parts of the ocean, OLR is high; these regions are relatively dry and cloud-free, and the radiation is coming from the warm surface.

Convection requires a warm surface *relative to the environmental air aloft*. This can be achieved by warming the surface by, for example, solar heating, leading to afternoon convection, especially in summer and, most of all, in the tropics where deep convection is very common. However, convection can also be achieved by cooling the air aloft. The latter occurs when winds bring cold air across a warm surface; for example, thunderstorms in middle latitudes are frequently associated with the passage of cold fronts. In middle latitudes, shallow convection is frequent (and is usually visible as cumulus clouds). Deep convection (always associated with heavy rain and often with thunderstorms) is intermittent.

4.7. RADIATIVE-CONVECTIVE EQUILIBRIUM

As we saw in Section 3.1.2, the thermal structure expected on the basis of radiative forcing alone has a temperature discontinuity at the ground, as illustrated in Fig. 3.4; the radiative equilibrium temperature of the ground is considerably warmer than that of the air above. This profile is unstable and convection will occur. Convective motions will transport heat upward from the surface; when the air parcels mix with the environment (as they will), they increase the environmental temperature until the environment itself approaches a state of neutrality, which, in the moist tropical atmosphere, is one of constant moist potential temperature. The tropical troposphere is indeed observed to be close to neutral for moist convection (cf. Fig. 4.9) where convection reaches up to the tropopause at height z_T. The whole tropical troposphere is in a state of *radiative-convective equilibrium*, with a convectively determined state below the tropopause and a radiative equilibrium state above, as sketched in Fig. 3.4.

How the temperature structure of the tropics is conveyed into middle latitudes is less certain. It seems unlikely that local convection is the primary control of the vertical temperature structure in middle latitudes. Here transport by larger scale systems plays an important role. In Chapter 8 we will discuss the role of mid-latitude weather systems in transporting heat, both vertically and horizontally.

4.8. FURTHER READING

The reader is referred to Holton (2004) and especially to Wallace and Hobbs (2006) for more detailed discussions of atmospheric thermodynamics, where many of the more exotic thermodynamic variables are defined and discussed. Emanuel (1994) gives a very thorough and advanced treatment of atmospheric convection.

4.9. PROBLEMS

1. Show that the buoyancy frequency, Eq. 4-22, may be written in terms of the environmental temperature profile thus:

$$N^2 = \frac{g}{T_E}\left(\frac{dT_E}{dz} + \Gamma_d\right),$$

where Γ_d is the dry adiabatic lapse rate.

2. From the temperature (T) profile shown in Fig. 4.9:

   (a) Estimate the tropospheric lapse rate and compare to the dry adiabatic lapse rate.

   (b) Estimate the pressure scale height RT_0/g, where T_o is the mean temperature over the 700 mbar to 300 mbar layer.

   (c) Estimate the period of buoyancy oscillations in mid-troposphere.

3. Consider the laboratory convection experiment described in Section 4.2.4. The thermodynamic equation (horizontally averaged over the tank) can be written:

$$\rho c_p \frac{dT}{dt} = \frac{\mathcal{H}}{h}, \qquad (4\text{-}31)$$

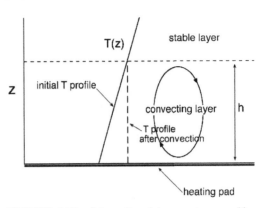

FIGURE 4.28. Schematic of temperature profiles before and after convection in the laboratory experiment GFD Lab II. The initial T profile, increasing linearly with height, is returned to the neutral state, one of uniform temperarture, by convection.

where h is the depth of the convection layer (see Fig. 4.28), $\mathcal{H}$ is the (constant) heat flux coming in at the bottom from the heating pad, ρ is the density, c_p is the specific heat, t is time, and T is temperature.

We observe that the temperature in the convection layer is almost homogeneous and "joins on" to the linear stratification into which the convection is burrowing, as sketched in Fig. 4.28. Show that if this is the case, Eq. 4-31 can be written thus:

$$\frac{\rho c_p \overline{T_z}}{2}\frac{d}{dt}h^2 = \mathcal{H},$$

where $\overline{T}$ is the initial temperature profile in the tank before the onset of convection, and $\overline{T_z} = \frac{d\overline{T}}{dz}$ is the initial stratification, assumed here to be constant.

   (a) Solve the above equation; show that the depth and temperature of the convecting layer increases by $\sqrt{t}$, and sketch the form of the solution.

   (b) Is your solution consistent with the plot of the observed temperature evolution from the laboratory experiment shown in Fig. 4.8?

(c) How would T have varied with time if initially the water in the tank had been of uniform temperature (i.e., was unstratified)? You may assume that the water remains well mixed at all times and so is of uniform temperature.

4. Consider an atmospheric temperature profile at dawn with a temperature discontinuity (inversion) at 1 km, and a tropopause at 11 km, such that

$$T(z) = \begin{cases} 10°C, & z < 1\,km \\ [15 - 8\,(z-1)]°C & 1 < z < 11\,km \\ -65°C & z > 11\,km \end{cases}$$

(where here z is expressed in km). Following sunrise at 6 a.m. until 1 p.m., the surface temperature steadily increases from its initial value of 10°C at a rate of 3°C per hour. Assuming that convection penetrates to the level at which air parcels originating at the surface and rising without mixing attain neutral buoyancy, describe the evolution during this time of convection

(a) if the surface air is completely dry.

(b) if the surface air is saturated.

You may assume the dry/wet adiabatic lapse rate is $10\,K\,km^{-1}/7\,K\,km^{-1}$, respectively.

5. For a perfect gas undergoing changes dT in temperature and dV in specific volume, the change ds in specific entropy, s, is given by

$$T\,ds = c_v dT + p\,dV.$$

(a) Hence, for unsaturated air, show that potential temperature θ

$$\theta = T\left(\frac{p_0}{p}\right)^\kappa,$$

is a measure of specific entropy; specifically, that

$$s = c_p \ln\theta + \text{constant},$$

where c_v and c_p are specific heats at constant volume and constant pressure, respectively.

(b) Show that if the environmental lapse rate is dry adiabatic (Eq. 4-14), it has constant potential temperature.

6. Investigate under what conditions we may approximate $\frac{L}{c_p T}dq$ by $d\left(Lq/c_p T\right)$ in the derivation of Eq. 4-30. Is this a good approximation in typical atmospheric conditions?

7. Assume the atmosphere is in hydrostatic balance and isothermal with temperature 280 K. Determine the potential temperature at altitudes of 5 km, 10 km, and 20 km above the surface. If an air parcel was moved adiabatically from 10 km to 5 km, what would its temperature be on arrival?

8. Somewhere (in a galaxy far, far away) there is a planet whose atmosphere is just like that of the Earth in all respects but one—it contains no moisture. The planet's troposphere is maintained by convection to be neutrally stable to vertical displacements. Its stratosphere is in radiative equilibrium, at a uniform temperature of −80°C, and temperature is continuous across the tropopause. If the surface pressure is 1000 mbar, and equatorial surface temperature is 32°C, what is the pressure at the equatorial tropopause?

9. Compare the dry-adiabatic lapse rate on Jupiter with that of Earth, given that the gravitational acceleration on Jupiter is $26\,m\,s^{-2}$ and its atmosphere is composed almost entirely of hydrogen and therefore has a different value of c_p.

10. In Section 3.3 we showed that the pressure of an isothermal atmosphere varies exponentially with height. Consider now an atmosphere with uniform *potential temperature*. Find how pressure varies with height, and show in particular that such an atmosphere has a discrete top (where $p \to 0$) at altitude $RT_o / (\kappa g)$, where R, κ, and g have their usual meanings, and T_o is the temperature at 1000 mbar pressure.

11. Consider the convective circulation shown in Fig. 4.29. Air rises in the center of the system; condensation occurs at altitude $z_B = 1$ km ($p_B = 880$ mbar), and the convective cell (cloud is shown by the shading) extends up to $z_T = 9$ km ($p_T = 330$ mbar), at which point the air diverges and descends adiabatically in the downdraft region. The temperature at the condensation level, T_B, is 20°C. Assume no entrainment and that all condensate falls out immediately as rain.

    (a) Determine the specific humidity at an altitude of 3 km within the cloud.

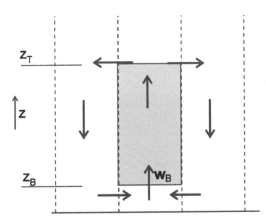

FIGURE 4.29. Air rises in the center of a convective cell with speed w_B at cloud base. Cloud is represented by shading; condensation occurs at altitude z_B and extends up to z_T, at which point the air diverges and descends adiabatically in the downdraft region.

    (b) The upward flux of air, per unit horizontal area, through the cloud at any level z is $w(z)\rho(z)$, where ρ is the density of dry air and w the vertical velocity. Mass balance requires that this flux be independent of height within the cloud. Consider the net upward flux of water vapor within the cloud, and hence show that the rainfall rate below the cloud (in units of mass per unit area per unit time) is $w_B \rho_B (q_{*B} - q_{*T})$, where the subscripts "B" and "T" denote the values at cloud base and cloud top, respectively. If $w_B = 5$ cm s^{-1}, and $\rho_B = 1.0$ kg m^{-3}, determine the rainfall rate in cm per day.

12. Observations show that, over the Sahara, air continuously subsides (hence the Saharan climate). Consider an air parcel subsiding in this region, where the environmental temperature T_e decreases with altitude at the constant rate of 7 K km^{-1}.

    (a) Suppose the air parcel leaves height z with the environmental temperature. Assuming the displacement to be adiabatic, show that, after a time δt, the parcel is warmer than its environment by an amount $w\Lambda_e \delta t$, where w is the subsidence velocity and

    $$\Lambda_e = \frac{dT_e}{dz} + \frac{g}{c_p} ,$$

    where c_p is the specific heat at constant pressure.

    (b) Suppose now that the displacement is not adiabatic, but that the parcel cools radiatively at such a rate that its temperature is *always the same as* its environment (so the circulation is in equilibrium). Show that the radiative rate of energy loss per unit volume must be

$\rho c_p w \Lambda_e$, and hence that the net radiative loss to space per unit horizontal area must be

$$\int_0^\infty \rho c_p w \Lambda_e \, dz = \frac{c_p}{g} \int_0^{p_s} w \Lambda_e \, dp \, ,$$

where p_s is surface pressure and ρ is the air density.

(c) Radiative measurements show that, over the Sahara, energy is being lost to space at a net, annually-averaged rate of $20 \, \text{W m}^{-2}$. Estimate the vertically-averaged (and annually-averaged) subsidence velocity.

5

The meridional structure of the atmosphere

5.1. Radiative forcing and temperature
    5.1.1. Incoming radiation
    5.1.2. Outgoing radiation
    5.1.3. The energy balance of the atmosphere
    5.1.4. Meridional structure of temperature
5.2. Pressure and geopotential height
5.3. Moisture
5.4. Winds
    5.4.1. Distribution of winds
5.5. Further reading
5.6. Problems

In previous chapters we considered those processes that play a role in setting the vertical distribution of atmospheric properties. Here we discuss how these properties vary horizontally, on the global scale. We shall see that geometrical effects play a major role in setting the observed horizontal distribution. The spherical Earth intercepts an essentially parallel beam of solar radiation, and so the incoming flux *per unit surface area* is greater at the equator than at the poles. An obvious and important consequence is that the atmosphere in the equatorial belt is warmer (and hence more moist) than the atmosphere over the polar caps. As we will discuss in this and subsequent chapters, these horizontal temperature gradients induce horizontal pressure gradients and hence motions, as sketched in Fig. 5.1. The resulting atmospheric wind patterns (along with ocean currents) act to transport heat from the warm tropics to the cool high latitudes, thereby playing a major role in climate.

In this chapter then we will describe the observed climatology[1] of atmospheric temperature, pressure, humidity, and wind.

---

[1] Here "climatology" implies some appropriate long-term average, such as the annual mean or seasonal mean, averaged over many years. In many of the figures shown here, the data are also averaged over longitude.

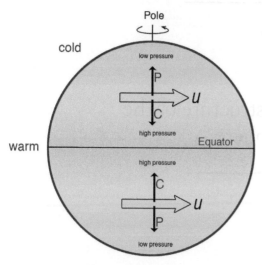

FIGURE 5.1. The atmosphere is warmer in the equatorial belt than over the polar caps. These horizontal temperature gradients induce, by hydrostatic balance, a horizontal pressure gradient force "P" that drive rings of air poleward. Conservation of angular momentum induces the rings to accelerate eastwards until Coriolis forces acting on them, "C," are sufficient to balance the pressure gradient force "P," as discussed in Chapters 6 and 7.

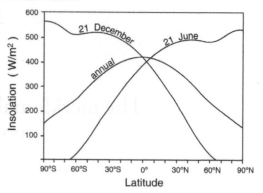

FIGURE 5.2. Distribution of annual mean and solstice (see Fig. 5.4) incoming solar radiation. The slight dip in the distribution at, for example, the winter solstice (December 21st) in the southern hemisphere corresponds to the edge of the polar day.

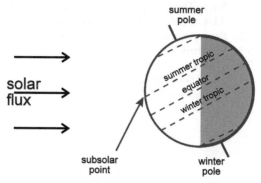

FIGURE 5.3. At the present time in history, the Earth's axis tilts at 23.5° and points towards the North Star. We sketch the incoming solar radiation at summer solstice when the Earth is tilted toward the Sun.

5.1. RADIATIVE FORCING AND TEMPERATURE

5.1.1. Incoming radiation

Annual mean

The latitudinal distribution of incoming solar radiation at the top of the atmosphere in the annual mean and at solstice is shown in Fig. 5.2. Its distribution is a consequence of the spherical geometry of the Earth and the tilt of the spin axis, depicted in Fig. 5.3. If the Earth's axis did not tilt with respect to the orbital plane, the average incident flux would maximize at a value of $S_{max} = S_0/\pi = 435W\,m^{-2}$ at the equator, and fall monotonically to zero at the poles. Because of the tilt, however, the poles do receive solar radiation during the summer half-year, and therefore the annual mean equator-to-pole difference is reduced, as Fig. 5.2 makes clear.

Seasonal

The daily averaged radiation received at any point on Earth varies through the year for two reasons. First, as illustrated in Fig. 5.4, the Earth's orbit around the Sun is not circular; the Earth is closest to the Sun—and the solar flux incident at the top of the atmosphere therefore maximizes—just after northern winter solstice. However, the variation of the Earth-Sun distance is less than ± 2%; although the corresponding variation in solar flux is not negligible, its contribution to the annual variation of the local solar flux per unit area at any given latitude is much less than that arising from

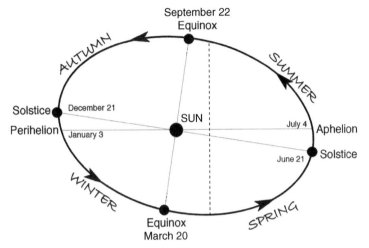

FIGURE 5.4. Earth describes an elliptical orbit around the Sun, greatly exaggerated in the figure. The longest (shortest) day occurs at the summer (winter) solstice when the Earth's spin axis points toward (away from) the Sun. The Earth is farthest from (closest to) the Sun at aphelion (perihelion). The seasons are labelled for the northern hemisphere.

the tilt of the rotation axis. At the present time in Earth history, the spin axis tilts from the vertical by 23.5°, the north pole pointing almost toward the North Star. At northern summer solstice, the north pole is tipped in the direction of the Sun, and the northern hemisphere has the longest day of the year. Conversely, at the northern winter solstice the north pole is tipped away from the Sun, and the northern hemisphere has the shortest day. At the equinoxes, daytime and nighttime are of equal length.

At solstice there is no incoming radiation at the winter pole (nor anywhere within "polar night"), but there is Sunlight 24 hours a day at the summer pole. It is for this reason that the incoming radiation actually maximizes (slightly) at the summer pole, when averaged over 24 hrs, as shown in Fig. 5.2. Nevertheless, the absorbed radiation at the summer pole is low because of the high albedo of snow and ice.

Before going on, we should emphasize that the Earth's tilt and its orbit around the Sun are not constant but change on very long time scales (of order 10^4–10^5 yr) in what are known as *Milankovitch cycles*. These changes are thought to play a role in climate change on very long time scales and, perhaps, in

pacing glacial-interglacial cycles, as will be discussed in Section 12.3.5.

5.1.2. Outgoing radiation

The net radiative budget of the Earth-atmosphere system, averaged over the year, is shown in Fig. 5.5. The absorbed solar radiation (incoming minus reflected) has a strong maximum in the tropics, where it is

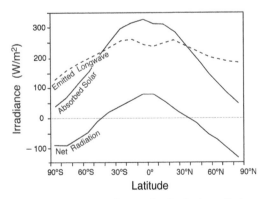

FIGURE 5.5. Annual mean absorbed solar radiation, emitted long-wave radiation, and net radiation, the sum of the two. The slight dip in emitted long-wave radiation at the equator is due to radiation from the (cold) tops of deep convecting clouds, as can be seen in Fig. 4.26.

about six times larger than at the poles. The latitudinal variation of emitted long-wave radiation, however, is much less, implying that the actual pole-to-equator temperature difference is smaller than it would be if the atmosphere was in thermodynamic balance at each latitude, column by column. Averaged over the year, there is a net surplus of incoming radiation in the tropics and a net deficit at high latitudes. Since local energy balance must be satisfied, Fig. 5.5 implies that there must be a transport of energy from low to high latitudes to maintain equilibrium (see Problem 1 at the end of this chapter).

5.1.3. The energy balance of the atmosphere

The required transport is quantified and plotted in Fig. 5.6 based on satellite

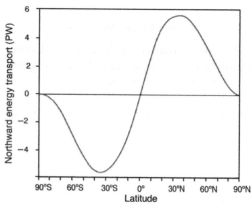

FIGURE 5.6. The northward energy transport deduced by top of the atmosphere measurements of incoming and outgoing solar and terrestrial radiation from the ERBE satellite. The units are in $PW = 10^{15}$W (see Trenberth and Caron, 2001). This curve is deduced by integrating the "net radiation" plotted in Fig. 5.5 meridionally. See Chapter 11 for a more detailed discussion.

measurements of incoming and outgoing solar and terrestrial radiation at the top of the atmosphere (see Section 11.5). In each hemisphere, the implied flux of energy is around 6×10^{15} W = 6 PW.[2] As will be discussed in the following chapters (particularly Chapters 8 and 11), the transport is achieved by fluid motions, especially in the atmosphere, but with the ocean also making a significant contribution.

5.1.4. Meridional structure of temperature

Troposphere

The observed structure of annual-mean temperature $\overline{T}$ (where the overbar implies zonal average[3]) and potential temperature $\overline{\theta}$ in the troposphere and lower stratosphere are shown in Figs. 5.7 and 5.8 respectively. Temperature decreases upward and (generally) poleward in the troposphere. The annual average surface temperature is below 0°C poleward of about 60° latitude, and reaches a maximum of 27°C just north of the equator. The annual-mean pole-to-equator temperature difference over the troposphere is typically 40°C.

As can be seen in Fig. 5.8, surfaces of constant potential temperature, often referred to as isentropic (constant θ implies constant entropy; see Problem 5 of Chapter 4) surfaces, slope upward toward the pole in the troposphere. Moreover $\overline{\theta}$ (unlike $\overline{T}$) always increases with height, reflecting the stability of the atmosphere to dry processes discussed in Section 4.3.2. The closely spaced contours aloft mark the stratosphere, the widely spaced contours below mark the troposphere. The transition between the two, the tropopause, is higher in the tropics than over the pole.

---

[2] 1 *PW* (petawatt) = 10^{15} W.

[3] The zonal average of a quantity X is conventionally written $\overline{X}$ (with an overbar) where:

$$\overline{X}(\varphi, z) = \frac{1}{2\pi} \int_0^{2\pi} X(\lambda, \varphi, z)\, d\lambda$$

(λ, φ) being (longitude, latitude).

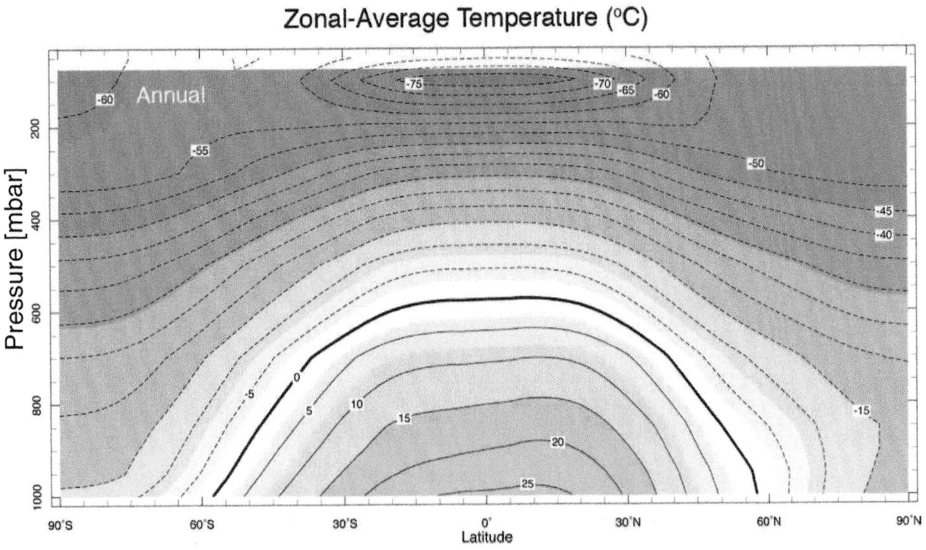

Zonal-Average Temperature (°C)

FIGURE 5.7.   The zonally averaged annual-mean temperature in °C.

Figure 5.9 shows the annual mean equivalent potential temperature, θ_e, defined by Eq. 4-30, and vividly displays the effects of vigorous convection in the tropics which remove vertical gradients of θ_e. This should be contrasted with the large vertical gradients of dry potential temperature, θ, seen in Fig. 5.8.

Stratosphere

The zonally averaged temperature is again shown in Fig. 5.10 (plotted here against height rather than pressure to emphasize upper altitudes) for solstice conditions. The features of the vertical temperature structure discussed in Chapter 3 are even clearer in Fig. 5.10: the temperature minima at the tropopause (at height 10–16 km) and mesopause (near 80 km), and the maximum at the stratopause (near 50 km). Note the latitudinal variation of these features, especially the variation of the tropopause, which is high and cold in the tropics, and much lower and warmer in high latitudes. In fact, there is something like a discontinuity of the tropopause in the subtropics (the "tropopause gap"),

which, as we will see in Chapter 8, is associated with the presence of strong winds in the jet stream. Air moving between the troposphere and stratosphere in a vertical direction (upward in the tropics, downward in the extratropics) does so very slowly, so that it has time to adjust its potential temperature to ambient values in response to weak diabatic processes. However, air can be exchanged more rapidly across the tropopause gap, since it can do so adiabatically by moving almost horizontally along isentropic surfaces between the tropical upper troposphere and the extratropical lower stratosphere.

The latitudinal temperature variation of the stratosphere is consistent with the incoming radiation budget; its temperature is greatest at the summer pole, where the averaged incoming radiation is most intense. However, in the troposphere the pole remains far colder than the tropics, even in summer. The polar regions, after a long cold winter, remain covered in highly reflective ice and snow (which do not have time to melt over the summer) and so have a high albedo (typically around 60%,

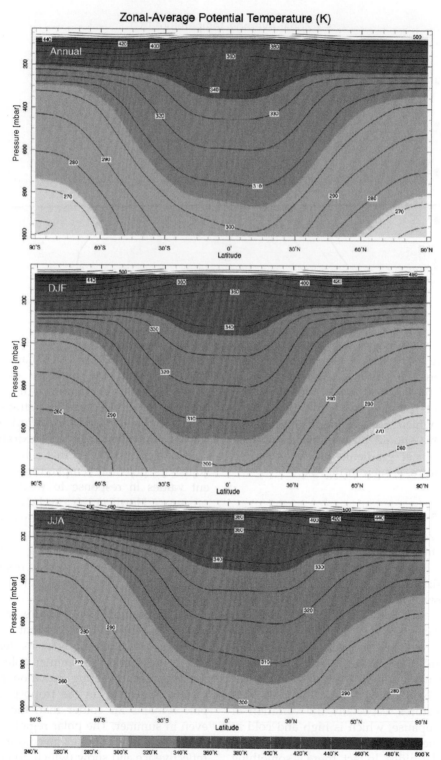

FIGURE 5.8.   The zonally averaged potential temperature in (top) the annual mean, averaged over (middle) December, January, and February (DJF), and (bottom) June, July, and August (JJA).

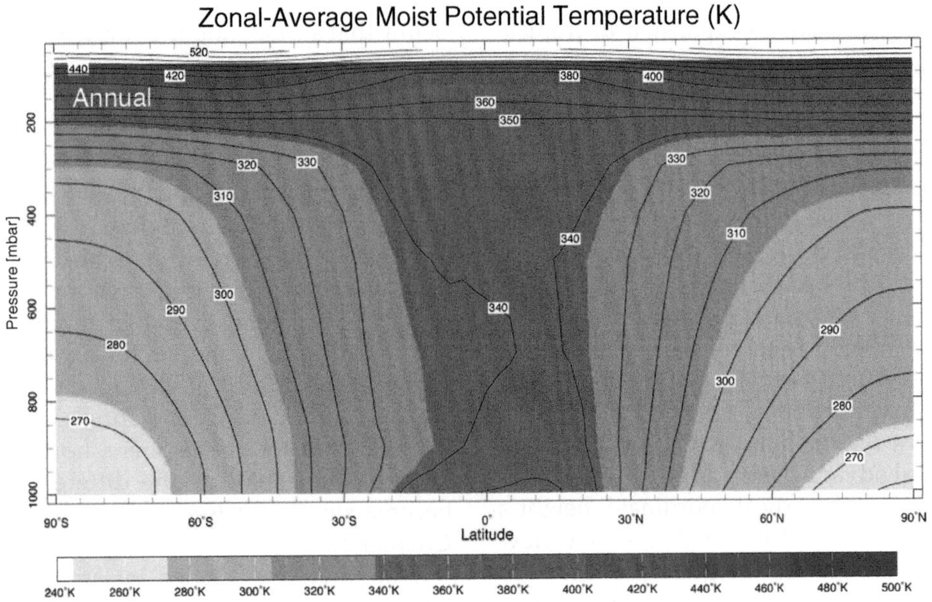

Zonal-Average Moist Potential Temperature (K)

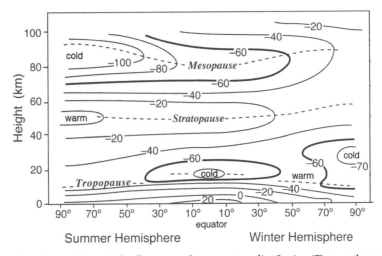

FIGURE 5.9. The zonal average, annual mean equivalent potential temperature, θ_e, Eq. 4-30.

FIGURE 5.10. The observed, longitudinally averaged temperature distribution (T) at northern summer solstice from the surface to a height of 100 km (after Houghton, 1986). Altitudes at which the vertical T gradient vanishes are marked by the dotted lines and correspond to the demarcations shown on the $T(z)$ profile in Fig. 3.1. The $-60°$C isopleth is thick. Note the vertical scale is in km compared to Fig. 5.7, which is in pressure. To convert between them, use Eq. 3-8.

compared to the global average of about 30%; cf. Fig. 2.5 and Table 2.2). Thus the solar radiation absorbed at the surface is substantially lower at the poles than in the tropics, even in summer.

5.2. PRESSURE AND GEOPOTENTIAL HEIGHT

We have seen that it is warmer in the tropics than at higher latitudes. We will now describe how, through hydrostatic balance, this warmth leads to expansion of tropical air columns relative to polar air columns and hence meridional pressure gradients. It is these pressure gradients that induce fluid accelerations and hence winds.

It is customary in meteorology to use pressure rather than height as the primary vertical coordinate. Some conceptual reasons will become clear in Chapter 6. Since, in hydrostatic balance, pressure is directly related to the overlying mass burden, pressure is actually a mass coordinate. In observations it is simpler to measure pressure *in situ* than height, so there are practical advantages also.[4]

With p as a vertical coordinate, height z becomes a *dependent variable*, so we now speak of $z(p)$—the height of a pressure surface—rather than $p(z)$. In principle, this is a trivial change: we can easily take a plot of $p(z)$, such as Fig. 3.6, and lay it on its side to give us $z(p)$. From Eq. 3-5 we may then write:

$$\frac{\partial z}{\partial p} = -\frac{RT}{gp}, \qquad (5\text{-}1)$$

or, noting that $p\frac{\partial}{\partial p} = \frac{\partial}{\partial \ln p}$,

$$\frac{\partial z}{\partial \ln p} = -\frac{RT}{g} = -H,$$

where H is the vertical scale height, Eq. 3-9. For an isothermal atmosphere (with constant scale height), z varies as $\ln p$, which of course is just another way of saying that p varies exponentially with z (see Eqs. 3-7 and 3-8). By integrating Eq. 5-1 vertically, we see that the height of a given pressure surface is dependent on the average temperature below that surface and the surface pressure, p_s, thus:

$$z(p) = R\int_p^{p_s} \frac{T}{g}\frac{dp}{p}, \qquad (5\text{-}2)$$

where we have set $z(p_s) = 0$. The *geopotential height* of the surface is defined by Eq. 5-2, but with g replaced by its (constant) surface value. In Chapter 1 we noted that g varies very little over the depth of the lower atmosphere and so, for most meteorological purposes, the difference between actual height and geopotential height is negligible. In the mesosphere, however, at heights above 100 km, the difference may become significant (see, e.g., Problem 6 in Chapter 3).

Note that, as sketched in Fig. 5.11, low height of a pressure surface corresponds to low pressure on a z surface.

The height of the 500 mbar surface (January monthly average) is plotted in Fig. 5.12. It has an average height of 5.5 km, as we deduced in Section 3.3, but is higher in the tropics (5.88 km) than over the pole (4.98 km), sloping down from equator to pole by about 900 m.

The zonally averaged geopotential height is plotted in Fig. 5.13 as a function of pressure and latitude for mean annual conditions. Note that the difference $\overline{z}(\varphi,p) - \langle z \rangle (p)$ is plotted where $\langle z \rangle (p)$ is the horizontal average at pressure level p. Except near the surface, pressure surfaces are generally high in the tropics and low at high latitudes, especially in winter. Since surface pressure does not vary much (a few tens of mbar at most), the height of a given

---

[4] Evangelista Torricelli (1608–1647) was the first person to create a sustained vacuum. He discovered the principle of the barometer which led to the development of the first instrument to measure pressure.

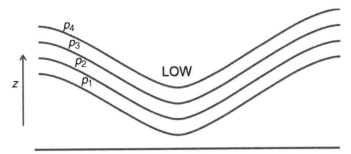

FIGURE 5.11. The geometry of pressure surfaces (surfaces of constants p_1, p_2, p_3, p_4, where $p_1 > p_2 > p_3 > p_4$) in the vicinity of a horizontal pressure minimum.

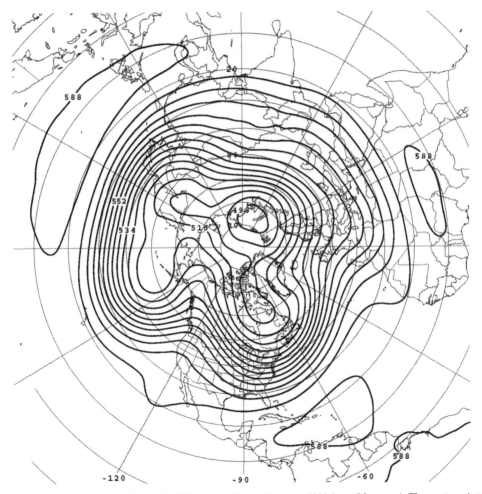

FIGURE 5.12. The mean height of the 500 mbar surface in January, 2003 (monthly mean). The contour interval is 6 decameters $\equiv$ 60 m. The surface is 5.88 km high in the tropics and 4.98 km high over the pole. Latitude circles are marked every 10°, longitude every 30°.

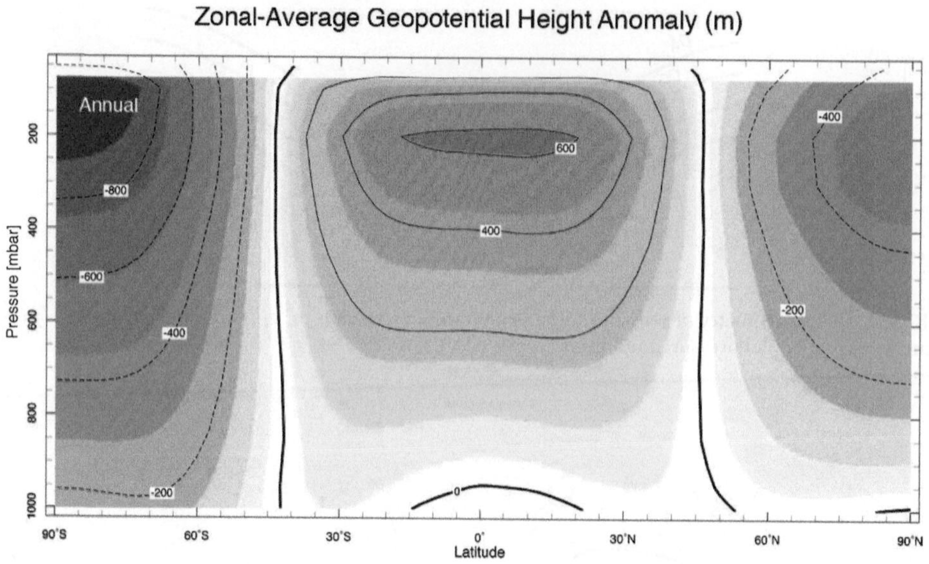

Zonal-Average Geopotential Height Anomaly (m)

FIGURE 5.13.   Zonal-mean geopotential height (m) for annual mean conditions. Values are departures from a horizontally uniform reference profile.

pressure surface is proportional to the mean temperature between that pressure surface and the ground. So where temperatures are cold (warm), air columns contract (expand) and geopotential heights are low (high).

We can estimate the expected tilt of a pressure surface as follows. For an atmosphere in which T varies in the horizontal but is vertically isothermal, the difference in the height of an isobaric surface, p, between warm and cold latitudes, $\Delta z_{\text{cold}}^{\text{warm}}$, is, using Eq. 3-8,

$$\Delta z_{\text{cold}}^{\text{warm}} = \frac{R \Delta T_{\text{cold}}^{\text{warm}}}{g} \ln\left(\frac{p_s}{p}\right), \qquad (5\text{-}3)$$

where p_s is the pressure at the surface. If $\Delta T_{\text{cold}}^{\text{warm}} = 40°C$ in the climatology, inserting numbers into the above, we find that the 500 mbar surface drops by $\Delta z_{\text{cold}}^{\text{warm}} = 811\,\text{m}$, as is evident in Figs. 5.12 and 5.13.

Finally, it is useful to define the *thickness* of an atmospheric layer, sandwiched between two pressure surfaces, such as p_1 and p_2 in Fig. 5.14. From Eq. 5-2 we have:

$$z_2 - z_1 = R \int_{p_2}^{p_1} \frac{T}{g} \frac{dp}{p}, \qquad (5\text{-}4)$$

which depends on T averaged over the layer. Atmospheric layers are "thick" in tropical regions, because they are warm, and "thin" in polar regions, because they are cold, leading to the large-scale slope of pressure surfaces seen in Fig. 5.12. Moreover, if tropical columns are warmer than polar columns at all levels, then the tilt of the pressure surfaces must increase with height, as sketched in Fig. 5.14 and seen in the observations, Fig. 5.13. We will see the importance of this fact when we discuss the distribution of atmospheric winds in Chapter 7.

5.3. MOISTURE

As discussed in Sections 1.3.2 and 4.5, the moisture distribution in the atmosphere is strongly controlled by the temperature distribution; the atmosphere is moist near the surface in the tropics where it is very warm and drier aloft and in polar latitudes where it is cold. As shown in Fig. 5.15, the specific humidity, defined in Eq. 4-23, reaches a maximum (of around $18\,\text{g kg}^{-1}$) at the surface near the equator and decreases

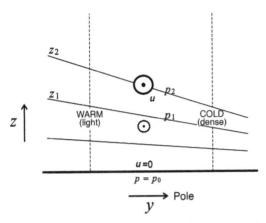

FIGURE 5.14. Warm columns of air expand, cold columns contract, leading to a tilt of pressure surfaces, a tilt which typically increases with height in the troposphere. In Section 7.3, we will see that the corresponding winds are out of the paper, as marked by $\odot$ in the figure.

to much lower values (around 1–$2\,\mathrm{g\,kg^{-1}}$) near the poles. At upper levels there is very little water vapor. This broad pattern can be understood by noting the striking correlation between q (Fig. 5.15) and T (Fig. 5.7). Air colder than $0\,^\circ\mathrm{C}$ can hold very little water vapor (see Fig. 1.5 and discussions in Sections 1.3 and 4.5).

The control by temperature of the specific humidity distribution can be seen more directly by comparing Fig. 5.15 with Fig. 5.16, which shows q_*, the specific humidity at saturation given by Eq. 4-24 with e_s given by Eq. 1-4. We see that q has the same spatial form as q_* but never reaches saturation even at the surface. As discussed in Section 4.5, U, the *relative humidity* defined in Eq. 4-25, is the ratio of q in Fig. 5.15 to q_* in Fig. 5.16. Zonal mean relative humidity, shown in Fig. 5.17, is (on average) 70–80% everywhere near the ground. The reason for the decrease of relative humidity with altitude is a little more subtle. Vertical transport of water vapor is effected mostly by convection, which (as we have seen) lifts the air to saturation. It may therefore seem odd that even the relative humidity decreases significantly with height through the troposphere. To understand this, we need to think about the entire circulation of a convective system, and not just the updraft. Consider Fig. 5.18.

The updraft in a convective cloud—the part considered in the parcel stability argument of Section 4.5.2—is rather narrow. Of course, the air must return and does so

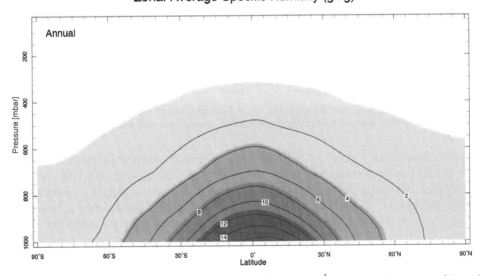

FIGURE 5.15. Zonally averaged specific humidity q, Eq. 4-23, in $\mathrm{g\,kg^{-1}}$ under annual mean conditions. Note that almost all the water vapor in the atmosphere is found where $T > 0\,^\circ\mathrm{C}$ (see Fig. 5.7).

Zonal-Average Saturated Specific Humidity (g/kg)

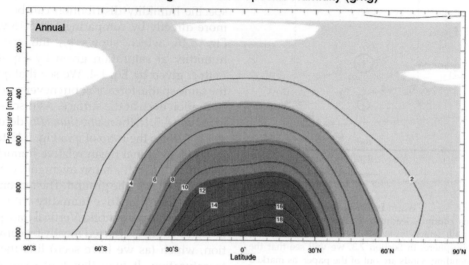

FIGURE 5.16. Zonally averaged saturated specific humidity, q_*, in $\mathrm{g\,kg^{-1}}$, for annual-mean conditions.

Zonal-Average Relative Humidity (%)

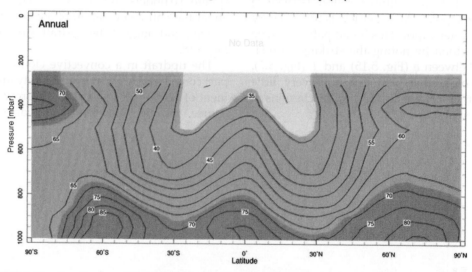

FIGURE 5.17. Zonal mean relative humidity (%), Eq. 4-25, under annual mean conditions. Note that data are not plotted above 300 mbar, where q is so small that it is difficult to measure accurately by routine measurements.

in a broad downdraft. Now, within the updraft, the air becomes saturated (whence the cloud) and will frequently produce precipitation: the excess water will rain out. Therefore, even though the air is saturated within the cloud, by the time the air flows out from the top of the cloud, it has lost most of its water (since the cloud top is at much lower temperature than the ground,

and hence its saturation specific humidity is very low). As this air descends and warms within the downdraft, it conserves its specific humidity. Since, once it has warmed, the saturation specific humidity at the air temperature has increased, the air becomes very dry in the sense that its relative humidity is very low. Hence, even though the air is saturated within the updraft, the average

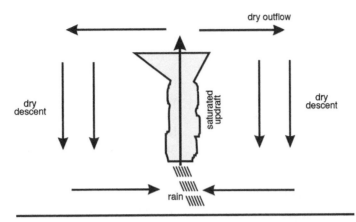

FIGURE 5.18.   Drying due to convection. Within the updraft, air becomes saturated and excess water is rained out. The descending air is very dry. Because the region of ascent is rather narrow and the descent broad, convection acts as a drying agent for the atmosphere as a whole.

(at a fixed height) over the system as a whole is low. Convection, by lifting air to saturation and thus causing precipitation of the air's water, acts as a drying agent for the atmosphere. This can be vividly seen in the satellite mosaic of the water vapor distribution over the globe between heights of 6–10 km shown in Fig. 3 of the Preface. The regions of relatively dry descent (dark regions) on either side of the equatorial moist band (light), mark the latitude of the deserts. (These issues will be discussed further in Chapter 8.)

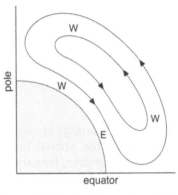

FIGURE 5.19.   The circulation envisaged by Hadley (1735) comprising one giant meridional cell stretching from equator to pole. Regions where Hadley hypothesized westerly (W) and easterly (E) winds are marked.

5.4. WINDS

We saw in Section 5.2 that because of the pole-equator temperature gradient, isobaric surfaces slope down from equator to pole, inducing a horizontal pressure gradient at upper levels. There is thus a pressure gradient force aloft, directed from high pressure to low pressure, which is from warm latitudes to cold latitudes, as seen in Fig. 5.1. We might expect air to move down this pressure gradient. Hadley[5] suggested one giant

meridional cell with rising motion in the tropics and descending motion at the pole, as sketched schematically in Fig. 5.19. One might expect rings of air circling the globe to be driven poleward by pressure gradient forces. As they contract, conserving the angular momentum imparted to them by the spinning Earth, westerly (W⟶E) winds will be induced (see detailed discussion in Section 8.2.1). At the poles Hadley imagined

---

[5]George Hadley (1685–1768). British meteorologist who was the first to recognize the relationship between the rotation of the Earth and the atmospheric circulation (in particular, the trade winds). Hadley presented his theory in 1735. The pattern of meridional circulation in tropical zones, called the Hadley circulation (or a Hadley cell—see Fig. 5.19) is named after him.

that the rings would sink and then expand outward as they flow equatorward below, generating easterly winds, as marked on the figure.

As attractive as this simple circulation may seem, we shall see that this picture of a single meridional cell extending from equator to pole is not in accord with observations.

5.4.1. Distribution of winds

Wind velocity is, of course, a vector with components $\mathbf{u} = (u, v, w)$ in the eastward, northward, and upward directions. The vertical component is, except in the most violent disturbances, very much smaller than the horizontal components (a consequence, among other things, of the thinness of the atmosphere), so much so that it cannot usually be directly measured but must be inferred from other measurements.

Mean zonal winds

The typical distribution of zonal-mean zonal wind, $\overline{u}$, in the annual mean and at the solstices (December, January, February [DJF]; and June, July, August [JJA]) is shown in Fig. 5.20. Except close to the equator, the zonal-mean winds are eastward (i.e., in meteorological parlance, westerly) almost everywhere. The stronger winds are found at the core of the *subtropical jets*, the strongest of which is located near 30° latitude in the winter hemisphere at 200 mbar, at a height of about 10 km with, on average, a speed[6] of around $30 \, \mathrm{m \, s^{-1}}$. A weaker jet of about $20 \, \mathrm{m \, s^{-1}}$ is located near 45°

in the summer hemisphere. The easterlies observed in the tropics are much weaker, especially in northern winter.

Note that the winds are much weaker near the ground, but show the same pattern; westerlies are poleward of about 30° and easterlies equatorward thereof. The low-level easterlies (which, as we shall see, are actually north-easterlies in the northern hemisphere and south-easterlies in the southern hemisphere) are known as the "trade winds", a term that comes from the days of sailing ships when, together with the westerlies and south-westerlies of higher latitudes, these winds allowed ships to complete a circuit of the North Atlantic and thus to trade easily between Europe and North America.[7] We shall see later, in Chapter 10, that this pattern of surface winds and the attendant stress is a primary driver of ocean circulation.

Mean meridional circulation

Figure 5.21 shows the zonal-mean circulation of the atmosphere in the meridional plane, known as the meridional overturning circulation, whose sense is marked by the arrows. Note the strong seasonal dependence. In DJF air rises just south of the equator and sinks in the subtropics of the northern hemisphere, around 30° N. (Conversely, in JJA air rises just north of the equator and sinks in the subtropics of the southern hemisphere.) We thus see strong upward motion on the summer side of the equator, where the warm surface triggers convection and rising motion, and strong descent on the winter side of the equator. In the annual mean we thus see two

---

[6]As we shall see, the jet actually wiggles around, both in longitude and in time, and so is smoothed out in the averaging process. Typical local, instantaneous maximum speeds are closer to $50 \, \mathrm{m \, s^{-1}}$.

[7] Matthew Fountaine Maury (1806–1873). U.S. Naval officer and oceanographer, the founder of the U.S. Naval Observatory, inventor of a torpedo, and pioneer of wind and current charts. Maury was the first to systematically study and map ocean currents and marine winds and recommend routes for sea clippers to take advantage of winds and currents.

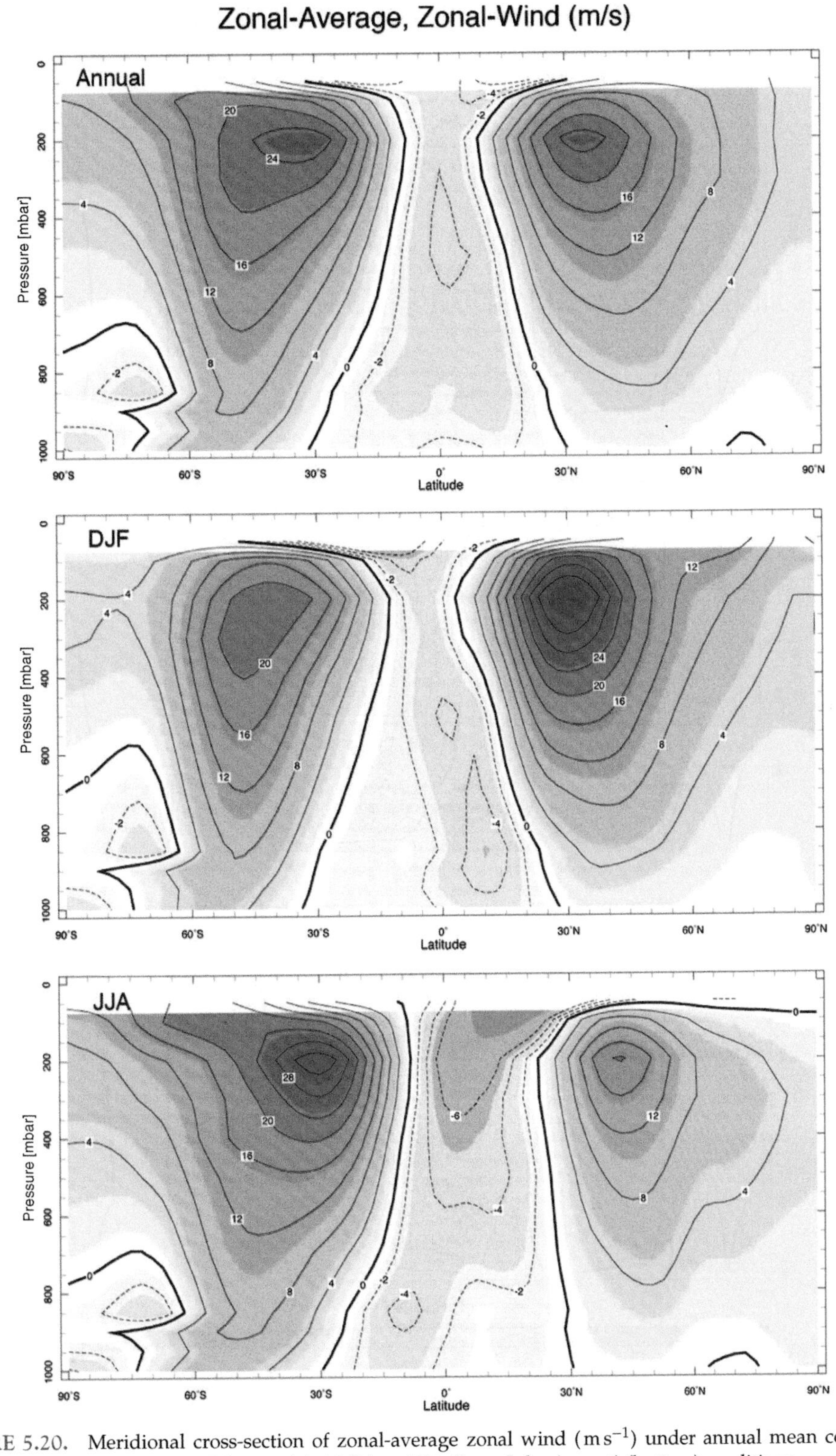

Zonal-Average, Zonal-Wind (m/s)

FIGURE 5.20. Meridional cross-section of zonal-average zonal wind ($\mathrm{m\,s^{-1}}$) under annual mean conditions (top), DJF (December, January, February) (middle) and JJA (June, July, August) (bottom) conditions.

Meridional Overturning Circulation (10⁹ kg/s)

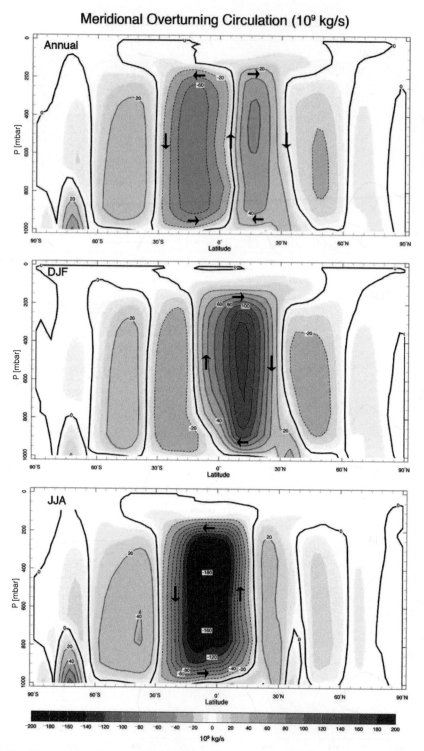

FIGURE 5.21. The meridional overturning streamfunction χ of the atmosphere in annual mean, DJF, and JJA conditions. [The meridional velocities are related to χ by $v = -(\rho a \cos \varphi)^{-1} \partial \chi / \partial z$; $w = (\rho a^2 \cos \varphi)^{-1} \partial \chi / \partial \varphi$.] Units are in $10^9 \, \text{kg s}^{-1}$, or Sverdrups, as discussed in Section 11.5.2. Flow circulates around positive (negative) centers in a clockwise (anticlockwise) sense. Thus in the annual mean, air rises just north of the equator and sinks around $\pm 30°$.

(weaker) cells, more or less symmetrically arranged about the equator, one branching north and the other south. Not surprisingly, the regions of mean upward motion coincide with the wet regions of the tropics, as evident, for example, in the presence of cold cloud tops in the OLR distribution shown in Fig. 4.26. In contrast, descending regions are very dry and cloud-free. The latter region is the desert belt and is also where the trade inversion discussed in the context of Fig. 4.16b is found.

This vertical motion is accompanied by meridional flow. Except in the tropics, mean northward winds are weak ($< 1 \, \mathrm{m \, s^{-1}}$) everywhere. In the tropical upper troposphere (near 200 mbar, between about 20° N and 20° S) we observe winds directed toward the winter hemisphere at speeds of up to $3 \, \mathrm{m \, s^{-1}}$. There is a return flow in the lower troposphere that is somewhat weaker, and which is directed mostly toward the summer hemisphere. Thus the "easterlies" we deduced from Fig. 5.20 are in fact north-easterlies, north of the equator in northern winter, and south-easterlies, south of the equator in southern winter. These are the *trade winds* mentioned earlier.

The overturning circulation of the tropical atmosphere evident in Fig. 5.21 is known as the *Hadley cell*; we will consider its dynamics in Chapter 8. The (much) weaker reverse cells in middle latitudes of each hemisphere are known as *Ferrel cells*, after William Ferrel (1817–1891), an American meteorologist who developed early theories of the atmospheric circulation.

Eddies and waves

Finally, in case the typical, zonally-averaged, cross sections presented here give the impression that the atmosphere actually looks like this at any given time, note that, in reality, the atmospheric structure is variable in time and three-dimensional. This is evident on any weather map (and from the fact that we need weather forecasts at all). A typical instantaneous 500 mbar geopotential height analysis is shown in Fig. 5.22 and should be compared to the (much smoother) monthly average shown in Fig. 5.12. Although the general features of the meridional structure are evident (in particular, the decrease of height of the pressure surface from low to high latitude) there are also many localized highs and lows in the instantaneous structure that, as we shall see, are indicative of the presence of *eddies* in the flow.[8] The atmosphere, especially in the extratropics, is full of eddying winds. As will be discussed in Chapter 8, the eddies are the key agency of meridional heat and moisture transport in the middle to high latitudes.

In summary, in this chapter we have discussed how warming of the tropical atmosphere and cooling over the poles leads, through hydrostatic balance, to a large-scale slope of the pressure surfaces and hence pressure gradient forces directed from equator to pole. It turns out that, as we go on to discuss in detail in Chapters 7 and 8, this pressure gradient force is balanced on the large-scale by Coriolis forces acting on the winds because of the Earth's rotation. In fact the temperature, pressure, and wind fields shown in this chapter are not independent of one another but are intimately connected through basic laws of physics. To take our discussion further and make it quantitative, we must next develop some of the underlying theory of atmospheric dynamics. This is interesting in itself because it involves applying the

---

[8] Vilhelm Bjerknes (1862–1951), Norwegian meteorologist. His father, Carl, was a professor of hydrodynamics, while his son, Jacob became a famous meteorologist in his own right (see Chapter 12). With Jacob, he created an early network of meteorological observations. Bjerknes was the founder of the "Bergen school", where the now-familiar synoptic concepts of cyclones, fronts, and air masses were first established.

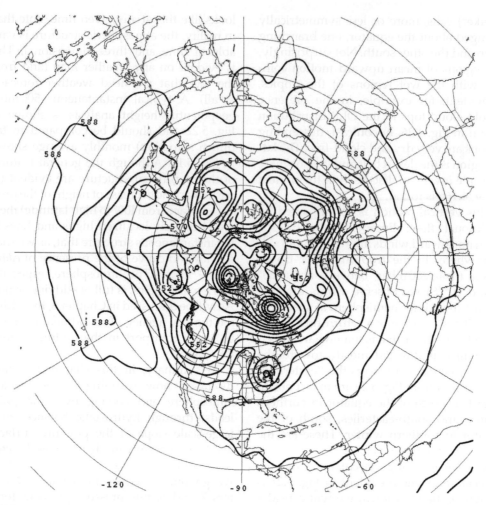

FIGURE 5.22. Typical 500 mbar height analysis: the height of the 500 mbar surface (in decameters) at 12 GMT on June 21, 2003. The contour interval is 6 decameters = 60 m. The minimum height is 516 decameters and occurs in the intense lows over the pole.

laws of mechanics and thermodynamics to a fluid on a rotating Earth; thinking about the importance, or otherwise, of rotation on the fluid motion; and contemplating the motion from different frames of reference, that of the rotating Earth itself and that of an inertial observer out in space looking back on the Earth.

5.5. FURTHER READING

A comprehensive survey of the observed climatological state of the atmosphere is given in Chapter 7 of Peixoto and Oort (1992).

5.6. PROBLEMS

1. Figure 5.5 shows the net incoming solar and outgoing long-wave irradiance at the top of the atmosphere. Note that there is a net gain of radiation in low latitudes and a net loss in high latitudes. By inspection of the figure, estimate the magnitude of the poleward energy flux that must

be carried by the atmosphere-ocean system across the 30° latitude circle, to achieve a steady state.

2. Suppose that the Earth's rotation axis were normal to the Earth-Sun line. The solar flux, measured per unit area in a plane normal to the Earth-Sun line, is S_0. By considering the solar flux incident on a latitude belt bounded by latitudes $(\varphi, \varphi + d\varphi)$, show that F, the 24-hr average of solar flux *per unit area of the Earth's surface*, varies with latitude as

$$F = \frac{S_0}{\pi} \cos \varphi.$$

   (a) Using this result, suppose that the atmosphere is completely transparent to solar radiation, but opaque to infrared radiation such that, *separately at each latitude*, the radiation budget can be represented by the "single slab" model shown in Fig. 2.7. Determine how surface temperature varies with latitude.

   (b) Calculate the surface temperature at the equator, 30°, and 60° latitude if Earth's albedo is 30% and $S_0 = 1367\,\mathrm{W\,m^{-2}}$. Compare your result with observations shown in Fig. 5.7.

3. Use the hydrostatic relation and the equation of state of an ideal gas to show that the 1000–500 mbar "thickness," $\Delta z = z(500\,\mathrm{mbar}) - z(1000\,\mathrm{mbar})$ is related to the mean temperature $\langle T \rangle$ of the 1000–500 mbar layer by

$$\Delta z = \frac{R \langle T \rangle}{g} \ln 2,$$

   where

$$\langle T \rangle = \frac{\int T\, d\ln p}{\int d\ln p},$$

where the integrals are from 500 mbar to 1000 mbar. (Note that $1000\,\mathrm{mbar} \equiv 1000\,\mathrm{hPa} \equiv 10^5\,\mathrm{Pa}$).

   (a) Compute the thickness of the surface to 500-mbar layer at 30° and 60° latitude, assuming that the surface temperatures computed in Problem 2b extend uniformly up to 500 mbar.

   (b) Figures 7.4 and 7.25 (of Chapter 7) show 500 mbar and surface pressure analyses for 12 GMT on June 21, 2003. Calculate $\langle T \rangle$ for the 1000-mbar to 500-mbar layer at the center of the 500-mbar trough at 50°N, 120°W, and at the center of the ridge at 40°N, 90°W. [N.B. You will need to convert from surface pressure, p_s, to height of the 1000 hPa surface, z_{1000}; to do so use the (approximate) formula

$$z_{1000} \cong 10\,(p_s - 1000),$$

   where z_{1000} is in meters and p_s is in hPa.]

   Is $\langle T \rangle$ greater in the ridge or the trough? Comment on and physically interpret your result.

4. Use the expression for saturated specific humidity, Eq. 4-24, and the empirical relation for saturated vapor pressure $e_s(T)$, Eq. 1-4 (where $A = 6.11$ mbar, $\beta = 0.067\,^\circ\mathrm{C^{-1}}$, and T is in °C), to compute from the graph of $T(p)$ in the tropical belt shown in Fig. 4.9, vertical profiles of saturated specific humidity, $q^*(p)$. You will need to look up values of R and R_v from Chapter 1.

   Compare your q^* profiles with observed profiles of q in the tropics shown in Fig. 5.15. Comment?

6

The equations of fluid motion

6.1. Differentiation following the motion
6.2. Equation of motion for a nonrotating fluid
    6.2.1. Forces on a fluid parcel
    6.2.2. The equations of motion
    6.2.3. Hydrostatic balance
6.3. Conservation of mass
    6.3.1. Incompressible flow
    6.3.2. Compressible flow
6.4. Thermodynamic equation
6.5. Integration, boundary conditions, and restrictions in application
6.6. Equations of motion for a rotating fluid
    6.6.1. GFD Lab III: Radial inflow
    6.6.2. Transformation into rotating coordinates
    6.6.3. The rotating equations of motion
    6.6.4. GFD Labs IV and V: Experiments with Coriolis forces on a parabolic rotating table
    6.6.5. Putting things on the sphere
    6.6.6. GFD Lab VI: An experiment on the Earth's rotation
6.7. Further reading
6.8. Problems

To proceed further with our discussion of the circulation of the atmosphere, and later the ocean, we must develop some of the underlying theory governing the motion of a fluid on the spinning Earth. A differentially heated, stratified fluid on a rotating planet cannot move in arbitrary paths. Indeed, there are strong constraints on its motion imparted by the angular momentum of the spinning Earth. These constraints are profoundly important in shaping the pattern of atmosphere and ocean circulation and their ability to transport properties around the globe. The laws governing the evolution of both fluids are the same and so our theoretical discussion will not be specific to either atmosphere or ocean, but can and will be applied to both. Because the properties of rotating fluids are often counterintuitive and sometimes difficult to grasp, alongside our theoretical development we will describe and carry out laboratory experiments with a tank of water on a rotating table (Fig. 6.1). Many of the laboratory

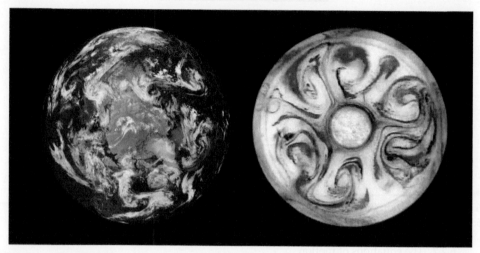

FIGURE 6.1. Throughout our text, running in parallel with a theoretical development of the subject, we study the constraints on a differentially heated, stratified fluid on a rotating planet (left), by using laboratory analogues to illustrate the fundamental processes at work (right). A complete list of the laboratory experiments can be found in Appendix A.4.

experiments we use are simplified versions of "classics" of geophysical fluid dynamics. They are listed in Appendix A.4. Furthermore we have chosen relatively simple experiments that, in the main, do not require sophisticated apparatus. We encourage you to "have a go" or view the attendant movie loops that record the experiments carried out in preparation of our text.

We now begin a more formal development of the equations that govern the evolution of a fluid. A brief summary of the associated mathematical concepts, definitions, and notation we employ can be found in Appendix A.2.

6.1. DIFFERENTIATION FOLLOWING THE MOTION

When we apply the laws of motion and thermodynamics to a fluid to derive the equations that govern its motion, we must remember that these laws apply to material elements of fluid that are usually mobile. We must learn, therefore, how to express the rate of change of a property of a fluid element, *following that element as it moves along*, rather than at a fixed point in space. It is useful to consider the following simple example.

Consider again the situation sketched in Fig. 4.13 in which a wind blows over a hill. The hill produces a pattern of waves in its lee. If the air is sufficiently saturated in water vapor, the vapor often condenses out to form a cloud at the "ridges" of the waves as described in Section 4.4 and seen in Figs. 4.14 and 4.15.

Let us suppose that a steady state is set up so the pattern of cloud does not change in time. If $C = C(x, y, z, t)$ is the cloud amount, where (x, y) are horizontal coordinates, z is the vertical coordinate, and t is time, then

$$\left(\frac{\partial C}{\partial t} \right)_{\substack{\text{fixed point} \\ \text{in space}}} = 0,$$

in which we keep at a fixed point in space, but at which, because the air is moving, there are constantly changing fluid parcels. The derivative $\left(\frac{\partial}{\partial t}\right)_{\text{fixed point}}$ is called the *Eulerian derivative* after Euler.[1]

But C is not constant *following along a particular parcel*; as the parcel moves upwards into the ridges of the wave, it cools, water condenses out, a cloud forms, and so C increases (recall GFD Lab 1, Section 1.3.3); as the parcel moves down into the troughs it warms, the water goes back in to the gaseous phase, the cloud disappears and C decreases. Thus

$$\left(\frac{\partial C}{\partial t}\right)_{\substack{\text{fixed}\\\text{particle}}} \neq 0,$$

even though the wave-pattern is fixed in space and constant in time.

So, how do we mathematically express "differentiation following the motion"? To follow particles in a continuum, a special type of differentiation is required. Arbitrarily small variations of $C(x, y, z, t)$, a function of position and time, are given to the first order by

$$\delta C = \frac{\partial C}{\partial t}\delta t + \frac{\partial C}{\partial x}\delta x + \frac{\partial C}{\partial y}\delta y + \frac{\partial C}{\partial z}\delta z,$$

where the partial derivatives $\partial/\partial t$ etc. are understood to imply that the other variables are kept fixed during the differentiation. The fluid velocity is the rate of change of position of the fluid element, following that element along. The variation of a property C *following an element of fluid* is thus derived by setting $\delta x = u\delta t$, $\delta y = v\delta t$, $\delta z = w\delta t$, where u is the speed in the x-direction, v is the speed in

the y-direction, and w is the speed in the z-direction, thus

$$(\delta C)_{\substack{\text{fixed}\\\text{particle}}} = \left(\frac{\partial C}{\partial t} + u\frac{\partial C}{\partial x} + v\frac{\partial C}{\partial y} + w\frac{\partial C}{\partial z}\right)\delta t,$$

where (u, v, w) is the velocity of the material element, which by definition is the fluid velocity. Dividing by δt and in the limit of small variations we see that

$$\left(\frac{\partial C}{\partial t}\right)_{\substack{\text{fixed}\\\text{particle}}} = \frac{\partial C}{\partial t} + u\frac{\partial C}{\partial x} + v\frac{\partial C}{\partial y} + w\frac{\partial C}{\partial z}$$

$$= \frac{DC}{Dt},$$

in which we use the symbol $\frac{D}{Dt}$ to identify the rate of change following the motion

$$\frac{D}{Dt} \equiv \frac{\partial}{\partial t} + u\frac{\partial}{\partial x} + v\frac{\partial}{\partial y} + w\frac{\partial}{\partial z} \equiv \frac{\partial}{\partial t} + \mathbf{u}.\nabla.$$

$$(6\text{-}1)$$

Here $\mathbf{u} = (u, v, w)$ is the velocity vector, and $\nabla \equiv \left(\frac{\partial}{\partial x}, \frac{\partial}{\partial y}, \frac{\partial}{\partial z}\right)$ is the gradient operator. D/Dt is called the Lagrangian derivative (after Lagrange; 1736–1813) (it is also called variously the *substantial*, the *total*, or the *material* derivative). Its physical meaning is *time rate of change of some characteristic of a particular element of fluid* (which in general is changing its position). By contrast, as introduced above, the Eulerian derivative $\partial/\partial t$ expresses the rate of change of some characteristic at a *fixed point* in space (but with constantly changing fluid element because the fluid is moving).

[1] Leonhard Euler (1707–1783). Euler made vast contributions to mathematics in the areas of analytic geometry, trigonometry, calculus and number theory. He also studied continuum mechanics, lunar theory, elasticity, acoustics, the wave theory of light, and hydraulics, and laid the foundation of analytical mechanics. In the 1750s Euler published a number of major works setting up the main formulas of fluid mechanics, the continuity equation, and the Euler equations for the motion of an inviscid, incompressible fluid.

Some writers use the symbol d/dt for the Lagrangian derivative, but this is better reserved for the ordinary derivative of a function of one variable, the sense it is usually used in mathematics. Thus for example the rate of change of the radius of a rain drop would be written dr/dt, with the identity of the drop understood to be fixed. In the same context D/Dt could refer to the motion of individual particles of water circulating within the drop itself. Another example is the vertical velocity, defined as $w = Dz/Dt$; if one sits in an air parcel and follows it around, w is the rate at which one's height changes.[2]

The term $\mathbf{u}.\nabla$ in Eq. 6-1 represents *advection* and is the mathematical representation of the ability of a fluid to carry its properties with it as it moves. For example, the effects of advection are evident to us every day. In the northern hemisphere, southerly winds (from the south) tend to be warm and moist because the air carries with it properties typical of tropical latitudes; northerly winds tend to be cold and dry because they advect properties typical of polar latitudes.

We will now use the Lagrangian derivative to help us apply the laws of mechanics and thermodynamics to a fluid.

6.2. EQUATION OF MOTION FOR A NONROTATING FLUID

The state of the atmosphere or ocean at any time is defined by five key variables:

$$\mathbf{u} = (u, v, w); \ p \text{ and } T,$$

(six if we include specific humidity in the atmosphere, or salinity in the ocean). Note that by using the equation of state, Eq. 1-1, we can infer ρ from p and T. To "tie" these variables down we need five independent equations. They are:

1. the laws of motion applied to a fluid parcel, yielding three independent

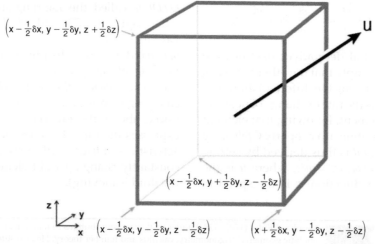

FIGURE 6.2.   An elementary fluid parcel, conveniently chosen to be a cube of sides $\delta x, \delta y, \delta z$, centered on (x, y, z). The parcel is moving with velocity $\mathbf{u}$.

---

[2]Meteorologists like working in pressure coordinates in which p is used as a vertical coordinate rather than z. In this coordinate an equivalent definition of "vertical velocity" is:

$$\omega = \frac{Dp}{Dt},$$

the rate at which pressure changes as the air parcel moves around. Since pressure varies much more quickly in the vertical than in the horizontal, this is still, for all practical purposes, a measure of vertical velocity, but expressed in units of hPa s^{-1}. Note also that upward motion has negative ω.

equations in each of the three orthogonal directions

2. conservation of mass

3. the law of thermodynamics, a statement of the thermodynamic state in which the motion takes place.

These equations, five in all, together with appropriate boundary conditions, are sufficient to determine the evolution of the fluid.

6.2.1. Forces on a fluid parcel

We will now consider the forces on an elementary fluid parcel, of infinitesimal dimensions $(\delta x, \delta y, \delta z)$ in the three coordinate directions, centered on (x, y, z) (see Fig. 6.2).

Since the mass of the parcel is $\delta M = \rho\, \delta x\, \delta y\, \delta z$, then, when subjected to a net force $\mathbf{F}$, Newton's Law of Motion for the parcel is

$$\rho\, \delta x\, \delta y\, \delta z \frac{D\mathbf{u}}{Dt} = \mathbf{F}, \qquad (6\text{-}2)$$

where $\mathbf{u}$ is the parcel's velocity. As discussed earlier we must apply Eq. 6-2 to the same material mass of fluid, which means we must follow the same parcel around. Therefore, the time derivative in Eq. 6-2 is the total derivative, defined in Eq. 6-1, which in this case is

$$\frac{D\mathbf{u}}{Dt} = \frac{\partial \mathbf{u}}{\partial t} + u \frac{\partial \mathbf{u}}{\partial x} + v \frac{\partial \mathbf{u}}{\partial y} + w \frac{\partial \mathbf{u}}{\partial z}$$

$$= \frac{\partial \mathbf{u}}{\partial t} + (\mathbf{u} \cdot \nabla)\, \mathbf{u}.$$

Gravity

The effect of gravity acting on the parcel in Fig. 6.2 is straightforward: the gravitational force is $g\, \delta M$, and is directed downward,

$$\mathbf{F}_{gravity} = -g\rho\hat{\mathbf{z}}\, \delta x\, \delta y\, \delta z, \qquad (6\text{-}3)$$

where $\hat{\mathbf{z}}$ is the unit vector in the upward direction and g is assumed constant.

Pressure gradient

Another force acting on a fluid parcel is the pressure force within the fluid. Consider Fig. 6.3. On each face of our parcel there is a force (directed inward) acting on the parcel equal to the pressure on that face multiplied by the area of the face. On face A, for example, the force is

$$F(A) = p(x - \frac{\delta x}{2}, y, z)\, \delta y\, \delta z,$$

directed in the positive x-direction. Note that we have used the value of p at the midpoint of the face, which is valid for small $\delta y, \delta z$. On face B, there is an x-directed force

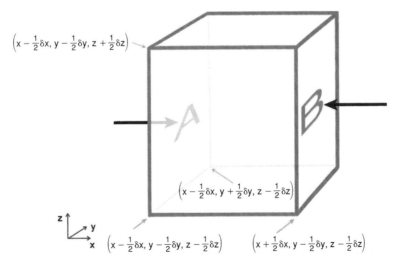

FIGURE 6.3. Pressure gradient forces acting on the fluid parcel. The pressure of the surrounding fluid applies a force to the right on face A and to the left on face B.

$$F(B) = -p(x + \frac{\delta x}{2}, y, z) \, \delta y \, \delta z,$$

which is negative (toward the left). Since these are the only pressure forces acting in the x-direction, the net x-component of the pressure force is

$$F_x = \left[p(x - \frac{\delta x}{2}, y, z) - p(x + \frac{\delta x}{2}, y, z) \right] \delta y \, \delta z.$$

If we perform a Taylor expansion (see Appendix A.2.1) about the midpoint of the parcel, we have

$$p(x + \frac{\delta x}{2}, y, z) = p(x, y, z) + \frac{\delta x}{2} \left(\frac{\partial p}{\partial x} \right),$$

$$p(x - \frac{\delta x}{2}, y, z) = p(x, y, z) - \frac{\delta x}{2} \left(\frac{\partial p}{\partial x} \right),$$

where the pressure gradient is evaluated at the midpoint of the parcel, and where we have neglected the small terms of $O(\delta x^2)$ and higher. Therefore the x-component of the pressure force is

$$F_x = -\frac{\partial p}{\partial x} \, \delta x \, \delta y \, \delta z.$$

It is straightforward to apply the same procedure to the faces perpendicular to the y- and z-directions, to show that these components are

$$F_y = -\frac{\partial p}{\partial y} \, \delta x \, \delta y \, \delta z,$$

$$F_z = -\frac{\partial p}{\partial z} \, \delta x \, \delta y \, \delta z.$$

In total, therefore, the net pressure force is given by the vector

$$\begin{aligned}
\mathbf{F}_{pressure} &= \left(F_x, F_y, F_z \right) \\
&= -\left(\frac{\partial p}{\partial x}, \frac{\partial p}{\partial y}, \frac{\partial p}{\partial z} \right) \delta x \, \delta y \, \delta z \\
&= -\nabla p \, \delta x \, \delta y \, \delta z.
\end{aligned} \qquad (6\text{-}4)$$

Note that the net force depends only on the *gradient* of pressure, ∇p; clearly, a uniform pressure applied to all faces of the parcel would not introduce any net force.

Friction

For typical atmospheric and oceanic flows, frictional effects are negligible except close to boundaries where the fluid rubs over the Earth's surface. The atmospheric boundary layer—which is typically a few hundred meters to 1 km or so deep—is exceedingly complicated. For one thing, the surface is not smooth; there are mountains, trees, and other irregularities that increase the exchange of momentum between the air and the ground. (This is the main reason why frictional effects are greater over land than over ocean.) For another, the boundary layer is usually *turbulent*, containing many small-scale and often vigorous eddies; these eddies can act somewhat like mobile molecules and diffuse momentum more effectively than molecular viscosity. The same can be said of oceanic boundary layers, which are subject, for example, to the stirring by turbulence generated by the action of the wind, as will be discussed in Section 10.1. At this stage, we will not attempt to describe such effects quantitatively but instead write the consequent frictional force on a fluid parcel as

$$\mathbf{F}_{fric} = \rho \, \mathcal{F} \, \delta x \, \delta y \, \delta z, \qquad (6\text{-}5)$$

where, for convenience, $\mathcal{F}$ is the frictional force *per unit mass*. For the moment we will not need a detailed theory of this term. Explicit forms for $\mathcal{F}$ will be discussed and employed in Sections 7.4.2 and 10.1.

6.2.2. The equations of motion

Putting all this together, Eq. 6-2 gives us

$$\rho \, \delta x \, \delta y \, \delta z \frac{D\mathbf{u}}{Dt} = \mathbf{F}_{gravity} + \mathbf{F}_{pressure} + \mathbf{F}_{fric}.$$

Substituting from Eqs. 6-3, 6-4, and 6-5, and rearranging slightly, we obtain

$$\frac{D\mathbf{u}}{Dt} + \frac{1}{\rho}\nabla p + g\hat{\mathbf{z}} = \mathcal{F}. \qquad (6\text{-}6)$$

This is our equation of motion for a fluid parcel.

Note that because of our use of vector notation, Eq. 6-6 seems rather simple. However, when written out in component form, as below, it becomes somewhat intimidating, even in Cartesian coordinates:

$$\frac{\partial u}{\partial t} + u\frac{\partial u}{\partial x} + v\frac{\partial u}{\partial y} + w\frac{\partial u}{\partial z} + \frac{1}{\rho}\frac{\partial p}{\partial x} = \mathcal{F}_x \quad (a)$$

$$\frac{\partial v}{\partial t} + u\frac{\partial v}{\partial x} + v\frac{\partial v}{\partial y} + w\frac{\partial v}{\partial z} + \frac{1}{\rho}\frac{\partial p}{\partial y} = \mathcal{F}_y \quad (b)$$

$$\frac{\partial w}{\partial t} + u\frac{\partial w}{\partial x} + v\frac{\partial w}{\partial y} + w\frac{\partial w}{\partial z} + \frac{1}{\rho}\frac{\partial p}{\partial z} + g = \mathcal{F}_z. \quad (c)$$

$$(6\text{-}7)$$

Fortunately we will often be able to make a number of simplifications. One such simplification, for example, is that, as discussed in Section 3.2, large-scale flow in the atmosphere and ocean is almost always close to hydrostatic balance, allowing Eq. 6-7c to be radically simplified as follows.

6.2.3. Hydrostatic balance

From the vertical equation of motion, Eq. 6-7c, we can see that if friction and the vertical acceleration Dw/Dt are negligible, we obtain

$$\frac{\partial p}{\partial z} = -\rho g, \qquad (6\text{-}8)$$

thus recovering the equation of hydrostatic balance, Eq. 3-3. For large-scale atmospheric and oceanic systems in which the vertical motions are weak, the hydrostatic equation is almost always accurate, though it may break down in vigorous systems of smaller horizontal scale such as convection.[3]

6.3. CONSERVATION OF MASS

In addition to Newton's laws there is a further constraint on the fluid motion: *conservation of mass*. Consider a fixed *fluid volume* as illustrated in Fig. 6.4. The volume has dimensions $(\delta x, \delta y, \delta z)$. The mass of the fluid occupying this volume, $\rho\,\delta x\,\delta y\,\delta z$, may change with time if ρ does so. However, mass continuity tells us that this can only

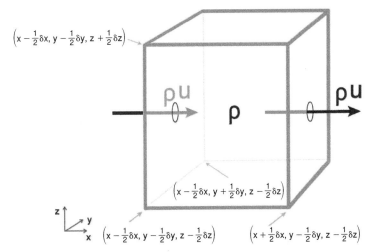

FIGURE 6.4.   The mass of fluid contained in the fixed volume, $\rho\delta x\,\delta y\,\delta z$, can be changed by fluxes of mass out of and into the volume, as marked by the arrows.

[3]It might appear from Eq. 6-7c that $|Dw/Dt| \ll g$ is a sufficient condition for the neglect of the acceleration term. This indeed is almost always satisfied. However, for hydrostatic balance to hold to sufficient accuracy to be useful, the condition is actually $|Dw/Dt| \ll g\Delta\rho/\rho$, where $\Delta\rho$ is a typical density variation on a pressure surface. Even in quite extreme conditions this more restrictive condition turns out to be very well satisfied.

occur if there is a flux of mass into (or out of) the volume, meaning that

$$\frac{\partial}{\partial t} \left(\rho \, \delta x \, \delta y \, \delta z \right) = \frac{\partial \rho}{\partial t} \, \delta x \, \delta y \, \delta z$$

$$= \left(\text{net mass flux into the volume} \right).$$

Now the volume flux in the x-direction per unit time into the left face in Fig. 6.4 is $u \left(x - 1/2 \, \delta x, y, z \right) \delta y \, \delta z$, so the corresponding mass flux is $[\rho u] \left(x - 1/2 \, \delta x, y, z \right) \delta y \, \delta z$, where $[\rho u]$ is evaluated at the left face. The flux out through the right face is $[\rho u] \left(x + 1/2 \, \delta x, y, z \right) \delta y \, \delta z$; therefore the net mass import in the x-direction into the volume is (again employing a Taylor expansion)

$$-\frac{\partial}{\partial x} \left(\rho u \right) \delta x \, \delta y \, \delta z.$$

Similarly the rate of net import of mass in the y-direction is

$$-\frac{\partial}{\partial y} \left(\rho v \right) \delta x \, \delta y \, \delta z,$$

and in the z-direction is

$$-\frac{\partial}{\partial z} \left(\rho w \right) \delta x \, \delta y \, \delta z.$$

Therefore the net mass flux into the volume is $-\nabla \cdot (\rho \mathbf{u}) \, \delta x \, \delta y \, \delta z$. Thus our *equation of continuity* becomes

$$\frac{\partial \rho}{\partial t} + \nabla \cdot (\rho \mathbf{u}) = 0. \qquad (6\text{-}9)$$

This has the general form of a physical conservation law:

$$\frac{\partial \, \text{Concentration}}{\partial t} + \nabla \cdot \left(\text{flux} \right) = 0$$

in the absence of sources and sinks.

Using the *total derivative D/Dt*, Eq. 6-1, and noting that $\nabla \cdot (\rho \mathbf{u}) = \rho \nabla \cdot \mathbf{u} + \mathbf{u} \cdot \nabla \rho$ (see the vector identities listed in Appendix A.2.2) we may therefore rewrite Eq. 6-9 in the alternative, and often very useful, form:

$$\frac{D\rho}{Dt} + \rho \nabla \cdot \mathbf{u} = 0. \qquad (6\text{-}10)$$

6.3.1. Incompressible flow

For incompressible flow (e.g., for a liquid such as water in our laboratory tank or in the ocean), the following simplified approximate form of the continuity equation almost always suffices:

$$\nabla \cdot \mathbf{u} = \frac{\partial u}{\partial x} + \frac{\partial v}{\partial y} + \frac{\partial w}{\partial z} = 0. \qquad (6\text{-}11)$$

Indeed this is the definition of incompressible flow: it is *nondivergent*—no bubbles allowed! Note that in any real fluid, Eq. 6-11 is never *exactly* obeyed. Moreover, despite Eq. 6-10, use of the incompressibility condition should not be understood as implying that $\frac{D\rho}{Dt} = 0$. On the contrary, the density of a parcel of water can be changed by internal heating and/or conduction (see, for example, Section 11.1). Although these density changes may be large enough to affect the buoyancy of the fluid parcel, they are too small to affect the mass budget. For example, the thermal expansion coefficient of water is typically $2 \times 10^{-4} \, \text{K}^{-1}$, and so the volume of a parcel of water changes by only 0.02% per degree of temperature change.

6.3.2. Compressible flow

A compressible fluid, such as air, is nowhere close to being nondivergent—ρ changes markedly as fluid parcels expand and contract. This is inconvenient in the analysis of atmospheric dynamics. However it turns out that, provided the hydrostatic assumption is valid (as it nearly always is), one can get around this inconvenience by adopting pressure coordinates. In pressure coordinates, (x, y, p), the elemental fixed "volume" is $\delta x \, \delta y \, \delta p$. Since $z = z \, (x, y, p)$, the vertical dimension of the elemental volume (in geometric coordinates) is $\delta z = \partial z / \partial p \, \delta p$, and so its mass is δM given by

$$\delta M = \rho \, \delta x \, \delta y \, \delta z$$

$$= \rho \left(\frac{\partial p}{\partial z} \right)^{-1} \delta x \, \delta y \, \delta p$$

$$= -\frac{1}{g} \, \delta x \, \delta y \, \delta p,$$

where we have used hydrostatic balance, Eq. 3-3. So the mass of an elemental fixed volume *in pressure coordinates* cannot change! In effect, comparing the top and bottom line of the previous equation, the equivalent of "density" in pressure coordinates—the mass per unit "volume"—is $1/g$, a constant. Hence, in the pressure-coordinate version of the continuity equation, there is no term representing rate of change of density; it is simply

$$\nabla_p \cdot \mathbf{u}_p = \frac{\partial u}{\partial x} + \frac{\partial v}{\partial y} + \frac{\partial \omega}{\partial p} = 0, \qquad (6\text{-}12)$$

where the subscript p reminds us that we are in pressure coordinates. The greater simplicity of this form of the continuity equation, as compared to Eqs. 6-9 or 6-10, is one of the reasons why pressure coordinates are favored in meteorology.

6.4. THERMODYNAMIC EQUATION

The equation governing the evolution of temperature can be derived from the first law of thermodynamics applied to a moving parcel of fluid. Dividing Eq. 4-12 by δt and letting $\delta t \longrightarrow 0$ we find:

$$\frac{DQ}{Dt} = c_p \frac{DT}{Dt} - \frac{1}{\rho} \frac{Dp}{Dt}. \qquad (6\text{-}13)$$

DQ/Dt is known as the *diabatic heating rate* per unit mass. In the atmosphere, this is mostly due to latent heating and cooling (from condensation and evaporation of H_2O) and radiative heating and cooling (due to absorption and emission of radiation). If the heating rate is zero then $DT/Dt = \frac{1}{\rho c_p} Dp/Dt$, and, as discussed in Section 4.3.1, the temperature of a parcel will decrease in ascent (as it moves to lower pressure) and increase in descent (as it moves to higher pressure). Of course this is why we introduced potential temperature in Section 4.3.2; in adiabatic motion, θ

is conserved. Written in terms of θ, Eq. 6-13 becomes

$$\frac{D\theta}{Dt} = \left(\frac{p}{p_0}\right)^{-\kappa} \frac{\dot{Q}}{c_p}, \qquad (6\text{-}14)$$

where $\dot{Q}$ (with a dot over the top) is a shorthand for $\frac{DQ}{Dt}$. Here θ is given by Eq. 4-17, the factor $\left(\frac{p}{p_0}\right)^{-\kappa}$ converts from T to θ, and $\frac{\dot{Q}}{c_p}$ is the diabatic heating in units of $K\,s^{-1}$. The analogous equations that govern the evolution of temperature and salinity in the ocean will be discussed in Chapter 11.

6.5. INTEGRATION, BOUNDARY CONDITIONS, AND RESTRICTIONS IN APPLICATION

The three equations in 6–7, together with 6–11 or 6–12, and 6–14 are our five equations in five unknowns. Together with initial conditions and boundary conditions, they are sufficient to determine the evolution of the flow.

Before going on, we make some remarks about restrictions in the application of our governing equations. The equations themselves apply very accurately to the detailed motion. In practice, however, variables are always averages over large volumes. We can only tentatively suppose that the equations are applicable to the average motion, such as the wind integrated over a 100-km square box. Indeed, the assumption that the equations do apply to average motion is often incorrect. This fact is associated with the representation of turbulent scales, both small scale and large scale. The treatment of turbulent motions remains one of the major challenges in dynamical meteorology and oceanography. Finally, our governing equations have been derived relative to a 'fixed' coordinate system. As we now go on to discuss, this is not really a restriction, but is usually an inconvenience.

6.6. EQUATIONS OF MOTION FOR A ROTATING FLUID

Equation 6-6 is an accurate representation of Newton's laws applied to a fluid observed from a fixed, inertial, frame of reference. However, we live on a rotating planet and observe winds and currents in its rotating frame. For example the winds shown in Fig. 5.20 are not the winds that would be observed by someone looking back at the Earth, as in Fig. 1. Rather, they are the winds measured by observers on the planet rotating with it. In most applications it is easier and more desirable to work with the governing equations in a frame rotating with the Earth. Moreover it turns out that rotating fluids have rather unusual properties and these properties are often most easily appreciated in the rotating frame. To proceed, then, we must write down our governing equations in a rotating frame. However, before going on to a formal "frame of reference" transformation of the governing equations, we describe a laboratory experiment that vividly illustrates the influence of rotation on fluid motion and demonstrates the utility of viewing and thinking about fluid motion in a rotating frame.

6.6.1. GFD Lab III: Radial inflow

We are all familiar with the swirl and gurgling sound of water flowing down a drain. Here we set up a laboratory illustration of this phenomenon and study it in rotating and nonrotating conditions. We rotate a cylinder about its vertical axis; the cylinder has a circular drain hole in the center of its bottom, as shown in Fig. 6.5. Water enters at a constant rate through a diffuser on its outer wall and exits through the drain. In so doing, the angular momentum imparted to the fluid by the rotating cylinder is conserved as it flows inwards, and paper dots floated on the surface acquire the swirling motion seen in Fig. 6.6 as the distance of the dots from the axis of rotation decreases.

The swirling flow exhibits a number of important principles of rotating fluid dynamics—conservation of angular momentum, geostrophic (and cyclostrophic) balance (see Section 7.1)—all of which will be used in our subsequent discussions. The experiment also gives us an opportunity to think about frames of reference because it is viewed by a camera co-rotating with the cylinder.

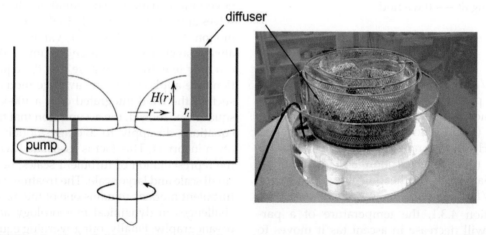

FIGURE 6.5. The radial inflow apparatus. A diffuser of 30-cm inside diameter is placed in a larger tank and used to produce an axially symmetric, inward flow of water toward a drain hole at the center. Below the tank there is a large catch basin, partially filled with water and containing a submersible pump whose purpose is to return water to the diffuser in the upper tank. The whole apparatus is then placed on a turntable and rotated in an anticlockwise direction. The path of fluid parcels is tracked by dropping paper dots on the free surface. See Whitehead and Potter (1977).

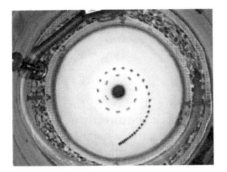
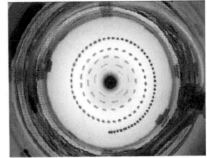

FIGURE 6.6. Trajectories of particles in the radial inflow experiment viewed in the rotating frame. The positions are plotted every 1/30 s. On the left $\Omega = 5$ rpm (revolutions per minute). On the right $\Omega = 10$ rpm. Note how the pitch of the particle trajectory increases as Ω increases, and how in both cases the speed of the particles increases as the radius decreases.

Observed flow patterns

When the apparatus is not rotating, water flows radially inward from the diffuser to the drain in the middle. The free surface is observed to be rather flat. When the apparatus is rotated, however, the water acquires a swirling motion: fluid parcels *spiral* inward, as can be seen in Fig. 6.6. Even at modest rotation rates of $\Omega = 10$ rpm (corresponding to a rotation period of around 6 seconds),[4] the effect of rotation is marked and parcels complete many circuits before finally exiting through the drain hole. The azimuthal speed of the particle increases as it spirals inwards, as indicated by the increase in the spacing of the particle positions in the figure. In the presence of rotation the free surface becomes markedly curved, high at the periphery and plunging downwards toward the hole in the center, as shown in the photograph, Fig. 6.7.

Dynamical balances

In the limit in which the tank is rotated rapidly, parcels of fluid circulate around many times before falling out through the drain hole (see the right hand frame of Fig. 6.6); the pressure gradient force directed radially inward (set up by the free surface tilt) is in large part balanced by a centrifugal force directed radially outward.

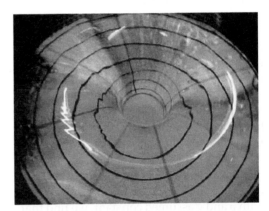

FIGURE 6.7. The free surface of the radial inflow experiment. The curved surface provides a pressure gradient force directed inward that is balanced by an outward centrifugal force due to the anticlockwise circulation of the spiraling flow.

If V_θ is the azimuthal velocity in the absolute frame (the frame of the laboratory) and v_θ is the azimuthal speed *relative* to the tank (measured using the camera co-rotating with the apparatus) then (see Fig. 6.8)

$$V_\theta = v_\theta + \Omega r, \qquad (6\text{-}15)$$

where Ω is the rate of rotation of the tank in radians per second. Note that Ωr is the azimuthal speed of a particle stationary relative to the tank at radius r from the axis of rotation.

---

[4]An Ω of 10 rpm (revolutions per minute) is equivalent to a rotation period $\tau = \frac{60}{10} = 6\,\mathrm{s}$. Various measures of rotation rate are set out in Table A.4 of the appendix.

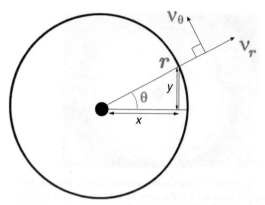

FIGURE 6.8.   The velocity of a fluid parcel viewed in the rotating frame of reference: $v_{rot} = (v_\theta, v_r)$ in polar coordinates (see Appendix A.2.3).

We now consider the balance of forces in the vertical and radial directions, expressed first in terms of the absolute velocity V_θ and then in terms of the relative velocity v_θ.

Vertical force balance   We suppose that hydrostatic balance pertains in the vertical, Eq. 3-3. Integrating in the vertical and noting that the pressure vanishes at the free surface (actually $p =$ atmospheric pressure at the surface, which here can be taken as zero), and with ρ and g assumed constant, we find that

$$p = \rho g(H - z),\qquad (6\text{-}16)$$

where $H(r)$ is the height of the free surface (where $p = 0$), and we suppose that $z = 0$ (increasing upwards) on the base of the tank (see Fig. 6.5, left).

Radial force balance in the non-rotating frame   If the pitch of the spiral traced out by fluid particles is tight (i.e., in the limit that $\frac{v_r}{v_\theta} \ll 1$, appropriate when Ω is sufficiently large) then the centrifugal force directed radially outward acting on a particle of fluid is balanced by the pressure gradient force directed inward associated with the tilt of

the free surface. This radial force balance can be written in the nonrotating frame thus:

$$\frac{V_\theta^2}{r} = \frac{1}{\rho}\frac{\partial p}{\partial r}.$$

Using Eq. 6-16, the radial pressure gradient can be directly related to the gradient of free surface height, enabling the force balance to be written[5]:

$$\frac{V_\theta^2}{r} = g\frac{\partial H}{\partial r}\qquad (6\text{-}17)$$

Radial force balance in the rotating frame   Using Eq. 6-15, we can express the centrifugal acceleration in Eq. 6-17 in terms of velocities in the rotating frame thus:

$$\frac{V_\theta^2}{r} = \frac{(v_\theta + \Omega r)^2}{r} = \frac{v_\theta^2}{r} + 2\Omega v_\theta + \Omega^2 r. \quad (6\text{-}18)$$

Hence

$$\frac{v_\theta^2}{r} + 2\Omega v_\theta + \Omega^2 r = g\frac{\partial H}{\partial r}.\qquad (6\text{-}19)$$

Equation 6.19 can be simplified by measuring the height of the free surface relative to that of a reference parabolic surface (see Section 6.6.4),[6] $\frac{\Omega^2 r^2}{2g}$, as follows:

$$\eta = H - \frac{\Omega^2 r^2}{2g}.\qquad (6\text{-}20)$$

Then, since $\partial\eta/\partial r = \partial H/\partial r - \Omega^2 r/g$, Eq. 6-19 can be written in terms of η thus:

$$\frac{v_\theta^2}{r} = g\frac{\partial\eta}{\partial r} - 2\Omega v_\theta.\qquad (6\text{-}21)$$

Equation 6-17 (nonrotating) and Eq. 6-21 (rotating) are equivalent statements of the balance of forces. The distinction between them is that the former is expressed in terms

---

[5]Note that the balance Eq. 6-17 cannot hold *exactly* in our experiment, because radial accelerations must be present associated with the flow of water inward from the diffuser to the drain. But if these acceleration terms are small the balance Eq. 6-17 is a good approximation.

[6]By doing so, we thus eliminate from Eq. 6-19 the centrifugal term $\Omega^2 r$ associated with the background rotation. We will follow a similar procedure for the spherical Earth in Section 6.6.3 (see also GFD Lab IV in Section 6.6.4)

of V_θ, the latter in terms of v_θ. Note that Eq. 6-21 has the same form as Eq. 6-17, except (i) η (measured relative to the reference parabola) appears rather than H (measured relative to a flat surface), and (ii) an extra term, $-2\Omega v_\theta$, appears on the right hand side of Eq. 6-21; this is called the *Coriolis acceleration*. It has appeared because we have chosen to express our force balance in terms of *relative*, rather than absolute, velocities. We shall see that the Coriolis acceleration plays a central role in the dynamics of the atmosphere and ocean.

Angular momentum

Fluid entering the tank at the outer wall will have angular momentum, because the apparatus is rotating. At r_1 the radius of the diffuser in Fig. 6.5, fluid has velocity Ωr_1 and hence angular momentum Ωr_1^2. As parcels of fluid flow inward, they will conserve this angular momentum (provided that they are not rubbing against the bottom or the side). Thus conservation of angular momentum implies that

$$V_\theta\, r = \text{constant} = \Omega r_1^2, \qquad (6\text{-}22)$$

where V_θ is the azimuthal velocity at radius r in the laboratory (inertial) frame given by Eq. 6-15. Combining Eqs. 6-22 and 6-15 we find

$$v_\theta = \Omega \frac{\left(r_1^2 - r^2\right)}{r}. \qquad (6\text{-}23)$$

We thus see that the fluid acquires a sense of rotation, which is the same as that of the rotating table but which is greatly magnified at small r. If $\Omega > 0$, meaning that the table rotates in an anticlockwise sense, then the fluid acquires an anticlockwise (cyclonic[7]) swirl. If $\Omega < 0$ the table rotates in a clockwise (anticyclonic) direction and the fluid acquires a clockwise (anticyclonic) swirl. This can be clearly seen in the trajectories plotted in Fig. 6.6. Equation 6-23 is, in fact, a rather good prediction for the azimuthal speed of the particles seen in Fig. 6.6.

We will return to this experiment later in Section 7.1.3 where we discuss the balance of terms in Eq. 6-21 and its relationship to atmospheric flows.

6.6.2. Transformation into rotating coordinates

In our radial inflow experiment we expressed the balance of forces in both the nonrotating and rotating frames. We have already written down the equations of motion of a fluid in a nonrotating frame, Eq. 6-6. Let us now formally transform it in to a rotating reference frame. The only tricky part is the acceleration term, $D\mathbf{u}/Dt$, which requires manipulations analogous to Eq. 6-18 but in a general framework. We need to figure out how to transform the operator D/Dt (acting on a vector) into a rotating frame. Of course, D/Dt of a scalar is the same in both frames, since this means "the rate of change of the scalar following a fluid parcel." The same fluid parcel is followed from both frames, and so scalar quantities (e.g., temperature or pressure) do not change when viewed from the different frames. However, a vector is not invariant under such a transformation, since the coordinate directions relative to which the vector is expressed are different in the two frames.

A clue is given by noting that the velocity in the absolute (inertial) frame $\mathbf{u}_{in}$ and the velocity in the rotating frame $\mathbf{u}_{rot}$, are related (see Fig. 6.9) through

$$\mathbf{u}_{in} = \mathbf{u}_{rot} + \Omega \times \mathbf{r}, \qquad (6\text{-}24)$$

where $\mathbf{r}$ is the position vector of a parcel in the rotating frame, Ω is the rotation vector of the rotating frame of reference, and $\Omega \times \mathbf{r}$ is the vector product of Ω and $\mathbf{r}$. This is just a generalization (to vectors) of the transformation used in Eq. 6-15 to express the absolute velocity in terms of the relative velocity in our radial inflow experiment. As we shall now go on to show, Eq. 6-24 is a special case of a general "rule" for transforming

---

[7]The term cyclonic (anticyclonic) means that the swirl is in the same (opposite) sense as the background rotation.

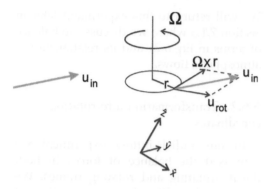

Inertial   Rotating

FIGURE 6.9.   On the left is the velocity vector of a particle $\mathbf{u}_{in}$ in the inertial frame. On the right is the view from the rotating frame. The particle has velocity $\mathbf{u}_{rot}$ in the rotating frame. The relation between $\mathbf{u}_{in}$ and $\mathbf{u}_{rot}$ is $\mathbf{u}_{in} = \mathbf{u}_{rot} + \Omega \times \mathbf{r}$, where $\Omega \times \mathbf{r}$ is the velocity of a particle fixed (not moving) in the rotating frame at position vector $\mathbf{r}$. The relationship between the rate of change of any vector $\mathbf{A}$ in the rotating frame and the change of $\mathbf{A}$ as seen in the inertial frame is given by: $(D\mathbf{A}/Dt)_{in} = (D\mathbf{A}/Dt)_{rot} + \Omega \times \mathbf{A}$.

the rate of change of vectors between frames, which we now derive.

Consider Fig. 6.9. In the rotating frame, any vector $\mathbf{A}$ may be written

$$\mathbf{A} = \hat{\mathbf{x}}A_x + \hat{\mathbf{y}}A_y + \hat{\mathbf{z}}A_z, \qquad (6\text{-}25)$$

where (A_x, A_y, A_z) are the components of $\mathbf{A}$ expressed instantaneously in terms of the three rotating coordinate directions, for which the unit vectors are $(\hat{\mathbf{x}}, \hat{\mathbf{y}}, \hat{\mathbf{z}})$. In the rotating frame, these coordinate directions are fixed, and so

$$\left(\frac{D\mathbf{A}}{Dt}\right)_{rot} = \hat{\mathbf{x}}\frac{DA_x}{Dt} + \hat{\mathbf{y}}\frac{DA_y}{Dt} + \hat{\mathbf{z}}\frac{DA_z}{Dt}.$$

However, *viewed from the inertial frame*, the coordinate directions in the rotating frame are not fixed, but are rotating at rate Ω, and so

$$\left(\frac{D\hat{\mathbf{x}}}{Dt}\right)_{in} = \Omega \times \hat{\mathbf{x}},$$

$$\left(\frac{D\hat{\mathbf{y}}}{Dt}\right)_{in} = \Omega \times \hat{\mathbf{y}},$$

$$\left(\frac{D\hat{\mathbf{z}}}{Dt}\right)_{in} = \Omega \times \hat{\mathbf{z}}.$$

Therefore, operating on Eq. 6-25,

$$\left(\frac{D\mathbf{A}}{Dt}\right)_{in} = \hat{\mathbf{x}}\frac{DA_x}{Dt} + \hat{\mathbf{y}}\frac{DA_y}{Dt} + \hat{\mathbf{z}}\frac{DA_z}{Dt}$$
$$+ \left(\frac{D\hat{\mathbf{x}}}{Dt}\right)_{in} A_x + \left(\frac{D\hat{\mathbf{y}}}{Dt}\right)_{in} A_y$$
$$+ \left(\frac{D\hat{\mathbf{z}}}{Dt}\right)_{in} A_z$$
$$= \hat{\mathbf{x}}\frac{DA_x}{Dt} + \hat{\mathbf{y}}\frac{DA_y}{Dt} + \hat{\mathbf{z}}\frac{DA_z}{Dt}$$
$$+ \Omega \times \left(\hat{\mathbf{x}}A_x + \hat{\mathbf{y}}A_y + \hat{\mathbf{z}}A_z\right),$$

whence

$$\left(\frac{D\mathbf{A}}{Dt}\right)_{in} = \left(\frac{D\mathbf{A}}{Dt}\right)_{rot} + \Omega \times \mathbf{A}, \qquad (6\text{-}26)$$

which yields our transformation rule for the operator $\frac{D}{Dt}$ acting on a vector.

Setting $\mathbf{A} = \mathbf{r}$, the position vector of the particle in the rotating frame, we arrive at Eq. 6-24. To write down the rate of change of velocity following a parcel of fluid in a rotating frame, $\left(\frac{D\mathbf{u}_{in}}{Dt}\right)_{in}$, we set $\mathbf{A} \longrightarrow \mathbf{u}_{in}$ in Eq. 6-26 using Eq. 6-24 thus:

$$\left(\frac{D\mathbf{u}_{in}}{Dt}\right)_{in} = \left[\left(\frac{D}{Dt}\right)_{rot} + \Omega \times\right](\mathbf{u}_{rot} + \Omega \times \mathbf{r})$$
$$= \left(\frac{D\mathbf{u}_{rot}}{Dt}\right)_{rot} + 2\Omega \times \mathbf{u}_{rot}$$
$$+ \Omega \times \Omega \times \mathbf{r}, \qquad (6\text{-}27)$$

since, by definition

$$\left(\frac{D\mathbf{r}}{Dt}\right)_{rot} = \mathbf{u}_{rot}.$$

Equation 6-27 is a more general statement of Eq. 6-18: we see that there is a one-to-one correspondence between the terms.

6.6.3. The rotating equations of motion

We can now write down our equation of motion in the rotating frame. Substituting from Eq. 6-27 into the inertial-frame equation of motion Eq. 6-6, we have, in the rotating frame,

$$\frac{D\mathbf{u}}{Dt} + \frac{1}{\rho}\nabla p + g\hat{\mathbf{z}}$$

$$= \underbrace{-2\Omega \times \mathbf{u}}_{\substack{\text{Coriolis}\\\text{accel}^n}} + \underbrace{-\Omega \times \Omega \times \mathbf{r}}_{\substack{\text{Centrifugal}\\\text{accel}^n}} + \mathcal{F},$$

$$(6\text{-}28)$$

where we have dropped the subscripts "rot" and it is now understood that $D\mathbf{u}/Dt$ and $\mathbf{u}$ refer to the rotating frame.

Note that Eq. 6-28 is the same as Eq. 6-6, except that $\mathbf{u} = \mathbf{u}_{rot}$ and "apparent" accelerations, introduced by the rotating reference frame, have been placed on the right-hand side of Eq. 6-28 (just as in Eq. 6-21). The apparent accelerations have been given names; the centrifugal acceleration ($-\Omega \times \Omega \times \mathbf{r}$) is directed radially outward (Fig. 6.9), and the Coriolis acceleration ($-2\Omega \times \mathbf{u}$) is directed "to the right" of the velocity vector if Ω is anticlockwise, sketched in Fig. 6.10. We now discuss these apparent accelerations in turn.

Centrifugal acceleration

As noted above, $-\Omega \times \Omega \times \mathbf{r}$ is directed radially outwards. If no other forces were acting on a particle, the particle would accelerate outwards. Because centrifugal acceleration can be expressed as the gradient of a potential,

$$-\Omega \times \Omega \times \mathbf{r} = \nabla\left(\frac{\Omega^2 r^2}{2}\right),$$

where r is the distance normal to the rotating axis (see Fig. 6.9) it is convenient to combine $\nabla\left(\frac{\Omega^2 r^2}{2}\right)$ with $g\hat{\mathbf{z}} = \nabla(gz)$, the gradient

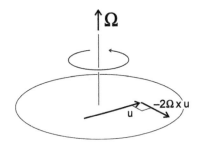

FIGURE 6.10. A fluid parcel moving with velocity u_{rot} in a rotating frame experiences a Coriolis acceleration, $-2\Omega \times u_{rot}$, directed "to the right" of u_{rot} if, as here, Ω is directed upwards, corresponding to anticlockwise rotation.

of the gravitational potential gz, and write Eq. 6-28 in the succinct form:

$$\frac{D\mathbf{u}}{Dt} + \frac{1}{\rho}\nabla p + \nabla\phi = -2\Omega \times \mathbf{u} + \mathcal{F}, \quad (6\text{-}29)$$

where

$$\phi = gz - \frac{\Omega^2 r^2}{2} \qquad (6\text{-}30)$$

is a modified (by centrifugal accelerations) gravitational potential "measured" in the rotating frame.[8] In this way gravitational and centrifugal accelerations can be conveniently combined in to a "measured" gravity, $\nabla\phi$. This is discussed in Section 6.6.4 at some length in the context of experiments with a parabolic rotating surface (GFD Lab IV).

Coriolis acceleration

The first term on the right hand side of Eq. 6-28 is the "Coriolis acceleration."[9] It describes a tendency for fluid parcels to turn, as shown in Fig. 6.10 and investigated

---

[8]Note that $\frac{\phi}{g} = z - \frac{\Omega^2 r^2}{2g}$ is directly analogous to Eq. 6-20 adopted in the analysis of the radial inflow experiment.

[9]  Gustave Gaspard Coriolis (1792–1843). French mathematician who discussed what we now refer to as the "Coriolis force" in addition to the already-known centrifugal force. The explanation of the effect sprang from problems of early 19th-century industry, for example, rotating machines, such as water-wheels.

in GFD Lab V. (Note that in this figure, the rotation is anticlockwise, meaning that $\Omega > 0$, like that of the northern hemisphere viewed from above the north pole; for the southern hemisphere, the effective sign of rotation is reversed; see Fig. 6.17.)

In the absence of any other forces acting on it, a fluid parcel would accelerate as

$$\frac{D\mathbf{u}}{Dt} = -2\mathbf{\Omega} \times \mathbf{u}. \qquad (6\text{-}31)$$

With the signs shown, the parcel would turn to the right in response to the Coriolis force (to the left in the southern hemisphere). Note that since, by definition, $(\mathbf{\Omega} \times \mathbf{u}) \cdot \mathbf{u} = 0$, the Coriolis force is *workless*: it does no work, but merely acts to change the flow direction.

To breathe some life in to these acceleration terms, we will now describe experiments with a parabolic rotating surface.

6.6.4. GFD Labs IV and V: Experiments with Coriolis forces on a parabolic rotating table

GFD Lab IV: *Studies of parabolic equipotential surfaces*

We fill a tank with water, set it turning, and leave it until it comes into solid body rotation, which is, the state in which fluid parcels have zero velocity in the rotating frame of reference. This is easily determined by viewing a paper dot floating on the free surface from a co-rotating camera. We note that the free surface of the water is not flat; it is depressed in the middle and rises up to its highest point along the rim of the tank, as sketched in Fig. 6.11. What's going on?

In solid body rotation, $\mathbf{u} = 0$, $\mathcal{F} = 0$, and so Eq. 6-29 reduces to $\frac{1}{\rho}\nabla p + \nabla\phi = 0$ (a generalization of hydrostatic balance to the rotating frame). For this to be true,

$$\frac{p}{\rho} + \phi = \text{constant}$$

everywhere in the fluid (note that here we are assuming $\rho = $ constant). Thus on surfaces where $p = $ constant, ϕ must be constant too; meaning that p and ϕ surfaces must be coincident with one another.

At the free surface of the fluid, $p = 0$. Thus, from Eq. 6-30

$$gz - \frac{\Omega^2 r^2}{2} = \text{constant}, \qquad (6\text{-}32)$$

the modified gravitational potential. We can determine the constant of proportionality by noting that at $r = 0$, $z = h(0)$, the height of the fluid in the middle of the tank (see

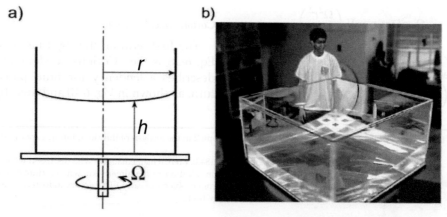

FIGURE 6.11. (a) Water placed in a rotating tank and insulated from external forces (both mechanical and thermodynamic) eventually comes into solid body rotation, in which the fluid does not move relative to the tank. In such a state the free surface of the water is not flat but takes on the shape of a parabola given by Eq. 6-33. (b) Parabolic free surface of water in a tank of 1 m square rotating at $\Omega = 20$ rpm.

Fig. 6.11a). Hence the depth of the fluid h is given by:

$$h(r) = h(0) + \frac{\Omega^2 r^2}{2g}, \qquad (6\text{-}33)$$

where r is the distance from the axis or rotation. Thus the free surface takes on a parabolic shape: it tilts so that it is always perpendicular to the vector $\mathbf{g}^*$ (gravity modified by centrifugal forces) given by $\mathbf{g}^* = -g\hat{\mathbf{z}} -\Omega \times \Omega \times \mathbf{r}$. If we hung a plumb line in the frame of the rotating table it would point in the direction of $\mathbf{g}^*$, which is slightly outward rather than directly down. The surface given by Eq. 6-33 is the reference to which H is compared to define η in Eq. 6-20.

Let us estimate the tilt of the free surface of the fluid by inserting numbers into Eq. 6-33 typical of our tank. If the rotation rate is 10 rpm (so $\Omega \simeq 1$ s^{-1}), and the radius of the tank is 0.30 m, then with $g = 9.81$ m s^{-2}, we find $\frac{\Omega^2 r^2}{2g} \sim 5$ mm, a noticeable effect but a small fraction of the depth to which the tank is typically filled. If one uses a large tank at high rotation, however (see Fig. 6.11b in which a 1 m square tank was rotated at a rate of 20 rpm) the distortion of the free surface can be very marked. In this case $\frac{\Omega^2 r^2}{2g} \sim 0.2$ m.

It is very instructive to construct a smooth parabolic surface on which one can roll objects. This can be done by filling a large flat-bottomed pan with resin on a turntable and letting the resin harden while the turntable is left running for several hours (this is how parabolic mirrors are made). The resulting parabolic surface can then be polished to create a low friction surface. The surface defined by Eq. 6-32 is an equipotential surface of the rotating frame, and so a body carefully placed on it at rest (in the rotating frame) should remain at rest. Indeed if we place a ball bearing on the rotating parabolic surface and make sure that the table is rotating at the same speed as was used to create the parabola, then we see that it does not fall into the center, but instead finds a state of rest in which the

component of gravitational force, g_H, resolved along the parabolic surface is exactly balanced by the outward-directed horizontal component of the centrifugal force, $(\Omega^2 r)_H$, as sketched in Fig. 6.12 and seen in action in Fig. 6.13.

GFD Lab V: Visualizing the Coriolis force

We can use the parabolic surface discussed in Lab IV in conjunction with a dry ice "puck" to help us visualize the Coriolis force. On the surface of the parabola, $\phi =$ constant and so $\nabla \phi = 0$. We can also assume that there are no pressure gradients acting on the puck because the air is so thin.

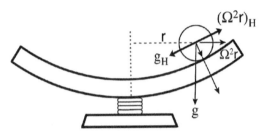

FIGURE 6.12. If a parabola of the form given by Eq. (6-33) is spun at rate Ω, then a ball carefully placed on it at rest does not fall in to the center but remains at rest: gravity resolved parallel to the surface, g_H, is exactly balanced by centrifugal accelerations resolved parallel to the surface, $(\Omega^2 r)_H$.

FIGURE 6.13. Studying the trajectories of ball bearings on a rotating parabola as described in Durran and Domonkos (1996). A co-rotating camera views and records the scene from above.

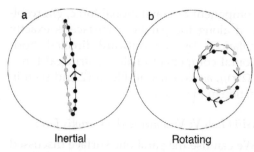

Inertial   Rotating

FIGURE 6.14.   Trajectory of the puck on the rotating parabolic surface during one rotation period of the table $2\pi/\Omega$ in (a) the inertial frame, and (b) the rotating frame of reference. The parabola is rotating in an anticlockwise (cyclonic) sense.

Furthermore, the gas sublimating off the bottom of the dry ice almost eliminates frictional coupling between the puck and the surface of the parabolic dish; thus we may also assume $\mathcal{F} = 0$. Hence the balance Eq. 6-31 applies.[10]

We can play games with the puck and study its trajectory on the parabolic turntable, both in the rotating and laboratory frames. It is useful to view the puck from the rotating frame using an overhead co-rotating camera. Fig. 6.14 plots the trajectory of the puck in the inertial (left) and in the rotating (right) frame. Notice that the puck is "deflected to the right" by the Coriolis force when viewed in the rotating frame, if the table is turning anticlockwise (cyclonically). The following are useful reference experiments:

1. We place the puck so that it is motionless in the rotating frame of reference. It follows a circular orbit around the center of the dish in the laboratory frame.

2. We launch the puck on a trajectory that crosses the rotation axis. Viewed from the laboratory the puck moves backward and forward along a straight line (the straight line expands out in to an ellipse if the frictional coupling between the puck and the rotating disc

is not negligible; see Fig. 6.14a). When viewed in the rotating frame, however, the particle is continuously deflected to the right and its trajectory appears as a circle as seen in Fig. 6.14b. This is the "deflecting force" of Coriolis. These circles are called *inertial circles*. (We will look at the theory of these circles below).

3. We place the puck on the parabolic surface again so that it appears stationary in the rotating frame, but is then slightly perturbed. In the rotating frame, the puck undergoes inertial oscillations consisting of small circular orbits passing through the initial position of the unperturbed puck.

Inertial circles   It is straightforward to analyze the motion of the puck in GFD Lab V. We adopt a Cartesian (x, y) coordinate in the rotating frame of reference whose origin is at the center of the parabolic surface. The velocity of the puck on the surface is $\mathbf{u}_{rot} = (u, v)$, where $u_{rot} = dx/dt$, $v_{rot} = dy/dt$, and we have reintroduced the subscript rot to make our frame of reference explicit. Further we assume that z increases upward in the direction of Ω.

Rotating frame   Let us write out Eq. 6-31 in component form (replacing D/Dt by d/dt, the rate of change of a property of the puck). Noting that (see VI, Appendix A.2.2)

$$2\Omega \times \mathbf{u}_{rot} = (0, 0, 2\Omega) \times (u_{rot}, v_{rot}, 0)$$
$$= (-2\Omega v_{rot}, 2\Omega u_{rot}, 0),$$

the two horizontal components of Eq. 6-31 are:

$$\frac{du_{rot}}{dt} - 2\Omega v_{rot} = 0; \quad \frac{dv_{rot}}{dt} + 2\Omega u_{rot} = 0 \quad (6\text{-}34)$$

$$u_{rot} = \frac{dx}{dt}; v_{rot} = \frac{dy}{dt}.$$

If we launch the puck from the origin of our coordinate system $x(0) = 0$; $y(0) = 0$

---

[10]A ball bearing can also readily be used for demonstration purposes, but it is not quite as effective as a dry ice puck.

(chosen to be the center of the rotating dish), with speed $u_{rot}(0) = 0$; $v_{rot}(0) = v_o$, the solution to Eq. 6-34 satisfying these boundary conditions is

$$u_{rot}(t) = v_o \sin 2\Omega t; \quad v_{rot}(t) = v_o \cos 2\Omega t$$

$$x(t) = \frac{v_o}{2\Omega} - \frac{v_o}{2\Omega} \cos 2\Omega t; \quad y(t) = \frac{v_o}{2\Omega} \sin 2\Omega t.$$
$$(6\text{-}35)$$

The puck's trajectory in the rotating frame is a circle (see Fig. 6.15) with a radius of $\frac{v_o}{2\Omega}$. It moves around the circle in a clockwise direction (anticyclonically) with a period π/Ω, known as the "inertial period." Note from Fig. 6.14 that in the rotating frame the puck is observed to complete two oscillation periods in the time it takes to complete just one in the inertial frame.

Inertial frame Now let us consider the same problem but in the nonrotating frame. The acceleration in a frame rotating at angular velocity Ω is related to the acceleration in an inertial frame of reference by Eq. 6-27. And so, if the balance of forces is $D\mathbf{u}_{rot}/Dt = -2\Omega \times \mathbf{u}_{rot}$ these two terms cancel out in Eq. 6-27, and it reduces to[11]:

$$\frac{d\mathbf{u}_{in}}{dt} = \Omega \times \Omega \times \mathbf{r} \qquad (6\text{-}36)$$

If the origin of our inertial coordinate system lies at the center of our dish, then the above can be written out in component form thus:

$$\frac{du_{in}}{dt} + \Omega^2 x = 0; \quad \frac{dv_{in}}{dt} + \Omega^2 y = 0 \qquad (6\text{-}37)$$

where the subscript $_{in}$ means inertial. This should be compared to the equation of motion in the rotating frame—see Eq. 6-34.

The solution of Eq. 6-37 satisfying our boundary conditions is:

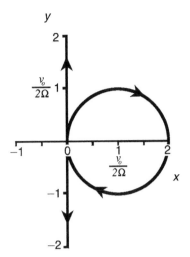

FIGURE 6.15. Theoretical trajectory of the puck during one complete rotation period of the table, $2\pi/\Omega$, in GFD Lab V in the inertial frame (straight line) and in the rotating frame (circle). We launch the puck from the origin of our coordinate system $x(0) = 0$; $y(0) = 0$ (chosen to be the center of the rotating parabola) with speed $u(0) = 0$; $v(0) = v_0$. The horizontal axes are in units of $v_0/2\Omega$. Observed trajectories are shown in Fig. 6.14.

$$u_{in}(t) = 0; \quad v_{in}(t) = v_o \cos \Omega t$$

$$x_{in}(t) = 0; \quad y_{in}(t) = \frac{v_o}{\Omega} \sin \Omega t. \qquad (6\text{-}38)$$

The trajectory in the inertial frame is a straight line shown in Fig. 6.15. Comparing Eqs. 6-35 and 6-38, we see that the length of the line marked out in the inertial frame is **twice** the diameter of the inertial circle in the rotating frame and the frequency of the oscillation is **one half** that observed in the rotating frame, just as observed in Fig. 6.14.

The above solutions go a long way to explaining what is observed in GFD Lab V and expose many of the curiosities of rotating versus nonrotating frames of reference. The deflection "to the right" by the Coriolis force is indeed a consequence of the rotation

---

[11]Note that if there are no frictional forces between the puck and the parabolic surface, then the *rotation* of the surface is of no consequence to the trajectory of the puck. The puck just oscillates back and forth according to:

$$\frac{d^2 r}{dt^2} = -g\frac{dh}{dr} = -\Omega^2 r,$$

where we have used the result that h, the shape of the parabolic surface, is given by Eq. 6-33. This is another (perhaps more physical) way of arriving at Eq. 6-37.

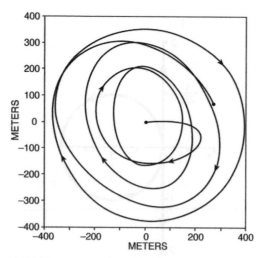

FIGURE 6.16.   Inertial circles observed by a current meter in the main thermocline of the Atlantic Ocean at a depth of 500 m; 28°N, 54°W. Five inertial periods are shown. The inertial period at this latitude is 25.6 h and 5 inertial periods are shown. Courtesy of Carl Wunsch, MIT.

of the frame of reference: the trajectory in the inertial frame is a straight line!

Before going on we note in passing that the theory of inertial circles discussed here is the same as that of the *Foucault pendulum*, named after the French experimentalist who in 1851 demonstrated the rotation of the Earth by observing the deflection of a giant pendulum swinging inside the Pantheon in Paris.

Observations of inertial circles   Inertial circles are not just a quirk of this idealized laboratory experiment. They are a common feature of oceanic flows. For example, Fig. 6.16 shows inertial motions observed by a current meter deployed in the main thermocline of the ocean at a depth of 500 m. The period of the oscillations is:

$$\text{Inertial Period} = \frac{\pi}{\Omega \sin \varphi}, \qquad (6\text{-}39)$$

where φ is latitude; the $\sin \varphi$ factor (not present in the theory developed here) is a geometrical effect due to the sphericity of the Earth, as we now go on to discuss. At the latitude of the mooring, 28° N, the period of the inertial circles is 25.6 hr.

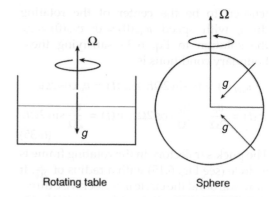

FIGURE 6.17.   In the rotating table used in the laboratory, Ω and g are always parallel or (as sketched here) antiparallel to one another. This should be contrasted with the sphere.

6.6.5. Putting things on the sphere

Hitherto, our discussion of rotating dynamics has made use of a laboratory turntable in which Ω and g are parallel or antiparallel to one another. But on the spherical Earth Ω and g are not aligned and we must take into account these geometrical complications, illustrated in Fig. 6.17. We will now show that, by exploiting the thinness of the fluid shell (Fig. 1.1) and the overwhelming importance of gravity, the equations of motion that govern the fluid on the rotating spherical Earth are essentially the same as those that govern the motion of the fluid of our rotating table if 2Ω is replaced by (what is known as) the *Coriolis parameter*, $f = 2\Omega \sin \varphi$. This is true because a fluid parcel on the rotating Earth "feels" a rotation rate of only $2\Omega \sin \varphi$—2Ω resolved in the direction of gravity, rather than the full 2Ω.

The centrifugal force, modified gravity, and geopotential surfaces on the sphere

Just as on our rotating table, so on the sphere the centrifugal term on the right of Eq. 6-28 modifies gravity and hence hydrostatic balance. For an inviscid fluid at rest *in the rotating frame*, we have

$$\frac{1}{\rho}\nabla p = -g\hat{\mathbf{z}} - \Omega \times \Omega \times \mathbf{r}.$$

Now, consider Fig. 6.18.

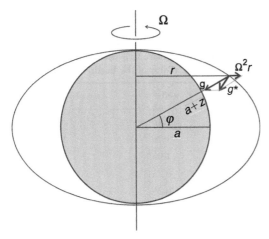

FIGURE 6.18. The centrifugal vector $\Omega \times \Omega \times r$ has magnitude $\Omega^2 r$, directed outward normal to the rotation axis. Gravity, g, points radially inward to the center of the Earth. Over geological time the surface of the Earth adjusts to make itself an equipotential surface—close to a reference ellipsoid—which is always perpendicular the the vector sum of $\Omega \times \Omega \times \mathbf{r}$ and **g**. This vector sum is "measured" gravity: $\mathbf{g}^* = -g\hat{\mathbf{z}} - \Omega \times \Omega \times \mathbf{r}$.

The centrifugal vector $\Omega \times \Omega \times \mathbf{r}$ has magnitude $\Omega^2 r$, directed outward normal to the rotation axis, where $r = (a + z) \cos\varphi \simeq a \cos\varphi$, a is the mean Earth radius, z is the altitude above the spherical surface with radius a, and φ is latitude, and where the "shallow atmosphere" approximation has allowed us to write $a + z \simeq a$. Hence on the sphere, Eq. 6-30 becomes:

$$\phi = gz - \frac{\Omega^2 a^2 \cos^2\varphi}{2},$$

defining the modified gravitational potential on the Earth. At the axis of rotation the height of a geopotential surface is geometric height, z (because $\varphi = \frac{\pi}{2}$). Elsewhere, geopotential surfaces are defined by:

$$z^* = z + \frac{\Omega^2 a^2 \cos^2\varphi}{2g}. \qquad (6\text{-}40)$$

We can see that Eq. 6-40 is exactly analogous to the form we derived in Eq. 6-33

for the free surface of a fluid in solid body rotation in our rotating table, when we realize that $r = a \cos\varphi$ is the distance normal to the axis of rotation. A plumb line is always perpendicular to z^* surfaces, and modified gravity is given by $\mathbf{g}^* = -\nabla z^*$.

Since (with $\Omega = 7.27 \times 10^{-5}\,\text{s}^{-1}$ and $a = 6.37 \times 10^6\,\text{m}$) $\Omega^2 a^2 / 2g \approx 11\,\text{km}$, geopotential surfaces depart only very slightly from a sphere, being 11 km higher at the equator than at the pole. Indeed, the figure of the Earth's surface—the geoid—adopts something like this shape, actually bulging more than this at the equator (by 21 km, relative to the poles).[12] So, if we adopt these (very slightly) aspherical surfaces as our basic coordinate system, then relative to these coordinates the centrifugal force disappears (being subsumed into the coordinate system) and hydrostatic balance again is described (to a very good approximation) by Eq. 6-8. This is directly analogous, of course, to adopting the surface of our parabolic turntable as a coordinate reference system in GFD Lab V.

Components of the Coriolis force on the sphere: the Coriolis parameter

We noted in Chapter 1 that the thinness of the atmosphere allows us (for most purposes) to use a local Cartesian coordinate system, neglecting the Earth's curvature. First, however, we must figure out how to express the Coriolis force in such a system. Consider Fig. 6.19.

At latitude φ, we define a local coordinate system such that the three coordinates in the (x, y, z) directions point (eastward, northward, upward), as shown. The components of Ω in these coordinates are $(0, \Omega \cos\varphi, \Omega \sin\varphi)$. Therefore, expressed in these coordinates,

$$\begin{aligned}
\Omega \times \mathbf{u} &= (0, \Omega \cos\varphi, \Omega \sin\varphi) \times (u, v, w) \\
&= (\Omega \cos\varphi\, w - \Omega \sin\varphi\, v, \Omega \sin\varphi\, u, \\
&\quad\ -\Omega \cos\varphi\, u).
\end{aligned}$$

---

[12]The discrepancy between the actual shape of the Earth and the prediction from Eq. 6-40 is due to the mass distribution of the equatorial bulge, which is not taken in the calculation presented here.

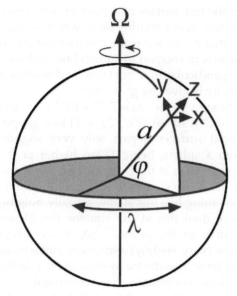

FIGURE 6.19. At latitude φ and longitude λ, we define a local coordinate system such that the three coordinates in the (x, y, z) directions point (eastward, northward, upward): $dx = a \cos \varphi \, d\lambda$; $dy = a \, d\varphi$; $dz = dz$, where a is the radius of the Earth. The velocity is $\mathbf{u} = (u, v, w)$ in the directions (x, y, z). See also Appendix A.2.3.

We can now make two (good) approximations. First, we note that the vertical component competes with gravity and so is negligible if $\Omega u \ll g$. Typically, in the atmosphere $|\mathbf{u}| \sim 10 \, \text{m s}^{-1}$, so $\Omega u \sim 7 \times 10^{-4} \, \text{m s}^{-2}$, which is negligible compared with gravity (we will see in Chapter 9 that ocean currents are even weaker, and so $\Omega u \ll g$ there too). Second, because of the thinness of the atmosphere and ocean, vertical velocities (typically $\leq 1 \, \text{cm s}^{-1}$) are much less than horizontal velocities; so we may neglect the term involving w in the x-component of $\Omega \times \mathbf{u}$. Hence we may write the Coriolis term as

$$2\Omega \times \mathbf{u} \simeq (-2\Omega \sin \varphi \, v, 2\Omega \sin \varphi \, u, 0) \quad (6\text{-}41)$$
$$= f \hat{\mathbf{z}} \times \mathbf{u} \,,$$

where

$$f = 2\Omega \sin \varphi \qquad (6\text{-}42)$$

is known as the *Coriolis parameter*. Note that $\Omega \sin \varphi$ is the *vertical component* of the Earth's

TABLE 6.1. Values of the Coriolis parameter, $f = 2\Omega \sin \varphi$ (Eq. 6-42), and its meridional gradient, $\beta = df/dy = 2\Omega/a \cos \varphi$ (Eq. 10-10), tabulated as a function of latitude. Here Ω is the rotation rate of the Earth and a is the radius of the Earth.

Latitude	$f \, (\times 10^{-4} \, \text{s}^{-1})$	$\beta \, (\times 10^{-11} \, \text{s}^{-1} \, \text{m}^{-1})$
90°	1.46	0
60°	1.26	1.14
45°	1.03	1.61
30°	0.73	1.98
10°	0.25	2.25
0°	0	2.28

rotation rate. This is the only component that matters (a consequence of the thinness of the atmosphere and ocean). For one thing this means that since $f \to 0$ at the equator, rotational effects are negligible there. Furthermore, $f < 0$ in the southern hemisphere (see Fig. 6.17, right). Values of f at selected latitudes are set out in Table 6.1.

We can now write Eq. 6-29 as (rearranging slightly),

$$\frac{D\mathbf{u}}{Dt} + \frac{1}{\rho}\nabla p + \nabla \phi + f \hat{\mathbf{z}} \times \mathbf{u} = \mathcal{F}, \quad (6\text{-}43)$$

where 2Ω has been replaced by $f \hat{\mathbf{z}}$. Writing this in component form for our *local* Cartesian system (see Fig. 6.19) and making the hydrostatic approximation for the vertical component, we have

$$\frac{Du}{Dt} + \frac{1}{\rho}\frac{\partial p}{\partial x} - fv = \mathcal{F}_x \,,$$
$$\frac{Dv}{Dt} + \frac{1}{\rho}\frac{\partial p}{\partial y} + fu = \mathcal{F}_y \,, \quad (6\text{-}44)$$
$$\frac{1}{\rho}\frac{\partial p}{\partial z} + g = 0 \,,$$

where $(\mathcal{F}_x, \mathcal{F}_y)$ are the (x, y) components of friction (and we have assumed the vertical component of $\mathcal{F}$ to be negligible compared with gravity).

The set, Eq. 6-44 is the starting point for discussions of the dynamics of a fluid in

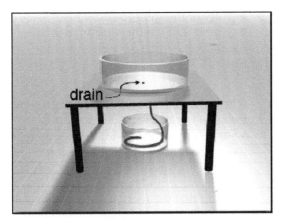

FIGURE 6.20. A leveled cylinder is filled with water, covered by a lid and left standing for several days. Attached to the small hole at the center of the cylinder is a hose (also filled with water and stopped by a rubber bung) which hangs down in to a pail of water. On releasing the bung the water flows out and, according to theory, should acquire a spin which has the same sense as that of the Earth.

a thin spherical shell on a rotating sphere, such as the atmosphere and ocean.

6.6.6. GFD Lab VI: An experiment on the Earth's rotation

A classic experiment on the Earth's rotation was carried out by Perrot in 1859.[13] It is directly analogous to the radial inflow experiment, GFD Lab III, except that the Earth's spin is the source of rotation rather than a rotating table. Perrot filled a large cylinder with water (the cylinder had a hole in the middle of its base plugged with a cork, as sketched in Fig. 6.20) and left it standing for two days. He returned and released the plug. As fluid flowed in toward the drain hole, it conserved angular momentum, thus "concentrating" the rotation of the Earth, and acquired a "spin" that was cyclonic (in the same sense of rotation as the Earth).

According to the theory below, we expect to see the fluid spiral in the same sense of rotation as the Earth. The close analogue with the radial inflow experiment is clear when one realizes that the container

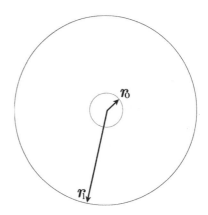

FIGURE 6.21. The Earth's rotation is magnified by the ratio $(r_1/r_0)^2 \gg 1$, if the drain hole has a radius r_0, very much less than the tank itself, r_1.

sketched in Fig. 6.20 is on the rotating Earth and experiences a rotation rate of $\Omega \times \sin lat$!

Theory We suppose that a particle of water initially on the outer rim of the cylinder at radius r_1 moves inward, conserving angular momentum until it reaches the drain hole at radius r_0 (see Fig. 6.21). The Earth's rotation Ω_{Earth} resolved in the

---

[13]Perrot's experiment can be regarded as the fluid-mechanical analogue of Foucault's 1851 experiment on the Earth's rotation using a pendulum.

direction of the local vertical is $\Omega_{Earth} \sin \varphi$ where φ is the latitude. Therefore a particle initially at rest relative to the cylinder at radius r_1, has a speed of $v_1 = r_1 \Omega_{Earth} \sin \varphi$ in the inertial frame. Its angular momentum is $A_1 = v_1 r_1$. At r_o what is the rate of rotation of the particle?

If angular momentum is conserved, then $A_o = \Omega_o r_o^2 = A_1$, and so the rate of rotation of the ball at the radius r_o is:

$$\Omega_o = \left(\frac{r_1}{r_o}\right)^2 \Omega_{Earth} \sin \varphi.$$

Thus if $r_1/r_o \gg 1$ the Earth's rotation can be "amplified" by a large amount. For example, at a latitude of $42°$ N, appropriate for Cambridge, Massachusetts, $\sin \varphi = 0.67$, $\Omega_{Earth} = 7.3 \times 10^{-5}\,\text{s}^{-1}$, and if the cylinder has a radius of $r_1 = 30\,\text{cm}$, and the inner hole has radius $r_o = 0.15\,\text{cm}$, we find that $\Omega_o = 1.96\,\text{rad s}^{-1}$, or a complete rotation in only 3 seconds!

Perrot's experiment, although based on sound physical ideas, is rather tricky to carry out. The initial (background) velocity has to be very tiny ($v \ll fr$) for the experiment to work, thus demanding great care in setup. Apparatus such as that shown in Fig. 6.20 can be used, but the experiment must be repeated many times. More often than not, the fluid does indeed acquire the spin of the Earth, swirling cyclonically as it exits the reservoir.

6.7. FURTHER READING

Discussions of the equations of motion in a rotating frame can be found in most texts on atmospheric and ocean dynamics, such as Holton (2004) and Cushman–Roisin (1994).

6.8. PROBLEMS

1. Consider the zonal-average zonal flow, u, shown in Fig. 5.20. Concentrate on the vicinity of the subtropical jet near $30°$ N in winter (DJF). If the x-component of the frictional force per unit mass in Eq. 6-44 is

$$\mathcal{F}_x = \nu \nabla^2 u,$$

where the kinematic viscosity coefficient for air is $\nu = 1.34 \times 10^{-5}\,\text{m}^2\,\text{s}^{-1}$, and $\nabla^2 \equiv \partial^2/\partial x^2 + \partial^2/\partial y^2 + \partial^2/\partial z^2$, compare the magnitude of this eastward force with the northward or southward Coriolis force, and thus convince yourself that the frictional force is negligible. [$10°$ of latitude $\simeq 1100\,\text{km}$; the jet is at an altitude of about $10\,\text{km}$. An order-of-magnitude calculation will suffice to make the point unambiguously.]

2. Using only the equation of hydrostatic balance and the rotating equation of motion, show that a fluid cannot be motionless unless its density is horizontally uniform. (Do *not* assume geostrophic balance, but you should assume that a motionless fluid is subjected to no frictional forces.)

3. a. What is the value of the centrifugal acceleration of a particle fixed to the Earth at the equator, and how does it compare to **g**? What is the deviation of a plumb line from the true direction to the center of the Earth at $45°$ N?

   b. By considering the centrifugal acceleration on a particle fixed to the surface of the Earth, obtain an order-of-magnitude estimate for the Earth's ellipticity $[(r_1 - r_0)/r_0]$, where r_1 is the equatorial radius and r_0 is the polar radius. As a simplifying assumption the gravitational contribution to **g** may be taken as constant and directed toward the center of the Earth. Discuss your estimate given that the ellipticity is observed to be $1/297$. You may assume that the mean radius of the Earth is $6000\,\text{km}$.

4. A punter kicks a football a distance of $60\,\text{m}$ on a field at latitude $45°$ N.

Assuming the ball, until being caught, moves with a constant forward velocity (horizontal component) of $15\,\mathrm{m\,s^{-1}}$, determine the lateral deflection of the ball from a straight line due to the Coriolis effect. [Neglect friction and any wind or other aerodynamic effects.]

5. Imagine that Concorde is (was) flying at speed u from New York to London along a latitude circle. The deflecting force due to the Coriolis effect is toward the south. By lowering the left wing ever so slightly, the pilot (or perhaps more conveniently the computer on board) can balance this deflection. Draw a diagram of the forces—gravity, uplift normal to the wings, and Coriolis—and use it to deduce that the angle of tilt, γ, of the aircraft from the horizontal required to balance the Coriolis force is

$$\tan\gamma = \frac{2\Omega\sin\varphi \times u}{g},$$

where Ω is the Earth's rotation, the latitude is φ and gravity is g. If $u = 600\,\mathrm{m\,s^{-1}}$, insert typical numbers to compute the angle. What analogies can you draw with atmospheric circulation? [Hint: cf. Eq. 7-8.]

6. Consider horizontal flow in circular geometry in a system rotating about a vertical axis with a steady angular velocity Ω. Starting from Eq. 6-29, show that the equation of motion for the azimuthal flow in this geometry is, in the rotating frame (neglecting friction and assuming 2-dimensional flow)

$$\frac{Dv_\theta}{Dt} + 2\Omega v_r \equiv \frac{\partial v_\theta}{\partial t} + v_r\frac{\partial v_\theta}{\partial r}$$
$$+ \frac{v_\theta}{r}\frac{\partial v_\theta}{\partial \theta} + \frac{v_\theta v_r}{r} + 2\Omega v_r$$
$$= -\frac{1}{\rho r}\frac{\partial p}{\partial \theta}, \qquad (6\text{-}45)$$

where (v_r, v_θ) are the components of velocity in the (r, θ) = (radial, azimuthal) directions (see Fig. 6.8). [Hint: write out Eq. 6-29 in cylindrical coordinates (see Appendix A.2.3) noting that $v_r = Dr/Dt$; $v_\theta = rD\theta/Dt$, and that the gradient operator is $\nabla = (\partial/\partial r, 1/r\,\partial/\partial\theta)$.]

(a) Assume that the flow is axisymmetric (i.e., all variables are independent of θ). For such flow, angular momentum (relative to an inertial frame) is conserved. This means, since the angular momentum per unit mass is

$$A = \Omega r^2 + v_\theta r, \qquad (6\text{-}46)$$

that

$$\frac{DA}{Dt} \equiv \frac{\partial A}{\partial t} + v_r\frac{\partial A}{\partial r} = 0. \qquad (6\text{-}47)$$

Show that Eqs. 6-45 and 6-47 are mutually consistent for axisymmetric flow.

(b) When water flows down the drain from a basin or a bath tub, it usually forms a vortex. It is often said that this vortex is anti-clockwise in the northern hemisphere, and clockwise in the southern hemisphere. Test this saying by carrying out the following experiment.

Fill a basin or a bath tub (preferably the latter—the bigger the better) to a depth of at least $10\,\mathrm{cm}$; let it stand for a minute or two, and then let it drain. When a vortex forms,[14] estimate as well as you can its angular velocity, direction, and radius (use small floats, such as pencil shavings, to help you see the flow). Hence

---

[14]A clear vortex (with a "hollow" center, as in Fig. 6.7) may not form. As long as there is an identifiable swirling motion, however, you will be able to proceed; if not, try repeating the experiment.

calculate the angular momentum per unit mass of the vortex.

Now, suppose that, at the instant you opened the drain, there was no motion (relative to the rotating Earth). If only the vertical component of the Earth's rotation matters, calculate the angular momentum density due to the Earth's rotation at the perimeter of the bath tub or basin. [Your tub or basin will almost certainly not be circular, but assume it is, with an effective radius R such that the area of your tub or basin is πR^2 to determine m.]

(c) Since angular momentum should be conserved, then if there was indeed no motion at the instant you pulled the plug, the maximum possible angular momentum per unit mass in the drain vortex should be the same as that at the perimeter at the initial instant (since that is where the angular momentum was greatest). Compare your answers and comment on the importance of the Earth's rotation for the drain vortex. Hence comment on the validity of the saying mentioned in (b).

(d) In view of your answer to (c), what are your thoughts on Perrot's experiment, GFD Lab VI?

7. We specialize Eq. 6-44 to two-dimensional, inviscid ($\mathcal{F} = 0$) flow of a homogeneous fluid of density ρ_{ref} thus:

$$\frac{Du}{Dt} + \frac{1}{\rho_{ref}}\frac{\partial p}{\partial x} - fv = 0$$

$$\frac{Dv}{Dt} + \frac{1}{\rho_{ref}}\frac{\partial p}{\partial y} + fu = 0,$$

where $D/Dt = \partial/\partial t + u\partial/\partial x + v\partial/\partial y$ and the continuity equation is

$$\frac{\partial u}{\partial x} + \frac{\partial v}{\partial y} = 0.$$

(a) By eliminating the pressure gradient term between the two momentum equations and making use of the continuity equation, show that the quantity $(\partial v/\partial x - \partial u/\partial y + f)$ is conserved following the motion, so that

$$\frac{D}{Dt}\left(\frac{\partial v}{\partial x} - \frac{\partial u}{\partial y} + f\right) = 0.$$

(b) Convince yourself that

$$\hat{\mathbf{z}}.\nabla \times \mathbf{u} = \frac{\partial v}{\partial x} - \frac{\partial u}{\partial y} \qquad (6\text{-}48)$$

(see Appendix A.2.2), that is $\partial v/\partial x - \partial u/\partial y$ is the vertical component of a vector quantity known as the vorticity, $\nabla \times \mathbf{u}$, the curl of the velocity field.

The quantity $\partial v/\partial x - \partial u/\partial y + f$ is known as the "absolute" vorticity, and is made up of "relative" vorticity (due to motion relative to the rotating planet) and "planetary" vorticity, f, due to the rotation of the planet itself.

(c) By computing the "circulation" —the line integral of $\mathbf{u}$ about the rectangular element in the (x, y) plane shown in Fig. 6.22—show that:

$$\frac{\text{circulation}}{\text{area enclosed}} = \begin{array}{l}\text{average normal}\\ \text{component of}\\ \text{vorticity}\end{array}$$

Hence deduce that if the fluid element is in solid body rotation,

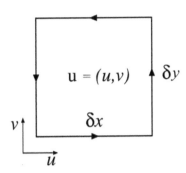

$$u = (u,v)$$

δy

v

δx

u

FIGURE 6.22.   Circulation integral schematic.

then the average vorticity is equal to twice the angular velocity of its rotation.

(d) If the tangential velocity in a hurricane varies like $v = 10^6/r\,\mathrm{m\,s^{-1}}$, where r is the radius, calculate the average vorticity between an inner circle of radius 300 km and an outer circle of radius 500 km. Express your answer in units of planetary vorticity f evaluated at 20° N. What is the average vorticity within the inner circle?

7

Balanced flow

7.1. Geostrophic motion
    7.1.1. The geostrophic wind in pressure coordinates
    7.1.2. Highs and lows; synoptic charts
    7.1.3. Balanced flow in the radial-inflow experiment
7.2. The Taylor-Proudman theorem
    7.2.1. GFD Lab VII: Taylor columns
7.3. The thermal wind equation
    7.3.1. GFD Lab VIII: The thermal wind relation
    7.3.2. The thermal wind equation and the Taylor-Proudman theorem
    7.3.3. GFD Lab IX: Cylinder "collapse" under gravity and rotation
    7.3.4. Mutual adjustment of velocity and pressure
    7.3.5. Thermal wind in pressure coordinates
7.4. Subgeostrophic flow: The Ekman layer
    7.4.1. GFD Lab X: Ekman layers: frictionally-induced cross-isobaric flow
    7.4.2. Ageostrophic flow in atmospheric highs and lows
    7.4.3. Planetary-scale ageostrophic flow
7.5. Problems

In Chapter 6 we derived the equations that govern the evolution of the atmosphere and ocean, setting our discussion on a sound theoretical footing. However, these equations describe myriad phenomena, many of which are not central to our discussion of the large-scale circulation of the atmosphere and ocean. In this chapter, therefore, we focus on a subset of possible motions known as *balanced flows*, which are relevant to the general circulation.

We have already seen that large-scale flow in the atmosphere and ocean is hydrostatically balanced in the vertical, in the sense that gravitational and pressure gradient forces balance one another rather than induce accelerations. It turns out that the atmosphere and ocean are also close to balance in the horizontal, in the sense that Coriolis forces are balanced by horizontal pressure gradients in what is known as *geostrophic motion*—from the Greek, *geo* for "Earth," and *strophe* for "turning." In this chapter we describe how the rather peculiar and counterintuitive properties of the geostrophic motion of a homogeneous fluid are encapsulated in the *Taylor-Proudman theorem*, which expresses in mathematical form the "stiffness" imparted to a fluid by rotation. This stiffness property will be repeatedly applied in later chapters to understand the large-scale circulation of the atmosphere and ocean. We go on to discuss how the Taylor-Proudman theorem is modified in a fluid in which the density is not homogeneous but varies from place to place, deriving the *thermal wind equation*. Finally we discuss so-called ageostrophic flow motion, which is not in geostrophic balance but is modified by friction in regions where the atmosphere and ocean rub against solid boundaries or at the atmosphere-ocean interface.

7.1. GEOSTROPHIC MOTION

Let us begin with the momentum equation, Eq. 6-43, for a fluid on a rotating Earth and consider the magnitude of the various terms. First, we restrict attention to the free atmosphere and ocean (by which we mean away from boundary layers), where friction is negligible.[1] Suppose each of the horizontal flow components u and v has a typical magnitude $\mathcal{U}$, and that each varies in time with a characteristic time scale $\mathcal{T}$ and with horizontal position over a characteristic length scale L. Then the first two terms on the left side of the horizontal components of Eq. 6-43 scale as[2]:

$$\underbrace{\frac{D\mathbf{u}}{Dt}} + f\hat{\mathbf{z}} \times \mathbf{u} = \underbrace{\frac{\partial \mathbf{u}}{\partial t}}_{\frac{\mathcal{U}}{\mathcal{T}}} + \underbrace{\mathbf{u} \cdot \nabla \mathbf{u}}_{\frac{\mathcal{U}^2}{L}} + \underbrace{f\hat{\mathbf{z}} \times \mathbf{u}}_{f\mathcal{U}}$$

For typical large-scale flows in the atmosphere, $\mathcal{U} \sim 10\,\mathrm{m\,s^{-1}}$, $L \sim 10^6\,\mathrm{m}$, and $\mathcal{T} \sim 10^5\,\mathrm{s}$, so $\mathcal{U}/\mathcal{T} \approx \mathcal{U}^2/L \sim 10^{-4}\,\mathrm{m\,s^{-2}}$. That $\mathcal{U}/\mathcal{T} \approx \mathcal{U}^2/L$ is no accident; the time scale on which motions change is intimately related to the time taken for the flow to traverse a distance L, viz., $L/\mathcal{U}$. So in practice the acceleration terms $\partial \mathbf{u}/\partial t$ and $\mathbf{u} \cdot \nabla \mathbf{u}$ are comparable in magnitude to one another and scale like $\mathcal{U}^2/L$. The ratio of these acceleration terms to the Coriolis term is known as the *Rossby number*[3]:

$$R_o = \frac{\mathcal{U}}{fL}. \qquad (7\text{-}1)$$

In middle latitudes (say near 45°; see Table 6.1), $f \simeq 2\Omega/\sqrt{2} = 1.03 \times 10^{-4}\,\mathrm{s^{-1}}$. So given our typical numbers, $R_o \simeq 0.1$ and we see that the Rossby number in the atmosphere is small. We will find in Section 9.3 that $R_o \simeq 10^{-3}$ for large-scale ocean circulation.

The smallness of R_o for large-scale motion in the free atmosphere and ocean[4] implies that the acceleration term in Eq. 6-43 dominates the Coriolis term, leaving

$$f\hat{\mathbf{z}} \times \mathbf{u} + \frac{1}{\rho}\nabla p = 0. \qquad (7\text{-}2)$$

---

[1]The effects of molecular viscosity are utterly negligible in the atmosphere and ocean, except *very* close to solid boundaries. Small-scale turbulent motions can in some ways act like viscosity, with an "effective eddy viscosity" that is much larger than the molecular value. However, even these effects are usually negligible away from the boundaries.

[2]We have actually anticipated something here that is evident only *a posteriori*: vertical advection makes a negligible contribution to $(\mathbf{u} \cdot \nabla)\mathbf{u}$.

[3]  Carl-Gustav Rossby (1898–1957). Swedish-born meteorologist, one of the major figures in the founding of modern dynamical study of the atmosphere and ocean. In 1928, he was appointed chair of meteorology in the Department of Aeronautics at MIT. This group later developed into the first Department of Meteorology in an academic institution in the United States. His name is recalled ubiquitously in Rossby waves, the Rossby number, and the Rossby radius of deformation, all ideas fundamental to the understanding of all planetary scale fluids.

[4]Near the equator, where $f \to 0$, the small Rossby number assumption breaks down, as will be seen, for example, in Section 12.2.2, below.

Equation 7-2 defines *geostrophic balance*, in which the pressure gradient is balanced by the Coriolis term. We expect this balance to be approximately satisfied for flows of small R_o. Another way of saying the same thing is that if we define the *geostrophic wind*, or *current*, to be the velocity $\mathbf{u}_g$ that *exactly* satisfies Eq. 7-2, then $\mathbf{u} \simeq \mathbf{u}_g$ in such flows. Since $\hat{\mathbf{z}} \times \hat{\mathbf{z}} \times \mathbf{u} = -\mathbf{u}$, Eq. 7-2 gives

$$\mathbf{u}_g = \frac{1}{f\rho}\hat{\mathbf{z}} \times \nabla p, \qquad (7\text{-}3)$$

or, in component form in the local Cartesian geometry of Fig. 6.19,

$$\left(u_g, v_g\right) = \left(-\frac{1}{f\rho}\frac{\partial p}{\partial y}, \frac{1}{f\rho}\frac{\partial p}{\partial x}\right). \qquad (7\text{-}4)$$

The geostrophic balance of forces described by Eq. 7-2 is illustrated in Fig. 7.1. The pressure gradient force is, of course, directed away from the high pressure system on the left, and towards the low pressure system on the right. The balancing Coriolis forces must be as shown, directed in the opposite sense, and consequently the geostrophically balanced flow must be *normal* to the pressure gradient, that is, along the contours of constant pressure, as Eq. 7-3 makes explicit. For the northern hemisphere cases ($f > 0$) illustrated in Fig. 7.1, the sense of the flow is clockwise around a high pressure system, and anticlockwise around a low. (The sense is opposite in the southern hemisphere.) The rule is summarized in Buys-Ballot's (the 19th-century Dutch meteorologist) law:

> If you stand with your back to the wind in the northern hemisphere, low pressure is on your left

("left" → "right" in the southern hemisphere).

We see from Eq. 7-3 that the geostrophic flow depends on the magnitude of the pressure gradient, and not just its direction.

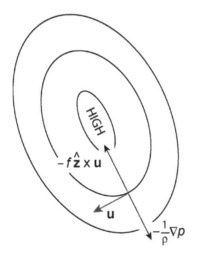 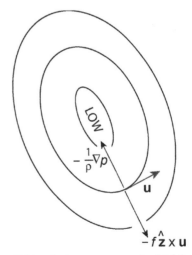

FIGURE 7.1. Geostrophic flow around a high pressure center (left) and a low pressure center (right). (Northern hemisphere case, $f > 0$.) The effect of Coriolis deflecting flow "to the right" (see Fig. 6.10) is balanced by the horizontal component of the pressure gradient force, $-1/\rho\nabla p$, directed from high to low pressure.

Consider Fig. 7.2, on which the curved lines show two isobars of constant pressure, p and $p + \delta p$, separated by the variable distance δs. From Eq. 7-3,

$$|\mathbf{u}_g| = \frac{1}{f\rho}|\nabla p| = \frac{1}{f\rho}\frac{\delta p}{\delta s}.$$

Since δp is constant along the flow, $|\mathbf{u}_g| \propto (\delta s)^{-1}$; the flow is strongest where the isobars are closest together. The geostrophic flow does not cross the pressure contours, and so the latter act like the banks of a river, causing the flow to speed up where the river is narrow and to slow down where it is wide. These characteristics explain in large part why the meteorologist is traditionally preoccupied with pressure maps: the pressure field determines the winds.

Note that the vertical component of the geostrophic flow, as defined by Eq. 7-3, is zero. This cannot be deduced directly from Eq. 7-2, which involves the horizontal components of the flow. However, consider for a moment an incompressible fluid (in the laboratory or the ocean) for which we can neglect variations in ρ. Further, while f varies on the sphere, it is almost constant

over scales of, for example, 1000 km or less.[5] Then Eq. 7-4 gives

$$\frac{\partial u_g}{\partial x} + \frac{\partial v_g}{\partial y} = 0. \qquad (7\text{-}5)$$

Thus the geostrophic flow is horizontally *nondivergent*. Comparison with the continuity Eq. 6-11 then tells us that $\partial w_g/\partial z = 0$; if $w_g = 0$ on, for example, a flat bottom boundary, then it follows that $w_g = 0$ everywhere, and so the geostrophic flow is indeed horizontal.

In a compressible fluid, such as the atmosphere, density variations complicate matters. We therefore now consider the equations of geostrophic balance in pressure coordinates, in which case such complications do not arise.

7.1.1. The geostrophic wind in pressure coordinates

To apply the geostrophic equations to atmospheric observations and particularly to upper air analyses (see below), we need to express them in terms of height gradients on a pressure surface, rather than, as in Eq. 7-4, of pressure gradients at constant height.

Consider Fig. 7.3. The figure depicts a surface of constant height z_0, and one of constant pressure p_0, which intersect at A, where of course pressure is $p_A = p_0$ and

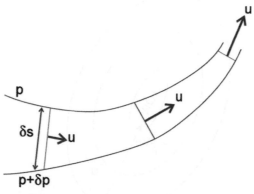

FIGURE 7.2. Schematic of two pressure contours (isobars) on a horizontal surface. The geostrophic flow, defined by Eq. 7-3, is directed along the isobars; its magnitude increases as the isobars become closer together.

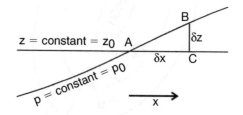

FIGURE 7.3. Schematic used in converting from pressure gradients on height surfaces to height gradients on pressure surfaces.

[5]Variations of f do matter, however, for motions of planetary scale, as will be seen, for example, in Section 10.2.1.

height is $z_A = z_0$. At constant height, the gradient of pressure in the x-direction is

$$\left(\frac{\partial p}{\partial x}\right)_z = \frac{p_C - p_0}{\delta x}, \qquad (7\text{-}6)$$

where δx is the (small) distance between C and A and subscript z means "keep z constant." Now, the gradient of height along the constant pressure surface is

$$\left(\frac{\partial z}{\partial x}\right)_p = \frac{z_B - z_0}{\delta x},$$

where subscript p means "keep p constant." Since $z_C = z_0$, and $p_B = p_0$, we can use the hydrostatic balance equation Eq. 3-3 to write

$$\frac{p_C - p_0}{z_B - z_0} = \frac{p_C - p_B}{z_B - z_C} = -\frac{\partial p}{\partial z} = g\rho,$$

and so $p_C - p_0 = g\rho\,(z_B - z_0)$.

Therefore from Eq. 7-6, and invoking a similar result in the y-direction, it follows that

$$\left(\frac{\partial p}{\partial x}\right)_z = g\rho \left(\frac{\partial z}{\partial x}\right)_p ;$$

$$\left(\frac{\partial p}{\partial y}\right)_z = g\rho \left(\frac{\partial z}{\partial y}\right)_p .$$

In pressure coordinates Eq. 7-3 thus becomes:

$$\mathbf{u}_g = \frac{g}{f}\widehat{\mathbf{z}}_p \times \nabla_p z , \qquad (7\text{-}7)$$

where $\widehat{\mathbf{z}}_p$ is the upward unit vector in pressure coordinates, and ∇_p denotes the gradient operator in pressure coordinates. In component form it is,

$$(u_g, v_g) = \left(-\frac{g}{f}\frac{\partial z}{\partial y}, \frac{g}{f}\frac{\partial z}{\partial x}\right). \qquad (7\text{-}8)$$

The wonderful simplification of Eq. 7-8 relative to Eq. 7-4 is that ρ does not explicitly appear and therefore, in evaluation from observations, we need not be concerned

about its variation. Just like p contours on surfaces of constant z, z contours on surfaces of constant p are streamlines of the geostrophic flow. The geostrophic wind is nondivergent in pressure coordinates if f is taken as constant:

$$\nabla_p \cdot \mathbf{u}_g = \frac{\partial u}{\partial x} + \frac{\partial v}{\partial y} = 0. \qquad (7\text{-}9)$$

Equation 7-9 enables us to define a streamfunction:

$$u_g = -\frac{\partial \psi_g}{\partial y}; \ v_g = \frac{\partial \psi_g}{\partial x}, \qquad (7\text{-}10)$$

which, as can be verified by substitution, satisfies Eq. 7-9 for any $\psi_g = \psi_g(x, y, p, t)$. Comparing Eq. 7-10 with Eq. 7-8 we see that:

$$\psi_g = \frac{g}{f}z. \qquad (7\text{-}11)$$

Thus height contours are *streamlines of the geostrophic flow* on pressure surfaces: the geostrophic flow streams along z contours, as can be seen in Fig. 7.4.

What does Eq. 7-8 imply about the magnitude of the wind? In Fig. 5.13 we saw that the 500-mbar pressure surface slopes down by a height $\Delta z = 800\,\text{m}$ over a meridional distance $L = 5000\,\text{km}$; then geostrophic balance implies a wind of strength $u = g/f\ \Delta z/L = \frac{9.81}{10^{-4}} \times \frac{800}{5\times10^6} \approx 15\,\text{m s}^{-1}$.

Thus Coriolis forces acting on a zonal wind of speed $\sim 15\,\text{m s}^{-1}$, are of sufficient magnitude to balance the poleward pressure gradient force associated with the pole-equator temperature gradient. This is just what is observed; see the strength of the midlevel flow shown in Fig. 5.20. Geostrophic balance thus "connects" Figs. 5.13 and 5.20.

Let us now look at some synoptic charts, such as those shown in Figs. 5.22 and 7.4, to see geostrophic balance in action.

7.1.2. Highs and lows; synoptic charts

Fig. 7.4 shows the height of the 500-mbar surface (contoured every 60 m) plotted with the observed wind vector (one full quiver represents a wind speed of 10 m s^{-1}) at an instant in time: 12 GMT on June 21, 2003, to be exact, the same time as the hemispheric map shown in Fig. 5.22. Note how

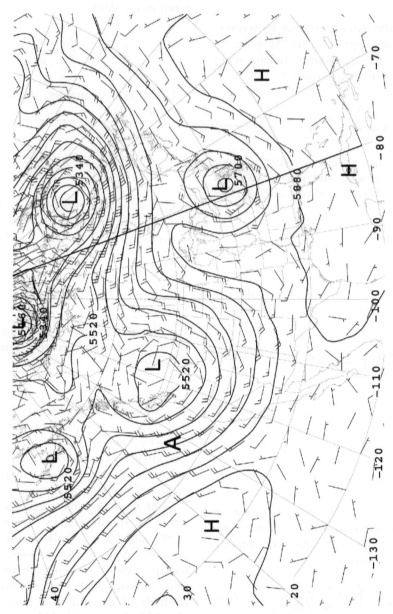

FIGURE 7.4. The 500-mbar wind and geopotential height field at 12 GMT on June 21, 2003. [Latitude and longitude (in degrees) are labelled by the numbers along the left and bottom edges of the plot.] The wind blows away from the quiver: one full quiver denotes a speed of 10 m s^{-1}, one half-quiver a speed of 5 m s^{-1}. The geopotential height is contoured every 60 m. Centers of high and low pressure are marked H and L. The position marked A is used as a check on geostrophic balance. The thick black line marks the position of the meridional section shown in Fig. 7.21 at 80° W extending from 20° N to 70° N. This section is also marked on Figs. 7.5, 7.20, and 7.25.

the wind blows along the height contours and is stronger the closer the contours are together. At this level, away from frictional effects at the ground, the wind is close to geostrophic.

Consider for example the point marked by the left "foot" of the "A" shown in Fig. 7.4, at 43° N, 133° W. The wind is blowing along the height contours to the SSE at a speed of 25 m s^{-1}. We estimate that the 500 mbar height surface slopes down at a rate of 60 m in 250 km here (noting that 1° of latitude is equivalent to a distance of 111 km and that the contour interval is 60 m). The geostrophic relation, Eq. 7-7, then implies a wind of speed $g/f \ \Delta z/L = \frac{9.81}{9.7 \times 10^{-5}} \times \frac{60}{2.5 \times 10^5} = 24$ m s^{-1}, close to that observed. Indeed the wind at upper levels in the atmosphere is very close to geostrophic balance.

In Fig. 7.5 we plot the R_o (calculated as $|\mathbf{u} \cdot \nabla \mathbf{u}| / |f \mathbf{u}|$) for the synoptic pattern shown in Fig. 7.4. It is about 0.1 over most of the region and so the flow is

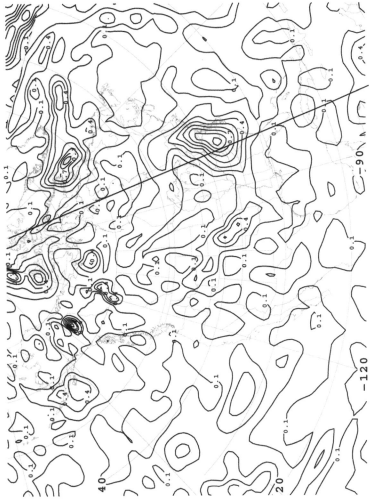

FIGURE 7.5. The Rossby number for the 500-mbar flow at 12 GMT on June 21, 2003, the same time as Fig. 7.4. The contour interval is 0.1. Note that $R_o \sim 0.1$ over most of the region but can approach 1 in strong cyclones, such as the low centered over 80° W, 40° N.

to a good approximation in geostrophic balance. However, R_o can approach unity in intense low pressure systems where the flow is strong and the flow curvature large, such as in the low centered over 80° W, 40° N. Here the Coriolis and advection terms are comparable to one another, and there is a three-way balance between Coriolis, inertial, and pressure gradient forces. Such a balance is known as *gradient wind balance* (see Section 7.1.3).

7.1.3. Balanced flow in the radial-inflow experiment

At this point it is useful to return to the radial inflow experiment, GFD Lab III, described in Section 6.6.1, and compute the Rossby number assuming that axial angular momentum of fluid parcels is conserved as they spiral into the drain hole (see Fig. 6.6). The Rossby number implied by Eq. 6-23 is given by:

$$R_o = \frac{v_\theta}{2\Omega r} = \frac{1}{2}\left(\frac{r_1^2}{r^2} - 1\right), \qquad (7\text{-}12)$$

where r_1 is the outer radius of the tank. It is plotted as a function of r/r_1 in Fig. 7.6 (right).

The observed R_o, based on tracking particles floating on the surface of the fluid (see Fig. 6.6) together with the theoretical prediction, Eq. 7-12, are plotted in Fig. 7.6. We see broad agreement, but the observations depart from the theoretical curve at small r and high R_o, perhaps because of the difficulty of tracking the particles in the high speed core of the vortex (note the blurring of the particles at small radius evident in Fig. 6.6).

According to Eq. 7-12 and Fig. 7.6, $R_o = 0$ at $r = r_1$, $R_o = 1$ at a radius $r_1/\sqrt{3} = 0.58r_1$, and rapidly increases as r decreases further. Thus the azimuthal flow is geostrophically balanced in the outer regions (small R_o) with the radial pressure gradient force balancing the Coriolis force in Eq. 6-21. In the inner regions (high R_o) the v_θ^2/r term in Eq. 6-21 balances the radial pressure gradient; this is known as *cyclostrophic balance*. In the middle region (where $R_o \sim 1$) all three terms in Eq. 6-21 play a role; this is known as *gradient wind balance*, of which geostrophic and cyclostrophic balance are limiting cases. As mentioned earlier, gradient wind balance can be seen in the synoptic chart shown in Fig. 7.4, in the low pressure regions where $R_o \sim 1$ (Fig. 7.5).

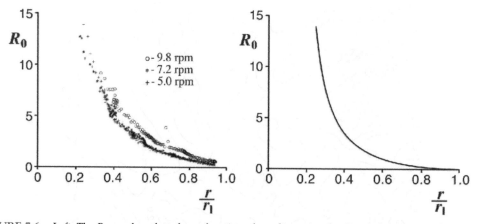

FIGURE 7.6.   Left: The R_o number plotted as a function of nondimensional radius (r/r_1) computed by tracking particles in three radial inflow experiments (each at a different rotation rate, quoted here in revolutions per minute [rpm]). Right: Theoretical prediction based on Eq. 7-12.

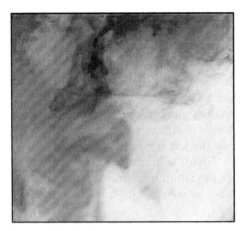 

FIGURE 7.7. Dye distributions from GFD Lab 0: on the left we see a pattern from dyes (colored red and green) stirred into a nonrotating fluid in which the turbulence is three-dimensional; on the right we see dye patterns obtained in a rotating fluid in which the turbulence occurs in planes perpendicular to the rotation axis and is thus two-dimensional.

7.2. THE TAYLOR-PROUDMAN THEOREM

A remarkable property of geostrophic motion is that if the fluid is homogeneous (ρ uniform) then, as we shall see, the geostrophic flow is two dimensional and does not vary in the direction of the rotation vector, Ω. Known as the Taylor-Proudman theorem, it is responsible for the glorious patterns observed in our dye shirring experiment, GFD Lab 0, shown again in Fig. 7.7. We discuss the theorem here and make much subsequent use of it, particularly in Chapters 10 and 11, to discuss the constraints of rotation on the motion of the atmosphere and ocean.

For the simplest derivation of the theorem, let us begin with the geostrophic relation written out in component form, Eq. 7-4. If ρ and f are constant, then taking the vertical derivative of the geostrophic flow components and using hydrostatic balance, we see that $\left(\partial u_g / \partial z, \partial v_g / \partial z\right) = 0$; therefore the geostrophic flow does not vary in the direction of $f\hat{\mathbf{z}}$.

A slightly more general statement of this result can be obtained if we go right back to the pristine form of the momentum equation (Eq. 6-29) in rotating coordinates. If the flow is sufficiently slow and steady ($R_o \ll 1$) and $\mathcal{F}$ is negligible, it reduces to:

$$2\Omega \times \mathbf{u} + \frac{1}{\rho}\nabla p + \nabla \phi = 0. \qquad (7\text{-}13)$$

The horizontal component of Eq. 7-13 yields geostrophic balance—Eq. 7-2 or Eq. 7-4 in component form—where now $\hat{\mathbf{z}}$ is imagined to point in the direction of Ω (see Fig. 7.8). The vertical component of Eq. 7-13 yields hydrostatic balance, Eq. 3-3. Taking the curl ($\nabla \times$) of Eq. 7-13, we find that if the fluid is *barotropic* [i.e., one in which $\rho = \rho(p)$], then[6]:

$$(\Omega \cdot \nabla)\,\mathbf{u} = 0, \qquad (7\text{-}14)$$

or (since $\Omega \cdot \nabla$ is the gradient operator in the direction of Ω, i.e., $\hat{\mathbf{z}}$)

$$\frac{\partial \mathbf{u}}{\partial z} = 0. \qquad (7\text{-}15)$$

---

[6]Using vector identities 2. and 6. of Appendix A.2.2, setting $\mathbf{a} \longrightarrow \Omega$ and $\mathbf{b} \longrightarrow \mathbf{u}$, remembering that $\nabla \cdot \mathbf{u} = 0$ and $\nabla \times \nabla$ (scalar)$= 0$.

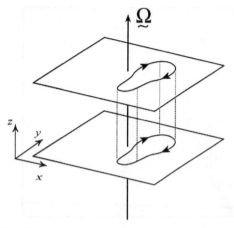

FIGURE 7.8. The Taylor-Proudman theorem, Eq. 7-14, states that slow, steady, frictionless flow of a barotropic, incompressible fluid is two-dimensional and does not vary in the direction of the rotation vector Ω.

Equation 7-14 is known as the Taylor-Proudman theorem (or T-P for short). T-P says that under the stated conditions—slow, steady, frictionless flow of a barotropic fluid—the velocity **u**, both horizontal and vertical components, cannot vary in the direction of the rotation vector Ω. In other words the flow is two-dimensional, as sketched in Fig. 7.8. Thus vertical columns of fluid remain vertical—they cannot be tilted over or stretched. We say that the fluid is made "stiff" in the direction of Ω. The columns are called *Taylor Columns* after G.I. Taylor, who first demonstrated them experimentally.[7]

Rigidity, imparted to the fluid by rotation, results in the beautiful dye patterns seen in experiment GFD Lab 0. On the right of Fig. 7.7, the rotating fluid, brought into gentle motion by stirring, is constrained to

move in two dimensions. Rich dye patterns emerge in planes perpendicular to Ω but with strong vertical coherence between the levels; flow at one horizontal level moves in lockstep with the flow at another level. In contrast, a stirred nonrotating fluid mixes in three dimensions and has an entirely different character with no vertical coherence; see the left frame of Fig. 7.7.

Taylor columns can readily be observed in the laboratory in a more controlled setting, as we now go on to describe.

7.2.1. GFD Lab VII: Taylor columns

Suppose a homogeneous rotating fluid moves in a layer of variable depth, as sketched in Fig. 7.9. This can easily be arranged in the laboratory by placing an obstacle (such as a bump made of a hockey puck) in the bottom of a tank of water rotating on a turntable and observing the flow of water past the obstacle, as depicted in Fig. 7.9. The T-P theorem demands that vertical columns of fluid move along contours of constant fluid depth, because they cannot be stretched.

At levels below the top of the obstacle, the flow must of course go around it. But Eq. 7-15 says that the flow must be the same at all z; so, at all heights, the flow must be deflected as if the bump on the boundary extended all the way through the fluid! We can demonstrate this behavior in the laboratory, using the apparatus sketched in Fig. 7.10 and described in the legend, by inducing flow past a submerged object.

We see the flow (marked by paper dots floating on the free surface) being diverted around the obstacle in a vertically

[7] Geoffrey Ingram Taylor (1886–1975). British scientist who made fundamental and long-lasting contributions to a wide range of scientific problems, especially theoretical and experimental investigations of fluid dynamics. The result, Eq. 7-14, was first demonstrated by Joseph Proudman in 1915 but is now called the Taylor-Proudman theorem. Taylor's name got attached because he demonstrated the theorem experimentally (see GFD Lab VII). In a paper published in 1921, he reported slowly dragging a cylinder through a rotating flow. The solid object all but immobilized an entire column of fluid parallel to the rotation axis.

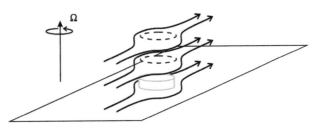

FIGURE 7.9. The T-P theorem demands that vertical columns of fluid move along contours of constant fluid depth because, from Eq. 7-14, they cannot be stretched in the direction of Ω. Thus fluid columns act as if they were rigid columns and move along contours of constant fluid depth. Horizontal flow is thus deflected as if the obstable extended through the whole depth of the fluid.

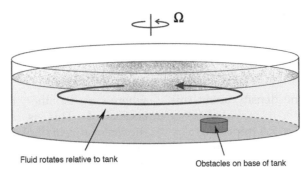

Fluid rotates relative to tank   Obstacles on base of tank

FIGURE 7.10. We place a cylindrical tank of water on a table turning at about 5 rpm. An obstacle such as a hockey puck is placed on the base of the tank whose height is a small fraction of the fluid depth; the water is left until it comes into solid body rotation. We now make a very small reduction in Ω (by 0.1 rpm or less). Until a new equilibrium is established (the "spin-down" process takes several minutes, depending on rotation rate and water depth), horizontal flow will be induced relative to the obstacle. Dots on the surface, used to visualize the flow (see Fig. 7.11), reveal that the flow moves around the obstacle as if the obstacle extended through the whole depth of the fluid.

coherent way (as shown in the photograph of Fig. 7.11) as if the obstacle extended all the way through the water, thus creating stagnant Taylor columns above the obstacle.

7.3. THE THERMAL WIND EQUATION

We saw in Section 5.2 that isobaric surfaces slope down from equator to pole. Moreover, these slopes increase with height, as can be seen, for example, in Fig. 5.13 and the schematic diagram, Fig. 5.14. Thus according to the geostrophic relation, Eq. 7-8, the geostrophic flow will increase with height, as indeed is observed

in Fig. 5.20. According to T-P, however, $\partial \mathbf{u}_g / \partial z = 0$. What's going on?

The Taylor-Proudman theorem pertains to a slow, steady, frictionless, barotropic fluid, in which $\rho = \rho(p)$. But in the atmosphere and ocean, density does vary on pressure surfaces, and so T-P does not strictly apply and must be modified to allow for density variations.

Let us again consider the water in our rotating tank, but now suppose that the density of the water varies thus:

$$\rho = \rho_{ref} + \sigma \text{ and } \frac{\sigma}{\rho_{ref}} \ll 1,$$

where ρ_{ref} is a constant reference density, and σ, called the density *anomaly*,

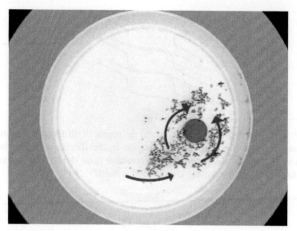

FIGURE 7.11. Paper dots on the surface of the fluid shown in the experiment described in Fig. 7.10. The dots move around, but not over, a submerged obstacle, in experimental confirmation of the schematic drawn in Fig. 7.9.

is the variation of the density about this reference.[8]

Now take $\partial/\partial z$ of Eq. 7-4 (replacing ρ by ρ_{ref} where it appears in the denominator) we obtain, making use of the hydrostatic relation Eq. 3-3:

$$\left(\frac{\partial u_g}{\partial z}, \frac{\partial v_g}{\partial z}\right) = \frac{g}{f\rho_{ref}}\left(\frac{\partial\sigma}{\partial y}, -\frac{\partial\sigma}{\partial x}\right) \quad (7\text{-}16)$$

or, in vector notation (see Appendix A.2.2, VI)

$$\frac{\partial \mathbf{u}_g}{\partial z} = -\frac{g}{f\rho_{ref}}\hat{\mathbf{z}} \times \nabla\sigma. \quad (7\text{-}17)$$

So if ρ varies in the horizontal then the geostrophic current will vary in the vertical. To express things in terms of temperature, and hence derive a connection (called the *thermal wind* equation) between the current and the thermal field, we can use our simplified equation of state for water, Eq. 4-4, which assumes that the density of water depends on temperature T in a linear fashion. Then Eq. 7-17 can be written:

$$\frac{\partial \mathbf{u}_g}{\partial z} = \frac{\alpha g}{f}\hat{\mathbf{z}} \times \nabla T, \quad (7\text{-}18)$$

where α is the thermal expansion coefficient. This is a simple form of the *thermal wind* relation connecting the vertical shear of the geostrophic current to horizontal temperature gradients. It tells us nothing more than the hydrostatic and geostrophic balances, Eq. 3-3 and Eq. 7-4, but it expresses these balances in a different way.

We see that there is an exactly analogous relationship between $\partial\mathbf{u}_g/\partial z$ and T as between $\mathbf{u}_g$ and p (compare Eqs. 7-18 and 7-3). So if we have horizontal gradients of temperature then the geostrophic flow will vary with height. The westerly winds increase with height because, through the thermal wind relation, they are associated with the poleward decrease in temperature. We now go on to study the thermal wind in the laboratory, in which we represent the cold pole by placing an ice bucket in the center of a rotating tank of water.

7.3.1. GFD Lab VIII: The thermal wind relation

It is straightforward to obtain a steady, axially-symmetric circulation driven by

---

[8]Typically the density of the water in the rotating tank experiments, and indeed in the ocean too (see Section 9.1.3), varies by only a few % about its reference value. Thus σ/ρ_{ref} is indeed very small.

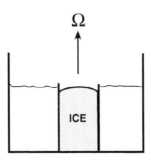
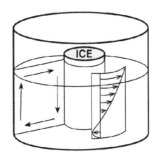

FIGURE 7.12. We place a cylindrical tank on a turntable containing a can at its center, fill the tank with water to a depth of 10 cm or so, and rotate the turntable very slowly (1 rpm or less). After solid-body rotation is achieved, we fill the can with ice/water. The can of ice in the middle induces a radial temperature gradient. A thermal wind shear develops in balance with it, which can be visualized with dye, as sketched on the right. The experiment is left for 5 minutes or so for the circulation to develop. The radial temperature gradient is monitored with thermometers and the currents measured by tracking paper dots floating on the surface.

FIGURE 7.13. Dye streaks being tilted over into a corkscrew pattern by an azimuthal current in thermal wind balance, with a radial temperature gradient maintained by the ice bucket at the center.

radial temperature gradients in our laboratory tank, which provides an ideal opportunity to study the thermal wind relation.

The apparatus is sketched in Fig. 7.12 and can be seen in Fig. 7.13. The cylindrical tank, at the center of which is an ice bucket, is rotated very slowly anticlockwise. The cold sides of the can cool the water adjacent to it and induce a radial temperature gradient. Paper dots sprinkled over the surface move in the same sense as, but more swiftly than, the rotating table—we have generated westerly (to the east) currents! We inject some dye. The dye streaks do not remain vertical but tilt over in an azimuthal direction, carried along by currents that increase in strength with height and are directed in the same sense as the rotating table (see the photograph in Fig. 7.13 and the schematic in Fig. 7.14. The Taylor columns have been tilted over by the westerly currents.

For our incompressible fluid in cylindrical geometry (see Appendix A.2.3), the azimuthal component of the thermal wind relation, Eq. 7-18, is:

$$\frac{\partial v_\theta}{\partial z} = \frac{\alpha g}{2\Omega}\frac{\partial T}{\partial r},$$

where v_θ is the azimuthal current (cf. Fig. 6.8) and f has been replaced by 2Ω. Since T increases moving outward from the cold center ($\partial T/\partial r > 0$) then, for positive Ω, $\partial v_\theta/\partial z > 0$. Since v_θ is constrained by friction to be weak at the bottom of the tank, we therefore expect to see $v_\theta > 0$ at the top, with the strongest flow at the radius of maximum density gradient. Dye streaks visible in Fig. 7.13 clearly show the thermal wind shear, especially near the cold can, where the density gradient is strong.

We note the temperature difference, ΔT, between the inner and outer walls a distance L apart (of order 1°C per 10 cm), and the speed of the paper dots at the surface relative to the tank (typically 1 cm s^{-1}). The

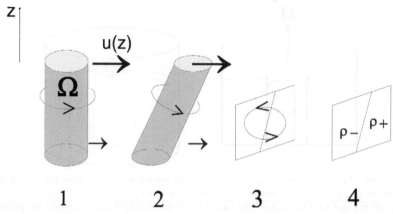

FIGURE 7.14. A schematic showing the physical content of the thermal wind equation written in the form Eq. 7-20: spin associated with the rotation vector 2Ω [1] is tilted over by the vertical shear (du/dz) [2]. Circulation in the transverse plane develops [3] creating horizontal density gradients from the stable vertical gradients. Gravity acting on the sloping density surfaces balances the overturning torque associated with the tilted Taylor columns [4].

tank is turning at 1 rpm, and the depth of water is $H \sim 10$ cm. From the above thermal wind equation we estimate that

$$\mathbf{u}_g \sim \frac{\alpha g}{2\Omega} \frac{\Delta T}{L} H \sim 1\,\mathrm{cm\,s^{-1}},$$

for $\alpha = 2 \times 10^{-4}$ K^{-1}, roughly as we observe.

This experiment is discussed further in Section 8.2.1 as a simple analogue of the tropical Hadley circulation of the atmosphere.

7.3.2. The thermal wind equation and the Taylor-Proudman theorem

The connection between the T-P theorem and the thermal wind equation can be better understood by noting that Eqs. 7-17 and 7-18 are simplified forms of a more general statement of the thermal wind equation, which we now derive.

Taking $\nabla \times$ (Eq. 7-13), but now relaxing the assumption of a barotropic fluid, we obtain [noting that the term of the left of Eq. 7-13 transforms as in the derivation of Eq. 7-14, and that $\nabla \times (1/\rho \nabla p) = -1/\rho^2 \nabla \rho \times \nabla p = 1/\rho^2 \nabla p \times \nabla \rho$]:

$$(2\Omega \cdot \nabla)\,\mathbf{u} = \frac{1}{\rho}\nabla p \times \frac{1}{\rho}\nabla \rho, \qquad (7\text{-}19)$$

which is a more general statement of the "thermal wind" relation. In the case of constant ρ, or more precisely in a barotropic fluid where $\rho = \rho(p)$ and so $\nabla \rho$ is parallel to ∇p, Eq. 7-19 reduces to 7-14. But now we are dealing with a baroclinic fluid in which density depends on temperature (see Eq. 4-4) and so ρ surfaces and p surfaces are no longer coincident. Thus the term on the right of Eq. 7-19, known as the baroclinic term, does not vanish. It can be simplified by noting that to a very good approximation, the fluid is in hydrostatic balance: $1/\rho \nabla p + g\hat{\mathbf{z}} = 0$, allowing it to be written:

$$(2\Omega \cdot \nabla)\,\mathbf{u} = 2\Omega\frac{\partial \mathbf{u}}{\partial z} = -\frac{g}{\rho}\hat{\mathbf{z}} \times \nabla \rho. \qquad (7\text{-}20)$$

When written in component form, Eq. 7-20 becomes Eq. 7-16 if $2\Omega \longrightarrow f\hat{\mathbf{z}}$ and $\rho \longrightarrow \rho_{ref} + \sigma$.

The physical interpretation of the right hand side of Eqs. 7-16 and 7-20 can now be better appreciated. It is the action of gravity on horizontal density gradients trying to return surfaces of constant density to the horizontal, the natural tendency of a fluid under gravity to find its own level. But on the large scale this tendency is counterbalanced by the rigidity of the Taylor columns, represented by the term on the left of Eq. 7-20. How this works is sketched in Fig. 7.14. The spin associated with the rotation vector 2Ω [1] is tilted over by the vertical shear (du/dz) of the current as time progresses [2]. Circulation in the transverse plane develops [3] and converts vertical stratification in to horizontal density gradients. If the environment is stably stratified, then the action of gravity acting on the sloping density surfaces is in the correct sense to balance the overturning torque associated with the tilted Taylor columns [4]. This is the torque balance at the heart of the thermal wind relation.

We can now appreciate how it is that gravity fails to return inclined temperature surfaces, such as those shown in Fig. 5.7, to the horizontal. It is prevented from doing so by the Earth's rotation.

7.3.3. GFD Lab IX: cylinder "collapse" under gravity and rotation

A vivid illustration of the role that rotation plays in counteracting the action of gravity on sloping density surfaces can be carried out by creating a density front in a rotating fluid in the laboratory, as shown in Fig. 7.15 and described in the legend. An initially vertical column of dense salty water is allowed to slump under gravity but is "held up" by rotation, forming a cone whose sides have a distinct slope. The photographs in Fig. 7.16 show the development of a cone. The cone acquires a definite sense of rotation, swirling in the same sense of rotation as the table. We measure typical speeds through the use of paper dots, measure the density of the dyed water and the slope of the side of the cone (the front), and interpret them in terms of the following theory.

Theory following Margules

A simple and instructive model of a front can be constructed as follows. Suppose that the density is ρ_1 on one side of the front and changes discontinuously to ρ_2 on the other, with $\rho_1 > \rho_2$ as sketched in Fig. 7.17. Let y be a horizontal axis and γ be the angle that the surface of discontinuity makes with the horizontal. Since the pressure must be the same on both sides of the front then the pressure change computed along paths [1]

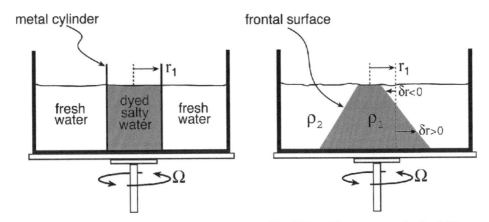

FIGURE 7.15. We place a large tank on our rotating table, fill it with water to a depth of 10 cm or so, and place in the center of it a hollow metal cylinder of radius $r_1 = 6$ cm, which protrudes slightly above the surface. The table is set into rapid rotation at 10 rpm and allowed to settle down for 10 minutes or so. While the table is rotating, the water within the cylinder is carefully and slowly replaced by dyed, salty (and hence dense) water delivered from a large syringe. When the hollow cylinder is full of colored saline water, it is rapidly removed to cause the least disturbance possible—practice is necessary! The subsequent evolution of the dense column is charted in Fig. 7.16. The final state is sketched on the right: the cylinder has collapsed into a cone whose surface is displaced a distance δr relative to that of the original upright cylinder.

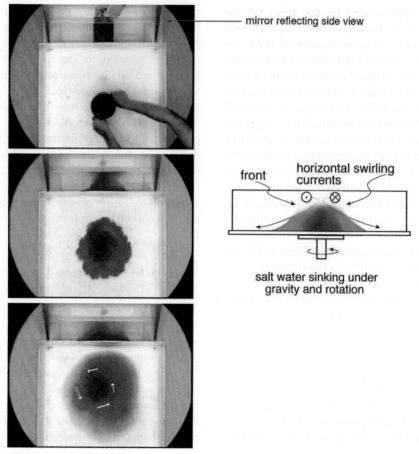

FIGURE 7.16. Left: Series of pictures charting the creation of a dome of salty (and hence dense) dyed fluid collapsing under gravity and rotation. The fluid depth is 10 cm. The white arrows indicate the sense of rotation of the dome. At the top of the figure we show a view through the side of the tank facilitated by a sloping mirror. Right: A schematic diagram of the dome showing its sense of circulation.

and [2] in Fig. 7.17 must be the same, since they begin and end at common points in the fluid:

$$\left(\frac{\partial p}{\partial z}\delta z + \frac{\partial p}{\partial y}\delta y\right)_{\text{path 2}} = \left(\frac{\partial p}{\partial y}\delta y + \frac{\partial p}{\partial z}\delta z\right)_{\text{path 1}}$$

for small δy, δz. Hence, using hydrostatic balance to express $\partial p/\partial z$ in terms of ρg on both sides of the front, we find:

$$\tan \gamma = \frac{dz}{dy} = \frac{\dfrac{\partial p_1}{\partial y} - \dfrac{\partial p_2}{\partial y}}{g\left(\rho_1 - \rho_2\right)}.$$

Using the geostrophic approximation to the current, Eq. 7-4, to relate the pressure

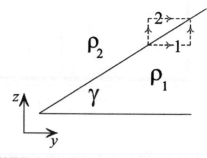

FIGURE 7.17. Geometry of the front separating fluid of differing densities used in the derivation of the Margules relation, Eq. 7-21.

gradient terms to flow speeds, we arrive at the following form of the thermal wind

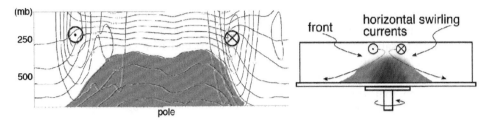

FIGURE 7.18. The dome of cold air over the north pole shown in the instantaneous slice across the pole on the left (shaded green) is associated with strong upper-level winds, marked ⊗ and ⊙ and contoured in red. On the right we show a schematic diagram of the column of salty water studied in GFD Lab IX (cf. Figs. 7.15 and 7.16). The column is prevented from slumping all the way to the bottom by the rotation of the tank. Differences in Coriolis forces acting on the spinning column provide a "torque" which balances that of gravity acting on salty fluid trying to pull it down.

equation (which should be compared to Eq. 7-16):

$$(u_2 - u_1) = \frac{g' \tan\gamma}{2\Omega}, \qquad (7\text{-}21)$$

where u is the component of the current parallel to the front, and $g' = g\Delta\rho/\rho_1$ is the "reduced gravity," with $\Delta\rho = \rho_1 - \rho_2$ (cf. parcel "buoyancy," defined in Eq. 4-3 of Section 4.2). Equation 7-21 is known as *Margules relation*, a formula derived in 1903 by the Austrian meteorologist Max Margules to explain the slope of boundaries in atmospheric fronts. Here it relates the swirl speed of the cone to g', γ, and Ω.

In the experiment we typically observe a slump angle γ of perhaps 30° for $\Omega \sim 1\,\mathrm{s}^{-1}$ (corresponding to a rotation rate of 10 rpm) and a $g' \sim 0.2\,\mathrm{m\,s}^{-2}$ (corresponding to a $\Delta\rho/\rho$ of 2%). Equation 7-21 then predicts a swirl speed of about $6\,\mathrm{cm\,s}^{-1}$, broadly in accord with what is observed in the high-speed core of the swirling cone.

Finally, to make the connection of the experiment with the atmosphere more explicit, we show in Fig. 7.18 the dome of cold air over the north pole and the strong upper-level wind associated with it. The horizontal temperature gradients and vertical wind shear are in thermal wind balance on the planetary scale.

7.3.4. Mutual adjustment of velocity and pressure

The cylinder collapse experiment encourages us to wonder about the adjustment between the velocity field and the pressure field. Initially (left frame of Fig. 7.15) the cylinder is not in geostrophic balance. The "end state," sketched in the right frame and being approached in Fig. 7.16 (bottom) and Fig. 7.18, is in "balance" and well described by the Margules formula, Eq. 7-21. How far does the cylinder have to slump sideways before the velocity field and the pressure field come in to this balanced state? This problem was of great interest to Rossby and is known as the *Rossby adjustment problem*. The detailed answer is in general rather complicated, but we can arrive at a qualitative estimate rather directly, as follows.

Let us suppose that as the column slumps it conserves angular momentum so that $\Omega r^2 + ur = $ constant, where r is the distance from the center of the cone and u is the velocity at its edge. If r_1 is the initial radius of a stationary ring of salty fluid (on the left of Fig. 7.15) then it will have an azimuthal speed given by:

$$u = -2\Omega\delta r, \qquad (7\text{-}22)$$

if it changes its radius by an amount δr (assumed small) as marked on the right of Fig. 7.15. In the upper part of the water column, $\delta r < 0$ and the ring will acquire a cyclonic spin; below, $\delta r > 0$ and the ring will spin anticyclonically.[9] This slumping will proceed until the resulting vertical shear is enough to satisfy Eq. 7-21. Assuming that $\tan \gamma \sim H / |\delta r|$, where H is the depth of the water column, combining Eqs. 7-22 and 7-21 we see that this will occur at a value of $\delta r \sim L_\rho = \sqrt{g'H}/2\Omega$. By noting that[10] $g'H \approx N^2 H^2$, L_ρ can be expressed in terms of the buoyancy frequency N of a continuously stratified fluid thus:

$$L_\rho = \frac{\sqrt{g'H}}{2\Omega} = \frac{NH}{2\Omega}, \qquad (7\text{-}23)$$

where now H is interpreted as the vertical scale of the motion. The horizontal length scale Eq. 7-23 is known as the *Rossby radius of deformation*. It is the scale at which the effects of rotation become comparable with those of stratification. More detailed analysis shows that on scales smaller than L_ρ, the pressure adjusts to the velocity field, whereas on scales much greater than L_ρ, the reverse is true and the velocity adjusts to the pressure.

For the values of g' and Ω appropriate to our cylinder collapse experiment, $L_\rho \sim 7\,\text{cm}$ if $H = 10\,\text{cm}$. This is roughly in accord with the observed slumping scale of our salty cylinder in Fig. 7.16. We shall see in Chapters 8 and 9 that $L_\rho \sim 1000\,\text{km}$ in the atmospheric troposphere, and $L_\rho \sim 30\,\text{km}$ in the main thermocline of the ocean; the respective deformation radii set the horizontal scale of the ubiquitous eddies observed in the two fluids.

7.3.5. Thermal wind in pressure coordinates

Equation 7-16 pertains to an incompressible fluid, such as water or the ocean. The thermal wind relation appropriate to the atmosphere is untidy when expressed with height as a vertical coordinate (because of

ρ variations). However, it becomes simple when expressed in pressure coordinates. To proceed in p coordinates, we write the hydrostatic relation:

$$\frac{\partial z}{\partial p} = -\frac{1}{g\rho}$$

and take, for example, the p-derivative of the x-component of Eq. 7-8 yielding:

$$\frac{\partial u_g}{\partial p} = -\frac{g}{f}\frac{\partial^2 z}{\partial p \partial y} = -\frac{g}{f}\left(\frac{\partial}{\partial y}\left[\frac{\partial z}{\partial p}\right]\right)_p$$
$$= \frac{1}{f}\frac{\partial}{\partial y}\left(\frac{1}{\rho}\right)_p.$$

Since $1/\rho = RT/p$, its derivative *at constant pressure* is

$$\frac{\partial}{\partial y}\left(\frac{1}{\rho}\right)_p = \frac{R}{p}\left(\frac{\partial T}{\partial y}\right)_p,$$

whence

$$\left(\frac{\partial u_g}{\partial p}, \frac{\partial v_g}{\partial p}\right) = \frac{R}{fp}\left(\left(\frac{\partial T}{\partial y}\right)_p, -\left(\frac{\partial T}{\partial x}\right)_p\right).$$
$$(7\text{-}24)$$

Equation 7-24 expresses the *thermal wind relationship* in pressure coordinates. By analogy with Eq. 7-8, just as height contours on a pressure surface act as streamlines for the geostrophic flow, then we see from Eq. 7-24 that temperature contours on a pressure surface act as streamlines for the thermal wind shear. We note in passing that one can obtain a relationship similar to Eq. 7-24 in height coordinates (see Problem 9 at end of chapter), but it is less elegant because of the ρ factors in Eq. 7-4. The thermal wind can also be written down in terms of potential temperature (see Problem 10, also at the end of this chapter).

The connection between meridional temperature gradients and vertical wind

---

[9]In practice, friction caused by the cone of fluid rubbing over the bottom brings currents there toward zero. The thermal wind shear remains, however, with cyclonic flow increasing all the way up to the surface.

[10]In a stratified fluid the buoyancy frequency (Section 4.4) is given by $N^2 = -g/\rho \, d\rho/dz$ or $N^2 \sim g'/H$ where $g' = g\Delta\rho/\rho$ and $\Delta\rho$ is a typical variation in density over the vertical scale H.

shear expressed in Eq. 7-24 is readily seen in the zonal-average climatology (see Figs. 5.7 and 5.20). Since temperature decreases poleward, $\partial T/\partial y < 0$ in the northern hemisphere, but $\partial T/\partial y > 0$ in the southern hemisphere; hence $f^{-1}\partial T/\partial y < 0$ in both.

Then Eq. 7-24 tells us that $\partial u/\partial p < 0$: so, with increasing height (decreasing pressure), winds must become increasingly eastward (westerly) in both hemispheres (as sketched in Fig. 7.19), which is just what we observe in Fig. 5.20.

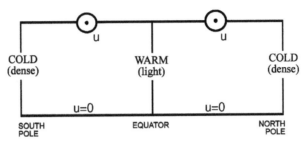

FIGURE 7.19. A schematic of westerly winds observed in both hemispheres in thermal wind balance with the equator-to-pole temperature gradient. (See Eq. 7-24 and the observations shown in Figs. 5.7 and 5.20.)

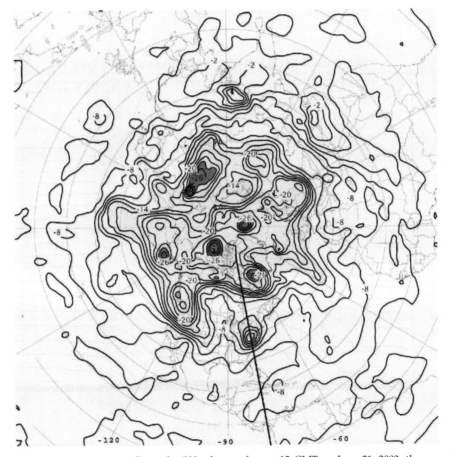

FIGURE 7.20. The temperature, T, on the 500-mbar surface at 12 GMT on June 21, 2003, the same time as Fig. 7.4. The contour interval is 2°C. The thick black line marks the position of the meridional section shown in Fig. 7.21 at 80° W extending from 20° N to 70° N. A region of pronounced temperature contrast separates warm air (pink) from cold air (blue). The coldest temperatures over the pole get as low as −32°C.

The atmosphere is also close to thermal wind balance on the large scale at any instant. For example, Fig. 7.20 shows T on the 500-mbar surface on 12 GMT on June 21, 2003, the same time as the plot of the 500-mbar height field shown in Fig. 7.4. Remember that by Eq. 7-24, the T contours are streamlines of the geostrophic shear, $\partial \mathbf{u}_g / \partial p$. Note the strong meridional gradients in middle latitudes associated with the strong meandering jet stream. These gradients are also evident in Fig. 7.21, a vertical cross section of temperature T and zonal wind u through the atmosphere at 80° W, extending from 20° N to 70° N at the same time as in Fig. 7.20. The vertical coordinate is pressure. Note that, in accord with Eq. 7-24, the wind increases with height

where T surfaces slope upward toward the pole and decreases with height where T surfaces slope downwards. The vertical wind shear is very strong in regions where the T surfaces steeply slope; the vertical wind shear is very weak where the T surfaces are almost horizontal. Note also the anomalously cold air associated with the intense low at 80° W, 40° N marked in Fig. 7.4.

In summary, then, Eq. 7-24 accounts quantitatively as well as qualitatively for the observed connection between horizontal temperature gradients and vertical wind shear in the atmosphere. As we shall see in Chapter 9, an analogous expression of thermal wind applies in the ocean too.

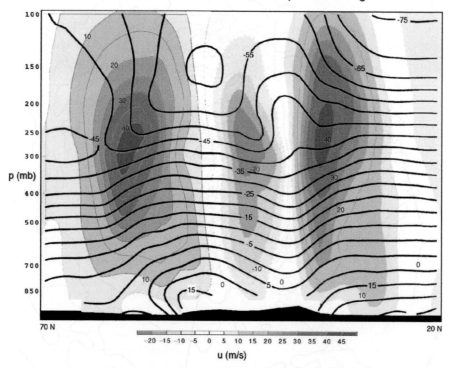

FIGURE 7.21. A cross section of zonal wind, u (color-scale, green indicating away from us and brown toward us, and thin contours every $5\,\mathrm{m\,s}^{-1}$), and temperature, T (thick contours every 5°C), through the atmosphere at 80° W, extending from 20° N to 70° N, on June 21, 2003, at 12 GMT, as marked on Figs. 7.20 and 7.4. Note that $\partial u / \partial p < 0$ in regions where $\partial T / \partial y < 0$ and vice versa.

7.4. SUBGEOSTROPHIC FLOW: THE EKMAN LAYER

Before returning to our discussion of the general circulation of the atmosphere in Chapter 8, we must develop one further dynamical idea. Although the large-scale flow in the free atmosphere and ocean is close to geostrophic and thermal wind balance, in boundary layers where fluid rubs over solid boundaries or when the wind directly drives the ocean, we observe marked departures from geostrophy due to the presence of the frictional terms in Eq. 6-29.

The momentum balance, Eq. 7-2, pertains if the flow is sufficiently slow ($Ro \ll 1$) and frictional forces $\mathcal{F}$ sufficiently small, which is when both $\mathcal{F}$ and $D\mathbf{u}/Dt$ in Eq. 6-43 can be neglected. Frictional effects are indeed small in the interior of the atmosphere and ocean, but they become important in boundary layers. In the bottom kilometer or so of the atmosphere, the roughness of the surface generates turbulence, which communicates the drag of the lower boundary to the free atmosphere above. In the top one hundred meters or so of the ocean the wind generates turbulence, which carries the momentum of the wind down into the interior. The layer in which $\mathcal{F}$ becomes important is called the *Ekman layer*, after the famous Swedish oceanographer who studied the wind-drift in the ocean, as will be discussed in detail in Chapter 10.

If the Rossby number is again assumed to be small, but $\mathcal{F}$ is now *not* negligible, then the horizontal component of the momentum balance, Eq. 6-43, becomes:

$$f\hat{\mathbf{z}} \times \mathbf{u} + \frac{1}{\rho}\nabla p = \mathcal{F}. \qquad (7\text{-}25)$$

To visualize these balances, consider Fig. 7.22. Let's start with **u**: the Coriolis force per unit mass, $-f\hat{\mathbf{z}} \times \mathbf{u}$, must be to the right of the flow, as shown. If the frictional force per unit mass, $\mathcal{F}$, acts as a "drag" it will be directed opposite to the prevailing flow. The sum of these two forces is

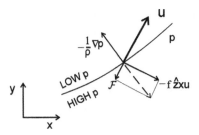

FIGURE 7.22. The balance of forces in Eq. 7-25: the dotted line is the vector sum $\mathcal{F} - f\hat{\mathbf{z}} \times u$ and is balanced by $-1/\rho\nabla p$.

depicted by the dashed arrow. This must be balanced by the pressure gradient force per unit mass, as shown. Thus the pressure gradient is no longer normal to the wind vector, or (to say the same thing) the wind is no longer directed along the isobars. Although there is still a tendency for the flow to have low pressure on its left, there is now a (frictionally-induced) component down the pressure gradient (toward low pressure).

Thus we see that in the presence of $\mathcal{F}$, the flow speed is subgeostrophic (less than geostrophic), and so the Coriolis force (whose magnitude is proportional to the speed) is not quite sufficient to balance the pressure gradient force. Thus the pressure gradient force "wins," resulting in an ageostrophic component directed from high to low pressure. The flow "falls down" the pressure gradient slightly.

It is often useful to explicitly separate the horizontal flow $\mathbf{u}_h$ in the geostrophic and ageostrophic components thus:

$$\mathbf{u}_h = \mathbf{u}_g + \mathbf{u}_{ag}, \qquad (7\text{-}26)$$

where $\mathbf{u}_{ag}$ is the *ageostrophic* current, the departure of the actual horizontal flow from its geostrophic value $\mathbf{u}_g$ given by Eq. 7-3. Using Eq. 7-25, Eq. 7-26, and the geostrophic relation Eq. 7-2, we see that:

$$f\hat{\mathbf{z}} \times \mathbf{u}_{ag} = \mathcal{F}. \qquad (7\text{-}27)$$

Thus the ageostrophic component is always directed "to the right" of $\mathcal{F}$ (in the northern hemisphere).

We can readily demonstrate the role of Ekman layers in the laboratory as follows.

7.4.1. GFD Lab X: Ekman layers: frictionally-induced cross-isobaric flow

We bring a cylindrical tank filled with water up to solid-body rotation at a speed of 5 rpm. A few crystals of potassium permanganate are dropped into the tank—they leave streaks through the water column as they fall and settle on the base of the tank—and float paper dots on the surface to act as tracers of upper level flow. The rotation rate of the tank is then reduced by 10% or so. The fluid continues in solid rotation, creating a cyclonic vortex (same sense of rotation as the table) implying, through the geostrophic relation, lower pressure in the center and higher pressure near the rim of the tank. The dots on the surface describe concentric circles and show little tendency toward radial flow. However at the bottom of the tank we see plumes of dye spiral inward to the center of the tank at about 45° relative to the geostrophic current (see Fig. 7.23, top panel). Now we increase the rotation rate. The relative flow is now anticyclonic with, via geostrophy, high pressure in the center and low pressure on the rim. Note how the plumes of dye sweep around to point outward (see Fig. 7.23, bottom panel).

In each case we see that the rough bottom of the tank slows the currents down there, and induces cross-isobaric, ageostrophic flow from high to low pressure, as schematized in Fig. 7.24. Above the frictional layer, the flow remains close to geostrophic.

7.4.2. Ageostrophic flow in atmospheric highs and lows

Ageostrophic flow is clearly evident in the bottom kilometer or so of the atmosphere, where the frictional drag of the rough underlying surface is directly felt by the flow. For example, Fig. 7.25 shows the surface pressure field and wind at the

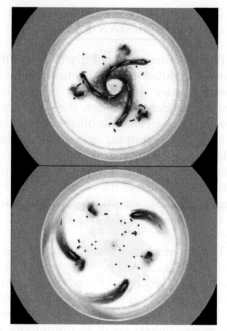

FIGURE 7.23. Ekman flow in a low-pressure system (top) and a high-pressure system (bottom), revealed by permanganate crystals on the bottom of a rotating tank. The black dots are floating on the free surface and mark out circular trajectories around the center of the tank directed anticlockwise (top) and clockwise (bottom).

surface at 12 GMT on June 21, 2003, at the same time as the upper level flow shown in Fig. 7.4. We see that the wind broadly circulates in the sense expected from geostrophy, anticyclonically around highs and cyclonically around the lows. But the surface flow also has a marked component directed down the pressure gradient, into the lows and out of the highs, due to frictional drag at the ground. The sense of the ageostrophic flow is exactly the same as that seen in GFD Lab X (cf. Fig. 7.23 and Fig. 7.24).

A simple model of winds in the Ekman layer

Equation 7-25 can be solved to give a simple expression for the wind in the Ekman layer. Let us suppose that the x-axis is directed along the isobars and that the surface stress decreases uniformly throughout the depth of the Ekman layer from its surface value to become small at

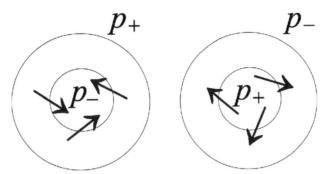

FIGURE 7.24. Flow spiraling in to a low-pressure region (left) and out of a high-pressure region (right) in a bottom Ekman layer. In both cases the ageostrophic flow is directed from high pressure to low pressure, or down the pressure gradient.

$z = \delta$, where δ is the depth of the Ekman layer such that

$$\mathcal{F} = -\frac{k}{\delta}\mathbf{u}, \qquad (7\text{-}28)$$

where k is a drag coefficient that depends on the roughness of the underlying surface. Note that the minus sign ensures that $\mathcal{F}$ acts as a drag on the flow. Then the momentum equations, Eq. 7-25, written out in component form along and across the isobars, become:

$$-fv = -k\frac{u}{\delta} \qquad (7\text{-}29)$$

$$fu + \frac{1}{\rho}\frac{\partial p}{\partial y} = -k\frac{v}{\delta}.$$

Note that v is entirely ageostrophic, being the component directed across the isobars. But u has both geostrophic and ageostrophic components.

Solving Eq. 7-29 gives:

$$u = -\frac{1}{\left(1 + \frac{k^2}{f^2\delta^2}\right)}\frac{1}{\rho f}\frac{\partial p}{\partial y}; \quad \frac{v}{u} = \frac{k}{f\delta}. \quad (7\text{-}30)$$

Note that the wind speed is *less* than its geostrophic value, and if $u > 0$, then $v > 0$ and vice versa; v is directed down the pressure gradient, from high to low pressure, just as in the laboratory experiment and in Fig. 7.24.

In typical meteorological conditions, $\delta \sim 1\,\text{km}$, k is $(1 \longrightarrow 1.5) \times 10^{-2}\,\text{m}\,\text{s}^{-1}$, and

$k/f\delta \sim 0.1$. So the wind speed is only slightly less than geostrophic, but the wind blows across the isobars at an angle of $6°$ to $12°$. The cross-isobaric flow is strong over land (where k is large), where the friction layer is shallow (δ small), and at low latitudes (f small). Over the ocean, where k is small, the atmospheric flow is typically much closer to its geostrophic value than over land.

Ekman developed a theory of the boundary layer in which he set $\mathcal{F} = \mu\partial^2\mathbf{u}/\partial z^2$ in Eq. 7-25, where μ was a constant eddy viscosity. He obtained what are now known as *Ekman spirals*, in which the current spirals from its most geostrophic to its most ageostrophic value (as will be seen in Section 10.1 and Fig. 10.5). But such details depend on the precise nature of $\mathcal{F}$, which in general is not known. Qualitatively, the most striking and important feature of the Ekman layer solution is that the wind in the boundary layer has a component directed toward lower pressure; this feature is independent of the *details* of the turbulent boundary layer.

Vertical motion induced by Ekman layers

Unlike geostrophic flow, ageostrophic flow is not horizontally nondivergent; on the contrary, its divergence drives vertical motion, because in pressure coordinates, Eq. 6-12 can be written (if f is constant,

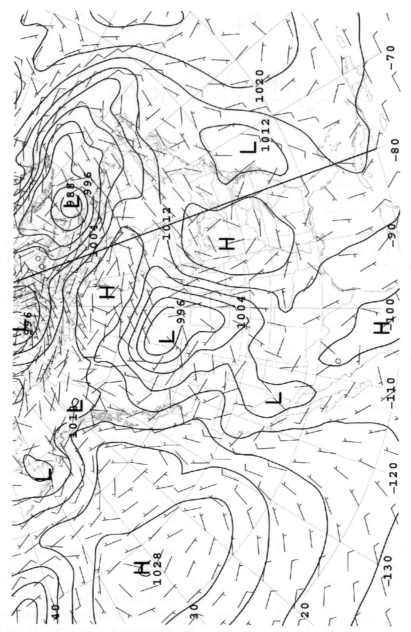

FIGURE 7.25. Surface pressure field and surface wind at 12 GMT on June 21, 2003, at the same time as the upper level flow shown in Fig. 7.4. The contour interval is 4 mbar. One full quiver represents a wind of $10 \, \text{m s}^{-1}$; one half quiver a wind of $5 \, \text{m s}^{-1}$. The thick black line marks the position of the meridional section shown in Fig. 7.21 at 80° W.

so that geostrophic flow is horizontally nondivergent):

$$\nabla_p \cdot \mathbf{u}_{ag} + \frac{\partial \omega}{\partial p} = 0.$$

This has implications for the behavior of weather systems. Fig. 7.26 shows schematics of a cyclone (low-pressure system) and an anticyclone (high-pressure system). In the free atmosphere, where the flow is

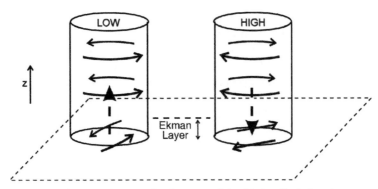

FIGURE 7.26. Schematic diagram showing the direction of the frictionally induced ageostrophic flow in the Ekman layer induced by low pressure and high pressure systems. There is flow into the low, inducing rising motion (the dotted arrow), and flow out of a high, inducing sinking motion.

Atmospheric Surface Pressure (mb)

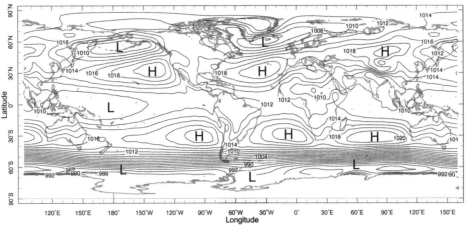

FIGURE 7.27. The annual-mean surface pressure field in mbar, with major centers of high and low pressure marked. The contour interval is 5 mbar.

geostrophic, the wind just blows around the system, cyclonically around the low and anticyclonically around the high. Near the surface in the Ekman layer, however, the wind deviates toward low pressure, inward in the low, outward from the high. Because the horizontal flow is convergent into the low, mass continuity demands a compensating vertical outflow. This *Ekman pumping* produces ascent, and, in consequence, cooling, clouds, and possibly rain in low pressure systems. In the high, the divergence of the Ekman layer flow demands subsidence (through *Ekman suction*). Therefore

high pressure systems tend to be associated with low precipitation and clear skies.

7.4.3. Planetary-scale ageostrophic flow

Frictional processes also play a central role in the atmospheric boundary layer on planetary scales. Fig. 7.27 shows the annual average surface pressure field in the atmosphere, p_s. We note the belt of high pressure in the subtropics (latitudes $\pm 30°$) of both hemispheres, more or less continuous in the southern hemisphere, confined mainly to the ocean basins in the northern hemisphere. Pressure is relatively low at the surface in the tropics and at high latitudes (±60°), particularly in the southern hemisphere. These features are readily seen in the zonal-average p_s shown in the top panel of Fig. 7.28.

To a first approximation, the surface wind is in geostrophic balance with the pressure field. Accordingly (see the top and middle panels of Fig. 7.28 since $\partial p_s / \partial y < 0$ in the latitudinal belt between 30° and 60° N, then from Eq. 7-4, $u_s > 0$ and we observe westerly winds there; between 0° and 30° N, p_s increases, $\partial p_s / \partial y > 0$ and we find easterlies, $u_s < 0$—the trade winds. A similar pattern is seen in the southern hemisphere (remember $f < 0$ here); note the particularly strong surface westerlies

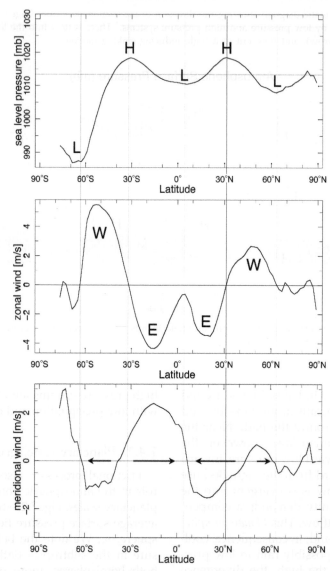

FIGURE 7.28.   Anually and zonally averaged (top) sea level pressure in mbar, (middle) zonal wind in $\mathrm{m\,s}^{-1}$, and (bottom) meridional wind in $\mathrm{m\,s}^{-1}$. The horizontal arrows mark the sense of the meridional flow at the surface.

around 50° S associated with the very low pressure observed around Antarctica in Figs. 7.27 and 7.28 (top panel).

Because of the presence of friction in the atmospheric boundary layer, the surface wind also shows a significant ageostrophic component directed from high pressure to low pressure. This is evident in the bottom panel of Fig. 7.28, which shows the zonal average of the meridional component of the surface wind, v_s. This panel shows the surface branch of the meridional flow in Fig. 5.21 (top panel). Thus in the zonal average we see v_s, which is entirely ageostrophic, feeding rising motion along the inter-tropical convergence zone at the equator, and being supplied by sinking of fluid into the subtropical highs of each hemisphere, around $\pm 30°$, consistent with Fig. 5.21.

We have now completed our discussion of balanced dynamics. Before going on to apply these ideas to the general circulation of the atmosphere and, in subsequent chapters, of the ocean, we summarize our key equations in Table 7.1.

7.5. PROBLEMS

1. Define a streamfunction ψ for non-divergent, two-dimensional flow in a vertical plane:

$$\frac{\partial u}{\partial x} + \frac{\partial v}{\partial y} = 0$$

and interpret it physically.

Show that the instantaneous particle paths (streamlines) are defined by $\psi = $ const, and hence in steady flow the contours $\psi = $ const are particle trajectories. When are trajectories and streamlines not coincident?

2. What is the pressure gradient required to maintain a geostrophic wind at a speed of $v = 10 \, \text{m s}^{-1}$ at 45° N? In the absence of a pressure gradient, show that air parcels flow around circles in an anticyclonic sense of radius v/f.

3. Draw schematic diagrams showing the flow, and the corresponding balance of forces, around centers of low and high

TABLE 7.1. Summary of key equations. Note that (x, y, p) is not a right-handed coordinate system. So although $\hat{z}$ is a unit vector pointing toward increasing z, and therefore upward, $\hat{z}_p$ is a unit vector pointing toward decreasing p, and therefore also upward.

(x, y, z) coordinates	(x, y, z) coordinates	(x, y, p) coordinates
$\nabla \equiv \left(\frac{\partial}{\partial x}, \frac{\partial}{\partial y}, \frac{\partial}{\partial z} \right)$	$\nabla \equiv \left(\frac{\partial}{\partial x}, \frac{\partial}{\partial y}, \frac{\partial}{\partial z} \right)$	$\nabla_p \equiv \left(\frac{\partial}{\partial x}, \frac{\partial}{\partial y}, \frac{\partial}{\partial p} \right)$
general	(incompressible—OCEAN)	(comp. perfect gas—ATMOS)
Continuity		
$\frac{\partial \rho}{\partial t} + \nabla \cdot (\rho u) = 0$	$\nabla \cdot u = 0$	$\nabla_p \cdot u = 0$
Hydrostatic balance		
$\frac{\partial p}{\partial z} = -g\rho$	$\frac{\partial p}{\partial z} = -g\rho$	$\frac{\partial z}{\partial p} = -\frac{1}{g\rho}$
Geostrophic balance		
$fu = \frac{1}{\rho}\hat{z} \times \nabla p$	$fu = \frac{1}{\rho_{ref}}\hat{z} \times \nabla p$	$fu = g\hat{z}_p \times \nabla_p z$
Thermal wind balance		
$(2\boldsymbol{\Omega} \cdot \nabla)\mathbf{u} = \frac{1}{\rho}\nabla p \times \frac{1}{\rho}\nabla \rho$	$f\frac{\partial u}{\partial z} = -\frac{g}{\rho_{ref}}\hat{z} \times \nabla \sigma$	$f\frac{\partial u}{\partial p} = -\frac{R}{p}\hat{z}_p \times \nabla T$

pressure in the midlatitude southern hemisphere. Do this for:

(a) the geostrophic flow (neglecting friction), and

(b) the subgeostrophic flow in the near-surface boundary layer.

4. Consider a low pressure system centered on 45° S, whose sea level pressure field is described by

$$p = 1000\,\text{hPa} - \Delta p\, e^{-r^2/R^2},$$

where r is the radial distance from the center. Determine the structure of the geostrophic wind around this system; find the maximum geostrophic wind, and the radius at which it is located, if $\Delta p = 20$ hPa, and $R = 500$ km. [Assume constant Coriolis parameter, appropriate to latitude 45° S, across the system.]

5. Write down an equation for the balance of radial forces on a parcel of fluid moving along a horizontal circular path of radius r at constant speed v (taken positive if the flow is in the same sense of rotation as the Earth).

Solve for v as a function of r and the radial pressure gradient, and hence show that:

(a) if $v > 0$, the wind speed is less than its geostrophic value,

(b) if $|v| \ll fr$, then the flow approaches its geostrophic value, and

(c) there is a limiting pressure gradient for the balanced motion when $v > -1/2fr$.

Comment on the asymmetry between clockwise and anticlockwise vortices.

6. (i) A typical hurricane at, say, 30° latitude may have low-level winds of 50 m s^{-1} at a radius of 50 km from its

center: do you expect this flow to be geostrophic?

(ii) Two weather stations near 45° N are 400 km apart, one exactly to the northeast of the other. At both locations, the 500-mbar wind is exactly southerly at 30 m s^{-1}. At the northeastern station, the height of the 500-mbar surface is 5510 m; what is the height of this surface at the other station?

What vertical displacement would produce the same pressure difference between the two stations? Comment on your answer. You may take $\rho_s = 1.2\,\text{kg m}^{-3}$.

7. Write down an expression for the centrifugal acceleration of a ring of air moving uniformly along a line of latitude with speed u relative to the Earth, which itself is rotating with angular speed Ω. Interpret the terms in the expression physically.

By hypothesizing that the relative centrifugal acceleration resolved parallel to the Earth's surface is balanced by a meridional pressure gradient, deduce the geostrophic relationship:

$$fu + \frac{1}{\rho}\frac{\partial p}{\partial y} = 0$$

(in our usual notation and where $dy = a\,d\varphi$).

If the gas is perfect and in hydrostatic equilibrium, derive the thermal wind equation.

8. The vertical average (with respect to log pressure) of atmospheric temperature below the 200-mbar pressure surface is about 265 K at the equator and 235 K at the winter pole. Calculate the equator-to-winter-pole height difference on the 200-mbar pressure surface, assuming surface pressure is 1000 mbar everywhere. Assuming that

this pressure surface slopes uniformly between 30° and 60° latitude and is flat elsewhere, use the geostrophic wind relationship (zonal component) in pressure coordinates,

$$u = -\frac{g}{f}\frac{\partial z}{\partial y}$$

to calculate the mean eastward geostrophic wind on the 200-mbar surface at 45° latitude in the winter hemisphere. Here $f = 2\Omega \sin\varphi$ is the Coriolis parameter, g is the acceleration due to gravity, z is the height of a pressure surface, and $dy = a \times d\varphi$, where a is the radius of the Earth is a northward pointing coordinate.

9. From the pressure coordinate thermal wind relationship, Eq. 7-24, and approximating

$$\frac{\partial u}{\partial p} \simeq \frac{\partial u/\partial z}{\partial p/\partial z},$$

show that, in geometric height coordinates,

$$f\frac{\partial u}{\partial z} \simeq -\frac{g}{T}\frac{\partial T}{\partial y}.$$

The winter polar stratosphere is dominated by the "polar vortex," a strong westerly circulation at about 60° latitude around the cold pole, as depicted schematically in Fig. 7.29. (This circulation is the subject of considerable interest, because it is within the polar vortices—especially that over Antarctica in southern winter and spring—that most ozone depletion is taking place.)

Assuming that the temperature at the pole is (at all heights) 50 K colder at 80° latitude than at 40° latitude (and that it varies uniformly in between), and that the westerly wind speed at 100 mbar pressure and 60° latitude is $10\,\mathrm{m\,s^{-1}}$, use the thermal wind relation to estimate the wind speed at 1 mbar pressure and 60° latitude.

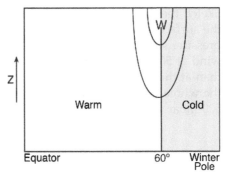

FIGURE 7.29. A schematic of the winter polar stratosphere dominated by the "polar vortex," a strong westerly circulation at ~60° around the cold pole.

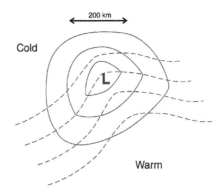

FIGURE 7.30. A schematic of surface pressure contours (solid) and mean 1000 mbar—500 mbar temperature contours (dashed), in the vicinity of a typical northern hemisphere depression (storm).

10. Starting from Eq. 7-24, show that the thermal wind equation can be written in terms of potential temperature, Eq. 4-17, thus:

$$\left(\frac{\partial u_g}{\partial p}, \frac{\partial v_g}{\partial p}\right) = \frac{1}{\rho\theta}\left(\left(\frac{\partial\theta}{\partial y}\right)_p, -\left(\frac{\partial\theta}{\partial x}\right)_p\right).$$

11. Fig. 7.30 shows, schematically, the surface pressure contours (solid) and mean 1000 mbar to 500 mbar temperature contours (dashed), in the vicinity of a typical northern hemisphere depression (storm). "L" indicates the low pressure center.

Sketch the directions of the wind near the surface, and on the 500-mbar pressure surface. (Assume that the wind at 500 mbar is significantly larger than at the surface.) If the movement of the whole system is controlled by the 500-mbar wind (i.e., it simply gets blown downstream by the 500-mbar wind), how do you expect the storm to move? [Use density of air at 1000 mbar = 1.2 kg m^{-3}; rotation rate of Earth = 7.27×10^{-5} s^{-1}; gas constant for air = 287 J kg^{-1}K.]

CHAPTER

8

The general circulation of the atmosphere

8.1. Understanding the observed circulation
8.2. A mechanistic view of the circulation
    8.2.1. The tropical Hadley circulation
    8.2.2. The extra tropical circulation and GFD Lab XI: Baroclinic instability
8.3. Energetics of the thermal wind equation
    8.3.1. Potential energy for a fluid system
    8.3.2. Available potential energy
    8.3.3. Release of available potential energy in baroclinic instability
    8.3.4. Energetics in a compressible atmosphere
8.4. Large-scale atmospheric energy and momentum budget
    8.4.1. Energy transport
    8.4.2. Momentum transport
8.5. Latitudinal variations of climate
8.6. Further reading
8.7. Problems

In this chapter we return to our discussion of the large-scale circulation of the atmosphere and make use of the dynamical ideas developed in the previous two chapters to enquire into its cause. We begin by reviewing the demands on the atmospheric circulation imposed by global energy and angular momentum budgets as depicted in Fig. 8.1. The atmosphere must transport energy from equator to pole to maintain the observed pole-equator temperature gradient. In addition, because there are westerly winds at the surface in middle latitudes and easterly winds in the tropics (see, e.g., Fig. 7.28), angular momentum must be transported from the tropics to higher latitudes. We discuss how this transfer is achieved by the Hadley circulation in the tropics and weather systems in middle latitudes. We will discover that weather systems arise from a hydrodynamic instability of the thermal wind shear associated with the pole-equator temperature gradient. Produced through a mechanism known as baroclinic instability, the underlying dynamics are illustrated through laboratory experiment and discussed from the perspective of the available potential energy "stored" by the horizontal temperature gradients.

8.1. UNDERSTANDING THE OBSERVED CIRCULATION

The simplest observed global characteristic of the atmosphere is that the tropics are much warmer than the poles. As discussed in Chapter 5, this is a straightforward consequence of the geometry of the Earth: the annually averaged incoming solar radiation per unit area of the Earth's surface is much greater at the equator than at the poles, a difference that is compounded by the fact that the polar regions are covered in ice and snow and therefore reflect much of the incoming radiation back to space. A little less obvious is the fact that the tropical regions actually receive more energy from the Sun than they emit back to space, whereas the converse is true in high latitudes. Since both regions are, on an annual average, in equilibrium, there must be a process acting to transport excess energy from the tropics to make up the deficit in high latitudes, as depicted schematically in Fig. 8.1 (left).

The implied transport of some 6×10^{15}W (see Fig. 5.6) must be effected by the atmospheric circulation, carrying warm air poleward and cold air equatorward. (In fact, the ocean circulation also contributes, as discussed in Chapter 11.) As a result, the tropics are cooler, and polar regions warmer, than they would be in the absence of such transport. Thus, in this as in other respects, the atmospheric general circulation plays a key role in climate.

What are the motions that deliver the required transport? The gross features of the observed atmospheric circulation discussed in Chapter 5 are depicted in Fig. 8.2. The zonal flow is strongly westerly aloft in middle latitudes, a fact that we can now, following Section 7.3, understand as a straightforward consequence of the decrease of temperature with latitude. Surface winds are constrained to be weak by the action of friction near the ground, and thermal wind balance, Eq. 7-24, implies that a poleward decrease of temperature is necessarily accompanied by increasing westerly winds with height. Taken together, these two facts require a zonal flow in middle latitudes that increases from near zero at the ground to strong westerlies at altitude, as schematized in Fig. 7.19.

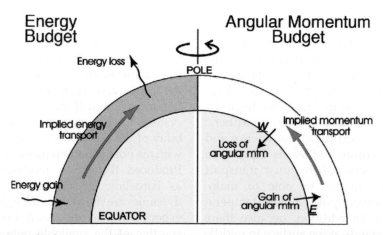

FIGURE 8.1.   Latitudinal transport of (left) energy and (right) angular momentum (mtm) implied by the observed state of the atmosphere. In the energy budget there is a net radiative gain in the tropics and a net loss at high latitudes; to balance the energy budget at each latitude, a poleward energy flux is implied. In the angular momentum budget the atmosphere gains angular momentum in low latitudes (where the surface winds are easterly) and loses it in middle latitudes (where the surface winds are westerly). A poleward atmospheric flux of angular momentum is thus implied.

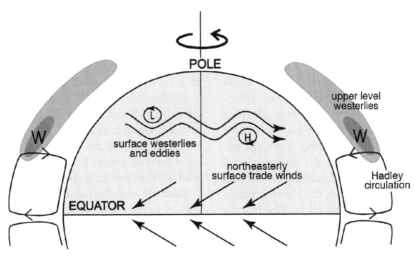

FIGURE 8.2. Schematic of the observed atmospheric general circulation for annual-averaged conditions. The upper level westerlies are shaded to reveal the core of the subtropical jet stream on the poleward flank of the Hadley circulation. The surface westerlies and surface trade winds are also marked, as are the highs and lows of middle latitudes. Only the northern hemisphere is shown. The vertical scale is greatly exaggerated.

Although the near-surface winds are weak—a few m s^{-1}—they nevertheless exhibit a distinct spatial distribution, with the zonal component being easterly in the tropics and westerly in middle latitudes (Fig. 7.28). In the presence of surface friction, the atmosphere must therefore be losing angular momentum to the ground in middle latitudes and gaining angular momentum in the tropics. As Fig. 8.1 (right) illustrates, the angular momentum balance of the atmosphere thus requires a transport of westerly angular momentum from low to middle latitudes to maintain equilibrium.

Even though the west-to-east circulation in the upper troposphere is the dominant component of the large-scale atmospheric flow, it cannot be responsible for the required poleward transports of energy and angular momentum, for which north-south flow is needed. As we saw in Chapter 5 (Fig. 5.21) there is indeed a mean circulation in the meridional plane, which is dominated by the Hadley circulation of the tropical atmosphere with, on an annual average, mean upwelling near the equator, poleward flow aloft, subsidence in the subtropics, and equatorward return flow near the surface. This circulation transports energy and angular momentum poleward, as required, within the tropics; however, the meridional circulation becomes much weaker in middle latitudes and so cannot produce much transport there. We saw in Section 5.4.1 that most of the north-south flow in the extratropical atmosphere takes the form of eddies, rather than a mean overturning as in the tropics. As we will see, it is these eddies that produce the poleward transports in the extratropics.

8.2. A MECHANISTIC VIEW OF THE CIRCULATION

To what extent can we explain the main features of the observed general circulation on the basis of the fluid dynamics of a simple representation of the atmosphere driven by latitudinal gradients in solar forcing? The emphasis here is on "simple." In reality, the Earth's surface is very inhomogeneous: there are large mountain ranges that disturb the flow and large contrasts (e.g., in temperature and in surface roughness) between oceans and continents. In the interests of simplicity, however, we shall neglect such variations in surface conditions. We also neglect seasonal and diurnal[1] variations, and assume maximum solar input at the equator, even though the subsolar point (Fig. 5.3) migrates between the two tropics through the course of a year. Thus we shall consider the response of an atmosphere on a longitudinally uniform, rotating planet (that is otherwise like the Earth) to a latitudinal gradient of heating. We shall see that the gross features of the observed circulation can indeed be understood on this basis; later, in Section 8.5, the shortcomings introduced by this approach will be briefly discussed.

Thus we ask how an axisymmetric atmosphere responds to an axisymmetric forcing, with stronger solar heating at the equator than at the poles. It seems reasonable to assume that the response (i.e., the induced circulation) will be similarly axisymmetric. Indeed, this assumption turns out to give us a qualitatively reasonable description of the circulation in the tropics, but it will become evident that it fails in the extratropical atmosphere, where the symmetry is broken by hydrodynamical instability.

8.2.1. The tropical Hadley circulation

If the Earth was not rotating, the circulation driven by the pole-equator temperature difference would be straightforward, with warm air rising in low latitudes and cold air sinking at high latitudes, as first suggested by Hadley and sketched in Fig. 5.19. But, as seen in Figs. 5.20 and 5.21, this is not quite what happens. We do indeed see a meridional circulation in the tropics, but the sinking motion is located in the subtropics around latitudes of $\pm 30°$. In fact, the following considerations tell us that one giant axisymmetric meridional cell extending from equator to pole is not possible on a rapidly rotating planet, such as the Earth.

Consider a ring of air encircling the globe as shown in Fig. 8.3, lying within and being advected by the upper level poleward flow of the Hadley circulation. Because this flow is by assumption axisymmetric, and also because friction is negligible in this upper level flow, well above the near-surface boundary layer, absolute angular momentum will be conserved by the ring as it moves around. The absolute angular momentum per unit mass is (recall our discussion of angular momentum in the radial inflow experiment in Section 6.6.1)

$$A = \Omega r^2 + ur,$$

the first term being the contribution from planetary rotation, and the second from the eastward wind u relative to the Earth, where r is the distance from the Earth's rotation axis. Since $r = a \cos \varphi$,

$$A = \Omega a^2 \cos^2 \varphi + ua \cos \varphi. \tag{8-1}$$

Now, suppose that $u = 0$ at the equator (Fig. 5.20 shows that this is a reasonable assumption in the equatorial upper troposphere). Then the absolute angular

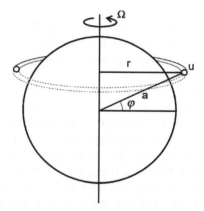

FIGURE 8.3. Schematic of a ring of air blowing west $\longrightarrow$ east at speed u at latitude φ. The ring is assumed to be advected by the poleward flow of the Hadley circulation conserving angular momentum.

---

momentum at the equator is simply $A_0 = \Omega a^2$. As the ring of air moves poleward, it retains this value, and so when it arrives at latitude φ, its absolute angular momentum is

$$A = \Omega a^2 \cos^2 \varphi + ua \cos \varphi = A_0 = \Omega a^2 .$$

Therefore the ring will acquire an eastward velocity

$$u(\varphi) = \frac{\Omega \left(a^2 - a^2 \cos^2 \varphi\right)}{a \cos \varphi} = \Omega a \frac{\sin^2 \varphi}{\cos \varphi} . \quad (8\text{-}2)$$

Note that this is directly analogous to Eq. 6-23 of our radial inflow experiment, when we realize that $r = a \cos \varphi$. Equation 8-2 implies unrealistically large winds far from the equator, at latitudes of $(10°, 20°, 30°)$, $u(\varphi) = (14, 58, 130) \text{ m s}^{-1}$, and of course the wind becomes infinite as $\varphi \to 90°$. On the grounds of physical plausibility, it is clear that such an axisymmetric circulation cannot extend all the way to the pole in the way envisaged by Hadley (and sketched in Fig. 5.19); it must terminate somewhere before it reaches the pole. Just how far the circulation extends depends (according to theory) on many factors.[2]

Consider the upper branch of the circulation, as depicted in Fig. 8.4. Near the equator, where f is small and the Coriolis effect is weak, angular momentum constraints are not so severe and the equatorial atmosphere acts as if the Earth were rotating slowly. As air moves away from the equator, however, the Coriolis parameter becomes increasingly large and in the northern hemisphere turns the wind to the right, resulting in a westerly component to the flow. At the poleward extent of the Hadley cell, then, we expect to find a strong westerly flow, as indeed we do (see Fig. 5.20). This subtropical jet is

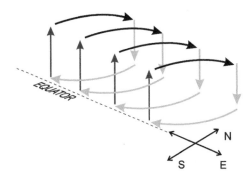

FIGURE 8.4. Schematic of the Hadley circulation (showing only the northern hemispheric part of the circulation; there is a mirror image circulation south of the equator). Upper level poleward flow induces westerlies; low level equatorward flow induces easterlies.

driven in large part by the advection from the equator of large values of absolute angular momentum by the Hadley circulation, as is evident from Eq. 8-2 and depicted in Fig. 8.4. Flow subsides on the subtropical edge of the Hadley cell, sinking into the subtropical highs (very evident in Fig. 7.27), before returning to the equator at low levels. At these low levels, the Coriolis acceleration, again turning the equatorward flow to its right in the northern hemisphere, produces the trade winds, northeasterly in the northern hemisphere (southeasterly in the southern hemisphere) (see Fig. 7.28). These winds are not nearly as strong as in the upper troposphere, because they are moderated by friction acting on the near-surface flow. In fact, as discussed above, there must be low-level westerlies somewhere. In equilibrium, the net frictional drag (strictly, torque) on the entire atmosphere must be zero, or the total angular momentum of the atmosphere could not be steady. (There would be a compensating change in the angular momentum of the solid Earth, and the length of the day would drift.[3])

---

[2] For example, if the Earth were rotating less (or more) rapidly, and other things being equal, the Hadley circulation would extend farther (or less far) poleward.

[3] In fact the length of the day does vary, ever so slightly, because of angular momentum transfer between the atmosphere and underlying surface. For example, on seasonal timescales there are changes in the length of the day of about a millisecond. The length of the day changes from day to day by about 0.1 ms! Moreover these changes can be attributed to exchanges of angular momentum between the Earth and the atmosphere.

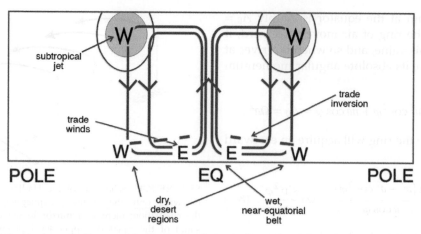

FIGURE 8.5. A schematic diagram of the Hadley circulation and its associated zonal flows and surface circulation.

So as shown in Fig. 8.5 the surface winds in our axisymmetric model would be westerly at the poleward edge of the circulation cell, and eastward near the equator. This is similar to the observed pattern (see Fig. 7.28, middle panel), but not quite the same. In reality the surface westerlies maximize near 50° N, S, significantly poleward of the subtropical jet, a point to which we return in Section 8.4.2.

As sketched in Fig. 8.5, the subtropical region of subsidence is warm (because of adiabatic compression) and dry (since the air aloft is much drier than surface air); the boundary formed between this subsiding air and the cooler, moister near-surface air is the "trade inversion" noted in Chapter 4, and within which the trade winds are located. Aloft, horizontal temperature gradients within the Hadley circulation are very weak (see Figs. 5.7 and 5.8), a consequence of very efficient meridional heat transport by the circulation.

Experiment on the Hadley circulation: GFD Lab VIII revisited

A number of aspects of the Hadley circulation are revealed in Expt VIII, whose experimental arrangement has already been described in Fig. 7.12 in the context of the thermal wind equation. The apparatus is just a cylindrical tank containing plain water, at the center of which is a metal can filled with ice. The consequent temperature gradient (decreasing "poleward") drives motions in the tank, the nature of which depends on the rotation rate of the apparatus. When slowly rotating, as in this experiment ($\Omega \lesssim 1$ rpm—yes, only 1 rotation of the table per minute: very slow!), we see the development of the thermal wind in the form of a strong "eastward" (i.e., super-rotating) flow in the upper part of the fluid, which can be revealed by paper dots floating on the surface and dye injected into the fluid as seen in Fig. 7.13.

The azimuthal current observed in the experiment, which is formed in a manner analogous to that of the subtropical jet by the Hadley circulation discussed previously, is maintained by angular momentum advection by the meridional circulation sketched in Fig. 7.12 (right). Water rises in the outer regions, moves inward in the upper layers, conserving angular momentum as it does, thus generating strong "westerly" flow, and rubs against the cold inner wall, becoming cold and descending. Potassium permanganate crystals, dropped into the fluid (not too many!), settle on the bottom and give an indication of the flow in the bottom boundary layer. In Fig. 8.6 we see flow moving radially outwards at the bottom and being

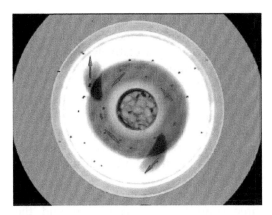

FIGURE 8.6. The Hadley regime studied in GFD Lab VIII, Section 7.3.1. Bottom flow is revealed by the two outward spiralling purple streaks showing anticyclonic (clockwise) flow sketched schematically in Fig. 7.12 (right); the black paper dots and green collar of dye mark the upper level flow and circulate cyclonically (anticlockwise).

deflected to the right, creating easterly flow opposite to that of the rotating table; note the two purple streamers moving outward and clockwise (opposite to the sense of rotation and the upper level flow). This bottom flow is directly analogous to the easterly and equatorward trade winds of the lower atmosphere sketched in Fig. 8.4 and plotted from observations in Fig. 7.28.

Quantitative study of the experiment (by tracking, for example, floating paper dots) shows that the upper level azimuthal flow does indeed conserve angular momentum, satisfying Eq. 6-23 quite accurately, just as in the radial inflow experiment. With $\Omega = 0.1$ s^{-1} (corresponding to a rotation rate of 0.95 rpm; see Table A.1, Appendix A.4.1) Eq. 6-23 implies a hefty $10\,\mathrm{cm\,s}^{-1}$ if angular momentum were conserved by a particle moving from the outer radius, $r_1 = 10\,\mathrm{cm}$, to an inner radius of 10 cm. Note, however, that if Ω were set 10 times higher, to 10 rpm, then angular momentum conservation would imply a speed of $1\,\mathrm{m\,s}^{-1}$, a very swift current. We will see in the next section that such swift currents are not observed if we turn up the rotation rate of the table. Instead the azimuthal current breaks down into eddying motions,

inducing azimuthal pressure gradients and breaking angular momentum conservation.

8.2.2. The extratropical circulation and GFD Lab XI: Baroclinic instability

Although the simple Hadley cell model depicted in Fig. 8.5 describes the tropical regions quite well, it predicts that little happens in middle and high latitudes. There, where the Coriolis parameter is much larger, the powerful constraints of rotation are dominant and meridional flow is impeded. We expect, however, (and have seen in Figs. 5.8 and 7.20), that there are strong gradients of temperature in middle latitudes, particularly in the vicinity of the subtropical jet. So, although there is little meridional circulation, there is a zonal flow in thermal wind balance with the temperature gradient. Since T decreases poleward, Eq. 7-24 implies westerly winds increasing with height. This state of a zonal flow with no meridional motion is a perfectly valid equilibrium state. Even though the existence of a horizontal temperature gradient implies horizontal pressure gradients (tilts of pressure surfaces), the associated force is entirely balanced by the Coriolis force acting on the thermal wind, as we discussed in Chapter 7.

Our deduction that the mean meridional circulation is weak outside the tropics is qualitatively in accord with observations (see Fig. 5.21) but leaves us with two problems:

1. Poleward heat transport is required to balance the energy budget. In the tropics, the overturning Hadley circulation transports heat poleward, but no further than the subtropics. How is heat transported further poleward if there is no meridional circulation?

2. Everyday observation tells us that a picture of the midlatitude atmosphere as one with purely zonal winds is very wrong. If it were true, weather would be very predictable (and very dull).

Our axisymmetric model is, therefore, only partly correct. The prediction of purely zonal flow outside the tropics is quite wrong. As we have seen, the midlatitude atmosphere is full of *eddies*, which manifest themselves as traveling weather systems.[4] Where do they come from? In fact, as we shall now discuss and demonstrate through a laboratory experiment, the extratropical atmosphere is hydrodynamically unstable, the flow spontaneously breaking down into eddies through a mechanism known as *baroclinic instability*.[5] These eddies readily generate meridional motion and, as we shall see, affect a meridional transport of heat.

Baroclinic instability

GFD Lab XI: Baroclinic instability in a dishpan To introduce our discussion of the breakdown of the thermal wind through baroclinic instability, we describe a laboratory experiment of the phenomenon. The apparatus, sketched in Fig. 7.12, is identical to that of Lab VIII used to study the thermal wind and the Hadley circulation. In the former experiments the table was rotated very slowly, at a rate of $\Omega \lesssim 1$ rpm. This time, however, the table is rotated much more rapidly, at $\Omega \sim 10$ rpm, representing the considerably greater Coriolis parameter found in middle latitudes. At this higher rotation rate something remarkable happens. Rather than a steady axisymmetric flow, as in the Hadley regime shown in Fig. 8.6, a strongly eddying flow is set up. The thermal wind remains, but

breaks down through instability, as shown in Fig. 8.7. We see the development of eddies that sweep (relatively) warm fluid from the periphery to the cold can in one sector of the tank (e.g., A in Fig. 8.7), and simultaneously carry cold fluid from the can to the periphery at another (e.g., B in Fig. 8.7). In this way a radially-inward heat transport is achieved, offsetting the cooling at the center caused by the melting ice.

For the experiment shown we observe three complete wavelengths around the tank in Fig. 8.7. By repeating the experiment but at different values of Ω, we observe that the scale of the eddies decreases, and the flow becomes increasingly irregular as Ω is increased. These eddies are produced by the same mechanism underlying the creation of atmospheric weather systems shown in Fig. 7.20 and discussed later.

Before going on we should emphasize that the flow in Fig. 8.7 does not conserve angular momentum. Indeed, if it did, as estimated at the end of Section 8.2.1, we would observe very strong horizontal currents near the ice bucket, of order $1 \, \mathrm{m \, s^{-1}}$ rather than a few $\mathrm{cm \, s^{-1}}$, as seen in the experiment. This, of course, is exactly what Eq. 8-2 is saying for the atmosphere: if rings of air conserved angular momentum as they move in axisymmetric motion from equator to pole, then we obtain unrealistically large winds. Angular momentum is not conserved because of the presence of zonal (or, in our tank experiment, azimuthal) pressure gradient associated with eddying motion (see Problem 6 of Chapter 6).

[4] Albert Defant (1884–1974), German Professor of Meteorology and Oceanography, who made important contributions to the theory of the general circulation of the atmosphere. He was the first to liken the meridional transfer of energy between the subtropics and pole to a turbulent exchange by large-scale eddies.

[5] In a "baroclinic" fluid, $\rho = \rho\,(p,T)$, and so there can be gradients of density (and therefore of temperature) along pressure surfaces. This should be contrasted to a "barotropic" fluid [$\rho = \rho\,(p)$] in which no such gradients exist.

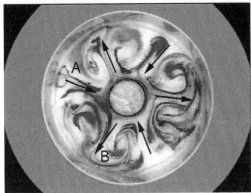

FIGURE 8.7. Top: Baroclinic eddies in the "eddy" regime viewed from the side. Bottom: View from above. Eddies draw fluid from the periphery in toward the centre at point A and vice versa at point B. The eddies are created by the instability of the thermal wind induced by the radial temperature gradient due to the presence of the ice bucket at the center of the tank. The diameter of the ice bucket is 15 cm.

Middle latitude weather systems The process of baroclinic instability studied in the laboratory experiment just described is responsible for the ubiquitous waviness of the midlatitude flow in the atmosphere. As can be seen in the observations shown on Figs. 5.22, 7.4, and 7.20, these waves

often form closed eddies, especially near the surface, where they are familiar as the high- and low-pressure systems associated with day-to-day weather. In the process, they also affect the poleward heat transport required to balance the global energy budget (see Fig. 8.1). The manner in which this is achieved on the planetary scale is sketched in Fig. 8.8. Eddies "stir" the atmosphere, carrying cold air equatorward and warm air poleward, thereby reducing the equator-to-pole temperature contrast. To the west of the low (marked L) in Fig. 8.8, cold air is carried into the tropics. To the east, warm air is carried toward the pole, but since poleward flowing air is ascending (remember the large-scale slope of the $\bar{\theta}$ surfaces shown in Fig. 5.8) it tends to leave the surface. Thus we get a concentrated gradient of temperature near point 1, where the cold air "pushes into" the warm air, and a second, less marked concentration where the warm air butts into cold at point 2. We can thus identify cold and warm fronts, respectively, as marked in the center panel of Fig. 8.8. Note that a triangle is used to represent the "sharp" cold front and a semicircle to represent the "gentler" warm front. In the bottom panel we present sections through cold fronts and warm fronts respectively.

Timescales and length scales We have demonstrated by laboratory experiment that a current in thermal wind balance can, and almost always does, become hydrodynamically unstable, spawning meanders and eddies. More detailed theoretical analysis[6] shows that the lateral scale of the eddies that form, L_{eddy} (as measured by, for example,

---

[6]Detailed analysis of the space-scales and growth rates of the instabilities that spontaneously arise on an initially zonal jet in thermal wind balance with a meridional temperature gradient shows that:

1. The Eady growth rate of the disturbance $e^{\sigma t}$ is given by $\sigma = 0.31 U/L\rho = 0.31\frac{f}{N}\frac{d\bar{u}}{dz}$ using Eq. (7-23) and $U = \frac{d\bar{u}}{dz}H$

   (see Gill, 1982). Inserting typical numbers for the troposphere, $N = 10^{-2}\,\text{s}^{-1}$, $\frac{d\bar{u}}{dz} = 2 \times 10^{-3}\,\text{s}^{-1}$, $f = 10^{-4}\,\text{s}^{-1}$, we find that $\sigma \simeq 10^{-5}\,\text{s}^{-1}$, an e-folding timescale of 1 day.

2. The wavelength of the fastest-growing disturbance is $4L\rho$ where $L\rho$ is given by Eq. 7-23. This yields a wavelength of 2800 km if $L\rho = 700$ km. The circumference of the Earth at 45° N is 21,000 km, and so about 7 synoptic waves can fit around the Earth at any one time. This is roughly in accord with observations; see, for example, Fig. 5.22.

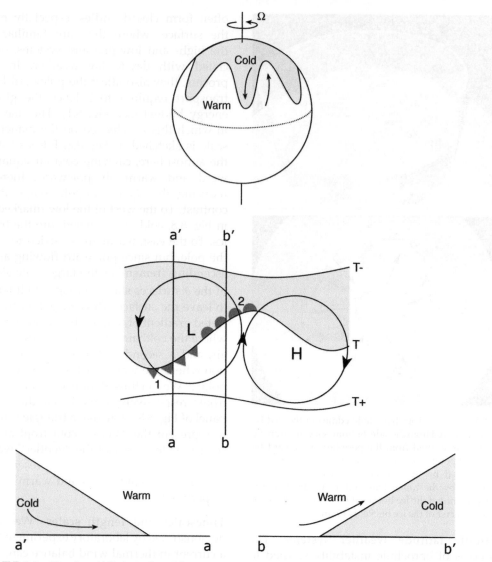

FIGURE 8.8.   Top: In middle latitudes eddies transport warm air poleward and upward and cold air equator-ward and downward. Thus the eddies tend to "stir" the atmosphere laterally, reducing the equator-to-pole temperature contrast. Middle: To the west of the "L," cold air is carried in to the tropics. To the east, warm air is carried toward the pole. The resulting cold fronts (marked by triangles) and warm fronts (marked by semicircles) are indicated. Bottom: Sections through the cold front, $a' \longrightarrow a$, and the warm front, $b \longrightarrow b'$, respectively.

the typical lateral scale of a low pressure system in the surface analysis shown in Fig. 7.25, or the swirls of dye seen in our tank experiment, Fig. 8.7) is proportional to the Rossby radius of deformation discussed in Section 7.3.4:

$$L_{eddy} \sim L_\rho, \qquad (8-3)$$

where L_ρ is given by Eq. 7-23.

The timescale of the disturbance is given by:

$$\tau_{eddy} \sim \frac{L_\rho}{U}, \qquad (8-4)$$

where $U = d\bar{u}/dz \, H$ is the strength of the upper level flow, $d\bar{u}/dz$ is the thermal wind given by Eq. 7-17, and H is the vertical scale of the flow. Eq. 8-4 is readily interpretable as the time it takes a flow moving at speed U to travel a distance L_ρ and is proportional to the (inverse of) the *Eady growth rate*, after Eric Eady[7], a pioneer of the theory of baroclinic instability.

In our baroclinic instability experiment, we estimate a deformation radius of some 10 cm, roughly in accord with the observed scale of the swirls in Fig. 8.7 (to determine a scale, note that the diameter of the ice can in the figure is 15 cm). Typical flow speeds observed by the tracking of paper dots floating at the surface of the fluid are about $1\,\text{cm s}^{-1}$. Thus Eq. 8-4 suggests a timescale of $\mathcal{T}_{eddy} \sim 7.0\,\text{s}$, or roughly one rotation period. Consistent with Eq. 8-3, the eddy scale decreases with increased rotation rate.

Applying the above formulae to the middle troposphere where $f \sim 10^{-4}\,\text{s}^{-1}$, $N \sim 10^{-2}\,\text{s}^{-1}$ (see Section 4.4), $H \sim 7\,\text{km}$ (see Section 3.3), and $U \sim 10\,\text{m s}^{-1}$, we find that $L_\rho = NH/f \sim 700\,\text{km}$ and $\mathcal{T}_{eddy} \sim 700\,\text{km}/10\,\text{m s}^{-1} \sim 1\,\text{d}$. These estimates are roughly in accord with the scales and growth rates of observed weather systems.

We have not yet discussed the underlying baroclinic instability mechanism and its energy source. To do so we now consider the energetics of and the release of potential energy from a fluid.

8.3. ENERGETICS OF THE THERMAL WIND EQUATION

The immediate source of kinetic energy for the eddying circulation observed in our baroclinic instability experiment and in the middle-latitude atmosphere, is the potential energy of the fluid. In the spirit of the energetic discussion of convection developed in Section 4.2.3, we now compute the potential energy available for conversion to motion. However, rather than, as there, considering the energy of isolated fluid parcels, here we focus on the potential energy of the whole fluid. It will become apparent that not all the potential energy of a fluid is available for conversion to kinetic energy. We need to identify that component of the potential energy—known as *available potential energy*—that can be released by a redistribution of mass of the system.

To keep things as simple as possible, we will first focus on an incompressible fluid, such as the water in our tank experiment. We will then go on to address the compressible atmosphere.

8.3.1. Potential energy for a fluid system

We assign to a fluid parcel of volume $dV = dx\,dy\,dz$ and density ρ, a potential energy of $gz \times (\text{parcel mass}) = gz\rho\,dV$. Then the potential energy of the entire fluid is

$$PE = g \int z\rho\,dV, \qquad (8\text{-}5)$$

where the integral is over the whole system. Clearly, PE is a measure of the vertical distribution of mass.

Energy can be released and converted to kinetic energy only if some rearrangement of the fluid results in a lower total potential energy. Such rearrangement is subject to certain constraints, the most obvious of which is that the total mass

[7] Eric Eady (1918–1967). A brilliant theorist who was a forecaster in the early part of his career, Eady expounded the theory of baroclinic instability in a wonderfully lucid paper—Eady (1949)—which attempted to explain the scale, growth rate, and structure of atmospheric weather systems.

$$M = \int \rho \, dV, \qquad (8\text{-}6)$$

cannot change. In fact, we may use Eq. 8-6 to rewrite Eq. 8-5 as

$$PE = gM \langle z \rangle, \qquad (8\text{-}7)$$

where

$$\langle z \rangle = \frac{\int z \, \rho \, dV}{\int \rho \, dV} = \frac{\int z \, \rho \, dV}{M} \qquad (8\text{-}8)$$

is the mass-weighted mean *height* of the fluid. We can think of $\langle z \rangle$ as the "height of the center of mass." It is evident from Eq. 8-7 that potential energy can be released only by lowering the center of mass of the fluid.

8.3.2. Available potential energy

Consider again the stratified incompressible fluid discussed in Section 4.2.2. Suppose, first, that $\partial \rho / \partial z < 0$ (stably stratified) and that there are no horizontal gradients of density, so ρ is a function of height *only*, as depicted in Fig. 4.5. In Section 4.2.3 we considered the energetic implications of adiabatically switching the position of two parcels, initially at heights z_1 and z_2, with densities ρ_1 and ρ_2, respectively, with $z_2 > z_1$ and $\rho_2 < \rho_1$. Since ρ is conserved under the displacement (remember we are considering an incompressible fluid here), the final state has a parcel with density ρ_1 at z_2, and one with ρ_2 at z_1. Thus density has increased at z_2 and decreased at z_1. The center of mass has therefore been raised, and so PE has been increased by the rearrangement; none can be released into kinetic energy. This of course is one way of understanding why the stratification is stable. Even though such a fluid has nonzero potential energy (as defined by Eq. 8-7), this energy is *unavailable* for conversion to kinetic energy. It cannot be reduced by any adiabatic rearrangement of fluid parcels. Therefore the fluid has *available* potential energy only if it has nonzero horizontal gradients of density.

Consider now the density distribution sketched in Fig. 8.9 (left); a two-layer fluid is shown with light fluid over heavy with the interface between the two sloping at angle γ. This can be considered to be a highly idealized representation of the radial density distribution in our tank experiment GFD Lab XI with heavier fluid (shaded) to the right (adjacent to the ice bucket) and light fluid to the left. Let us rearrange the fluid such that the interface becomes horizontal as shown. Now all fluid below the dashed horizontal line is dense, and all fluid above is light. The net effect of the rearrangement has been to exchange heavy fluid downward and light fluid upward such that, in the wedge B, heavy fluid has been replaced by light fluid, while the opposite has happened in the wedge A. (Such a rearrangement can be achieved in more than one way, as we shall see.) The center of mass has thus been lowered, and the final potential energy is therefore less than that in the initial state: potential energy has been released. Since no further energy can be released once the interface is horizontal, the difference in the potential energy between these two states defines the *available potential energy* (*APE*) of the initial state.

The important conclusions here are that available potential energy is associated with horizontal gradients of density and that release of that energy (and, by implication, conversion to kinetic energy of the motion) is effected by a reduction in those gradients.

The previous discussion can be made quantitative as follows. Consider again Fig. 8.9, in which the height of the interface between the two fluids is given by $h(y) = \frac{1}{2} H + \gamma y$, and to keep things simple we restrict the slope of the interface $\gamma < \frac{1}{2} \frac{H}{L}$ to ensure that it does not intersect the lower and upper boundaries at $z = 0$, $z = H$ respectively. Direct integration of Eq. 8-5 for the density distribution shown in Fig. 8.9, shows that the potential energy per unit length in the x–direction is given by (as derived in Appendix A.1.2)

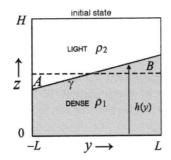
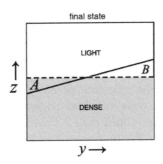

FIGURE 8.9. Reduction in the available potential energy of a two-layer fluid moving from an initial state in which the interface is tilted (left) to the final state in which the interface is horizontal (right). Dense fluid is shaded. The net effect of the rearrangement is to exchange heavy fluid downward and light fluid upward such that, in the wedge B, heavy fluid is replaced by light fluid, whereas the opposite occurs in the wedge A.

$$P = \int_0^H \int_{-L}^L g\rho z \, dy \, dz$$

$$= H^2 L \rho_1 \left(g + \frac{g'}{4} \right) + \frac{1}{3}\rho_1 g'\gamma^2 L^3,$$

where $g' = g(\rho_1 - \rho_2)/\rho_1$ is the reduced gravity of the system. P is at its absolute minimum, P_{min}, when $\gamma = 0$, and given by the first term on the rhs of the above. When $\gamma = 0$, no rearrangement of fluid can lower the center of mass, since all dense fluid lies beneath all light fluid. When $\gamma \neq 0$, however, P exceeds P_{min}, and the fluid has available potential energy in the amount

$$APE = P - P_{min} = \frac{1}{3}\rho_1 g'\gamma^2 L^3 . \qquad (8-9)$$

Note that APE is independent of depth H, because all the APE is associated with the interfacial slope; only fluid motion in the vicinity of the interface changes potential energy, as is obvious by inspection of Fig. 8.9.

It is interesting to compare the APE associated with the tilted interface in Fig. 8.9 with the kinetic energy associated with the thermal wind current, which is in balance with it. Making use of the Margules formula, Eq. 7-21, the geostrophic flow of the upper layer is given by $u_2 = g'\gamma/2\Omega$ if the lower layer is at rest. Thus the kinetic energy of the flow per unit length is, on evaluation:

$$KE = \int_h^H \int_{-L}^L \frac{1}{2}\rho_2 u_2^2 \, dy \, dz = \frac{1}{2}\rho_2 \frac{g'^2\gamma^2}{(2\Omega)^2}HL \qquad (8-10)$$

The ratio of APE to KE is thus:

$$\frac{APE}{KE} = \frac{2}{3}\left(\frac{L}{L_\rho} \right)^2 \qquad (8-11)$$

where we have set $\rho_1/\rho_2 \longrightarrow 1$ and used Eq. 7-23 to express the result in terms of the deformation radius L_ρ.

Equation 8-11 is a rather general result for balanced flows—the ratio of APE to KE is proportional to the square of the ratio of L, the lateral scale over which the flow changes, to the deformation radius L_ρ. Typically, on the largescale, L is considerably larger than L_ρ, and so the potential energy stored in the sloping density surfaces greatly exceeds the kinetic energy associated with the thermal wind current in balance with it. For example, T varies on horizontal scales of $L \sim 5000\,\text{km}$ in Fig. 5.7. In Section 8.2.2 we estimated that $L_\rho \sim 700\,\text{km}$; thus $(L/L_\rho)^2 \simeq 50$. This has important implications for atmospheric and (as we shall see in Chapter 10) oceanic circulation.[8]

---

[8]In fact in the ocean $(L/L\rho)^2 \simeq 400$, telling us that the available potential energy stored in the main thermocline exceeds the kinetic energy associated with ocean currents by a factor of 400. (See Section 10.5.)

8.3.3. Release of available potential energy in baroclinic instability

In a nonrotating fluid, the release of *APE* is straightforward and intuitive. If, at some instant in time, the interface of our two-component fluid were tilted as shown in Fig. 8.10 (left), we would expect the interface to collapse in the manner depicted by the heavy arrows in the figure. In reality the interface would overshoot the horizontal position, and the fluid would slosh around until friction brought about a motionless steady state with the interface horizontal. The *APE* released by this process would first have been converted to kinetic energy of the motion, and ultimately lost to frictional dissipation. In this case there would be no circulation in the long term, but if external factors such as heating at the left of Fig. 8.10 (left) and cooling at the right were to sustain the tilt in the interface, there would be a steady circulation (shown by the light arrows in the figure) in which the interface-flattening effects of the circulation balance the heating and cooling.

In a rotating fluid, things are not so obvious. There is no necessity for a circulation of the type shown in Fig. 8.10 (left) to develop since, as we saw in Section 7.3, an equilibrium state can be achieved in which forces associated with the horizontal density gradient are in thermal wind balance

with a vertical shear of the horizontal flow. In fact, rotation actively suppresses such circulations. We have seen that if Ω is sufficiently small, as in the Hadley circulation experiment GFD Lab VIII (Section 8.2.1), such a circulation is indeed established. At high rotation rates, however, a Hadley regime is not observed; nevertheless, *APE* can still be released, albeit by a very different mechanism. As we discussed in Section 8.2.2, the zonal state we have described is *unstable* to *baroclinic instability*. Through this instability, azimuthally asymmetric motions are generated, within which fluid parcels are exchanged along sloping surfaces as they oscillate from side to side, as sketched in Fig. 8.8. The azimuthal current is therefore not purely zonal, but wavy, as observed in the laboratory experiment shown in Fig. 8.7. The only way light fluid can be moved upward in exchange for downward motion of heavy fluid is for the exchange to take place at an angle that is steeper than the horizontal, but less steep than the density surfaces, or within the *wedge of instability*, as shown in Fig. 8.10 (right). Thus *APE* is reduced, not by overturning in the radial plane, but by fluid parcels oscillating back and forth within this wedge; light (warm) fluid moves upward and radially inward at one azimuth, while, simultaneously heavy (cold) fluid moves downward and

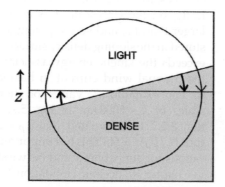
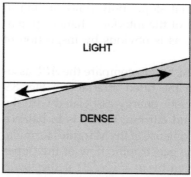

FIGURE 8.10.   Release of available potential energy in a two-component fluid. Left: Nonrotating (or very slowly rotating) case: azimuthally uniform overturning. Right: Rapidly rotating case: sloping exchange in the wedge of instability by baroclinic eddies.

outward at another, as sketched in Fig. 8.8. This is the mechanism at work in GFD Lab XI.[9]

8.3.4. Energetics in a compressible atmosphere

So far we have discussed energetics in the context of an incompressible fluid, such as water. However, it is not immediately clear how we can apply the concept of available potential energy to a compressible fluid, such as the atmosphere. Parcels of incompressible fluid can do no work on their surroundings and T is conserved in adiabatic rearrangement of the fluid; therefore potential energy is all one needs to consider. However, air parcels expand and contract during the redistribution of mass, doing work on their surroundings and changing their T and hence internal energy. Thus we must also consider changes in *internal energy*. It turns out that for an atmosphere in hydrostatic balance and with a flat lower boundary it is very easy to extend the definition Eq. 8-5 to include the internal energy. The internal energy (IE) of the entire atmosphere (assuming it to be a perfect gas) is, if we neglect surface topography (see Appendix A.1.3)

$$IE = c_v \int \rho T \, dV = \frac{c_v}{R} g \int z\rho \, dV . \quad (8\text{-}12)$$

Note the similarity to Eq. 8-5. In fact, for a diatomic gas like the atmosphere, $c_v/R = (1-\kappa)/\kappa = 5/2$, so its internal energy (neglecting topographic effects) is 2.5 times greater than its potential energy. The sum of internal and potential energy is conventionally (if somewhat loosely)

referred to as *total potential energy*, and given by

$$TPE = PE + IE = \frac{g}{\kappa} \int z\rho \, dV = \frac{g}{\kappa} M \langle z \rangle ,$$
$$(8\text{-}13)$$

where, as before, $\langle z \rangle$ is defined by Eq. 8-8 and M by Eq. 8-6.

We can apply the ideas discussed in the previous section to a compressible fluid if, as discussed in Chapter 4, we think in terms of the distribution of potential temperature, rather than density, since the former is conserved under adiabatic displacement.

Consider Fig. 8.11, in which the distribution of θ in the atmosphere, θ increasing upward and equatorward as in Fig. 5.8, is schematically shown. Again, we suppose

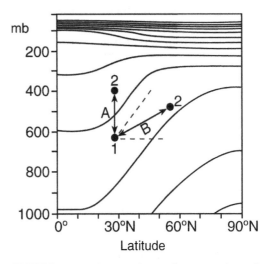

FIGURE 8.11. Air parcels 1 and 2 are exchanged along paths marked A and B, conserving potential temperature θ. The continuous lines are observed θ surfaces (see Fig. 5.8). The tilted dotted line is parallel to the local θ.

 [9] Jule Gregory Charney (1917–1981), American meteorologist and MIT Professor. He made wide-ranging contributions to the modern mathematical theory of atmospheric and oceanic dynamics, including the theory of large-scale waves, and of baroclinic instability. He was also one of the pioneers of numerical weather prediction, producing the first computer forecast in 1950.

that two air parcels are exchanged adiabatically. Suppose first that we exchange parcel 1, with $\theta = \theta_1$, with a second parcel with $\theta = \theta_2 > \theta_1$, vertically along path A. It is clear that the final potential energy will be greater than the initial value because the parcel moving upward is colder than the parcel it replaces. This is just the problem considered in our study of stability conditions for dry convection in Section 4.3. If parcels are exchanged along a horizontal path or along a path in a surface of constant θ (dotted lines in the figure), there will be no change in potential energy. But consider exchanging parcels along path B in Fig. 8.11, which has a slope *between* that of the θ surface and the horizontal. The parcel moving upward is now warmer than the parcel moving downward, and so potential energy has been reduced by the parcel rearrangement. In other words the center of gravity of the fluid is lowered. In the process, a lateral transport of heat is achieved from warm to cold. It is this release of *APE* that powers the eddying motion. As discussed in Section 8.3.3, fluid parcels are exchanged in this wedge of instability as they oscillate from side to side, as sketched in Fig. 8.8.

8.4. LARGE-SCALE ATMOSPHERIC ENERGY AND MOMENTUM BUDGET

8.4.1. Energy transport

We have seen two basic facts about the atmospheric energy budget:

1. There must be a *conversion* of energy into kinetic energy, from the available potential energy gained from solar heating, which requires that motions develop that transport heat upward (thereby tending to lower the atmospheric center of mass).

2. Accompanying this upward transport, there must be a *poleward transport* of heat from low latitudes; this transport

cools the tropics and heats the polar regions, thus balancing the observed radiative budget.

It is fairly straightforward to estimate the rate of poleward heat transport by atmospheric motions. Consider the northward flow at latitude φ across an elemental area dA, of height dz and longitudinal width $d\lambda$, so $dA = a \cos \varphi \, d\lambda \, dz$. The rate at which mass is flowing northward across dA is $\rho v \, dA$, and the associated flux of energy is $\rho v E \, dA$, where

$$E = c_p T + gz + Lq + \frac{1}{2}\mathbf{u} \cdot \mathbf{u} \qquad (8\text{-}14)$$

is the moist static energy, Eq. 4-27 (made up of the "dry static energy" $c_p T + gz$, and the latent heat content Lq) and the kinetic energy density $\frac{1}{2}\mathbf{u} \cdot \mathbf{u}$. If we now integrate across the entire atmospheric surface at that latitude, the net energy flux is

$$\overline{\mathcal{H}}^{\lambda}_{atmos} = \iint \rho v E \, dA$$
$$= a \cos \varphi \int_0^{2\pi} \int_0^{\infty} \rho v E \, dz \, d\lambda$$
$$= \frac{a}{g} \cos \varphi \int_0^{2\pi} \int_0^{p_s} v E \, dp \, d\lambda, \qquad (8\text{-}15)$$

(using hydrostatic balance to replace $\rho \, dz$ by $-dp/\rho$ where p_s is surface pressure. It turns out that the kinetic energy component of the net transport is negligible (typically less than 1% of the total), so we only need to consider the first three terms in Eq. 8-14.

Let us first consider energy transport in the tropics by the Hadley circulation which (in the annual mean) has equatorward flow in the lower troposphere and poleward flow in the upper troposphere. We may write

$$\overline{\mathcal{H}}^{\lambda}_{tropics} = \frac{2\pi a}{g} \cos \varphi \int_0^{p_s} v \left(c_p T + gz + Lq \right) \, dp, \qquad (8\text{-}16)$$

where we have assumed that the tropical atmosphere is independent of longitude λ.

Since temperature decreases with altitude, the heat carried poleward by the upper tropospheric flow is less than that carried equatorward at lower altitudes. Therefore the net heat transport—the first term in Eq. 8-16—is equatorward. The Hadley circulation carries heat toward the hot equator from the cooler subtropics! This does not seem very sensible. However, now let us add in the second term in Eq. 8-16, the potential energy density gz, to obtain the flux of dry static energy $v\left(c_pT + gz\right)$. Note that the vertical gradient of dry static energy is $\partial\left(c_pT + gz\right)/\partial z = c_p\partial T/\partial z + g$. For an atmosphere (like ours) that is generally stable to dry convection, $\partial T/\partial z + g/c_p > 0$ (as discussed in Section 4.3.1) and so the dry static energy increases with height. Therefore the poleward flow at high altitude carries more dry static energy poleward than the low-level flow takes equatorward, and the net transport is poleward, as we might expect.

To compute the total energy flux we must include all the terms in Eq. 8-16. The latent heat flux contribution is equatorward, since (see Fig. 5.15) q decreases rapidly with height, and so the equatorward flowing air in the lower troposphere carries more moisture than the poleward moving branch aloft. In fact, this term almost cancels the dry static energy flux. This is because the gross vertical gradient of moist static energy $c_pT + gz + Lq$, a first integral of Eq. 4-26 under hydrostatic motion, is weak in the tropics. This is just another way of saying that the tropical atmosphere is almost neutral to moist convection, a state that moist convection itself helps to bring about. This is clearly seen in the profiles of moist potential temperature shown in Fig. 5.9. Thus the equatorward and poleward branches of the circulation carry almost the same amount in each direction. In the net, then, the annually averaged energy flux by the Hadley cell is poleward, but weakly so. In fact, as can be seen in Fig. 8.13 and discussed further in Section 11.5, the heat transport by the ocean exceeds that of the atmosphere in the tropics up to 15° or so, particularly in the northern hemisphere.

The relative contributions of the various components of the energy flux are different in the extratropics, where the mean circulation is weak and the greater part of the transport is carried out by midlatitude eddies. In these motions, the poleward and equatorward flows occur at almost the same altitude, and so the strong vertical gradients of gz and Lq are not so important. Although these components of the energy flux are not negligible (and must be accounted for in any detailed calculation), we can use the heat flux alone to obtain an order of magnitude estimate of the net flux. So in middle latitudes we can represent Eq. 8-15 by

$$\overline{\mathcal{H}}^{\lambda}_{mid-lat} \sim 2\pi\frac{ac_p}{g}\cos\varphi\, p_s\, [v][T],$$

where $[v]$ is a typical northward wind velocity (dominated by the eddy component in middle latitudes), and $[T]$ is the typical magnitude of the temperature *fluctuations* in the presence of the eddies—the temperature difference between equatorward and poleward flowing air. Given $a = 6371$ km, $c_p = 1005$ J kg^{-1} K^{-1}, $g = 9.81$ m s^{-2}, $p_s \simeq 10^5$ Pa, then if we take typical values of $[v] \simeq 10$ m s^{-1} and $[T] \simeq 3$ K at a latitude of 45°, we estimate $\overline{\mathcal{H}}^{\lambda}_{mid-lat} \sim 8$ PW. As discussed earlier in Section 5.1.3 and in more detail later, this is of the same order as implied by the radiative imbalance.

In cartoon form, our picture of the low- and high-latitude energy balance is therefore as shown in Fig. 8.12 (left). In the tropics energy is transported poleward by the Hadley circulation; in higher latitudes, eddies are the principal agency of heat transport.

The results of more complete calculations, making use of top of the atmosphere radiation measurements and analyzed atmospheric fields, of heat fluxes in the atmosphere and ocean are shown in Fig. 8.13. The bulk of the required transport is carried by the atmosphere in middle and high latitudes, but the ocean makes up a considerable fraction, particularly in the

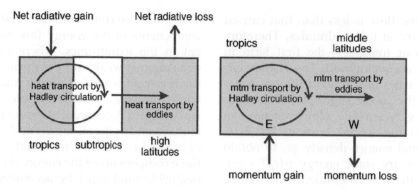

FIGURE 8.12. Schematic of the transport of (left) energy and (right) momentum by the atmospheric general circulation. Transport occurs through the agency of the Hadley circulation in the tropics, and baroclinic eddies in middle latitudes (see also Fig. 8.1).

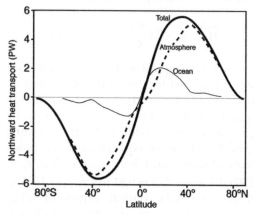

FIGURE 8.13. The ocean (thin) and atmospheric (dotted) contributions to the total northwards heat flux (thick) based on the NCEP reanalysis and top of the atmosphere radiation measurements (in PW = 10^{15} W) by (i) estimating the net surface heat flux over the ocean, (ii) the associated oceanic contribution, correcting for heat storage associated with global warming and constraining the ocean heat transport to be −0.1 PW at 68° S, and (iii) deducing the atmospheric contribution as a residual. The total meridional heat flux, as in Fig. 5.6, is also plotted (thick). From Trenberth and Caron (2001).

tropics where (as we have seen) atmospheric energy transport is weak. The role of the ocean in meridional heat transport, and the partition of heat transport between the atmosphere and ocean, are discussed in some detail in Section 11.5.2.

8.4.2. Momentum transport

In addition to transporting heat, the general circulation also transports angular momentum. In the Hadley circulation, as we have seen, the upper, poleward-flowing, branch is associated with strong westerly winds. Thus westerly momentum is carried poleward. The lower branch, on the other hand, is associated with easterlies that are weakened by surface friction; the equatorward flow carries weak easterly momentum equatorward. The net effect is a poleward transport of westerly angular momentum. Midlatitude eddies also transport angular momentum (albeit for less obvious and, in fact, quite subtle reasons, as sketched in Fig. 8.14), again mostly transporting eastward angular momentum poleward. This has the effect of modifying the surface winds from that sketched in Fig. 8.5, by shifting the low latitude westerlies into middle latitudes.

The atmospheric angular momentum budget may therefore be depicted as in Fig. 8.12 (right). Because there is a net export of momentum out of low latitudes, there must be a supply of momentum into this region; the only place it can come from is the surface, through friction acting on the low level winds. For the friction to supply angular momentum to the atmosphere (and yet act as a brake on the low level winds), the low level winds in the tropics must be easterly, consistent with that deduced for the Hadley circulation here. The balancing loss of westerly momentum from the atmosphere—which must be associated

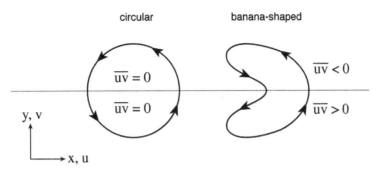

FIGURE 8.14. Circular eddies on the left are unable to affect a meridional transfer of momentum. The eddies on the right, however, by virtue of their "banana shape," transfer westerly momentum northwards south of, and southwards north of, the midline. Weather systems transport westerly momentum from the tropics ($\overline{uv} > 0$) toward higher latitudes, as required in Fig. 8.12 (right), by "trailing" their troughs down in to the tropics (see, e.g., the H_2O distribution shown in Fig. 3), as in the southern half of the "banana-shaped" eddy sketched on the right.

with drag on the near-surface westerlies—is located in middle latitudes and balances the supply of westerly momentum via eddy transport.

8.5. LATITUDINAL VARIATIONS OF CLIMATE

Putting everything together, we can depict the atmospheric wind systems in the upper and lower troposphere schematically as in Fig. 8.15.

As we have remarked at various points in this chapter, the structure of the circulation dictates more than just the pattern of winds. In the near-equatorial regions, the convergence of the trade winds is associated with frequent and intense rainfall, as is characteristic of the deep tropics. It is here, for example, that the world's great tropical rainforests are located. Around the subtropics (20°–30° latitude), in the warm, dry, descending branch of the Hadley circulation, the climate is hot and dry: this is the subtropical desert belt. Poleward of about 30°, where baroclinic eddies dominate the meteorology, local winds vary from day to day, though they are predominantly westerly, and the weather is at the mercy of the passing eddies: usually calm and fine in anticyclonic high pressure systems, and frequently wet and stormy in cyclonic, low pressure systems.

At least from a "big picture," we have thus been able to deduce the observed pattern of atmospheric winds, and the division of the world into major climate zones, through straightforward consideration of the properties of a dynamic atmosphere on a longitudinally uniform rotating planet subject to longitudinally independent differential heating. We cannot proceed much further without taking account of seasonal variations, and of the longitudinal asymmetry of the Earth, which we do not have the space to develop here. However, we will briefly note some of the major differences between our simple picture and observed circulation.

During the course of a year, the pattern of solar forcing migrates, north in northern summer, south in southern summer, and this has a significant impact on the atmospheric response. Our picture of the "typical" Hadley circulation (Fig. 8.5) inevitably shows symmetry about the equator. In reality, the whole circulation tends to shift seasonally such that the upwelling branch and associated rainfall are found on the summer side of the equator (see Fig. 5.21). The degree to which this happens is strongly controlled by local geography. Seasonal variations over the oceans, whose temperatures vary relatively little through the year, are weak, while they are much stronger over land. The migration of the main area of rainfall is most dramatic in the region of

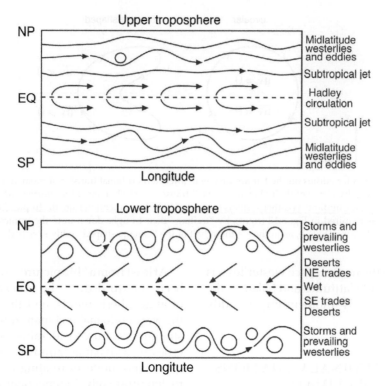

FIGURE 8.15.   Schematic of the global distributions of atmospheric winds in the upper (top) and lower (bottom) troposphere together with the primary latitudinal climate zones.

the Indian Ocean, where intense rain moves onto the Asian continent during the summer monsoon.

Although prevailing westerlies and migrating storms characterize most of the extratropical region throughout the year, there are seasonal variations (generally with more storms in winter, when the temperature gradients are greater, than in summer) and strong longitudinal variations, mostly associated with the distribution of land and sea. Longitudinal variations in climate can be dramatic: consider the contrast across the Atlantic Ocean, where the January mean temperature at Goose Bay, Newfoundland (at about 53° N) is about 30° C colder than that at Galway, at almost the same latitude on the west coast of Ireland. We cannot explain this with our longitudinally symmetric model of the world, but the explanation is nevertheless straightforward. Prevailing winds bring cold continental air

to the coast of Newfoundland, but mild oceanic air to the west coast of Ireland. The North Atlantic Ocean is much warmer than the continents at the same latitude in winter, partly because of the large heat capacity of the ocean, and partly because of heat transport by the ocean circulation. The ocean circulation and its role in climate is the subject of the following chapters.

8.6. FURTHER READING

An interesting historical account of evolving perspectives on the general circulation of the atmosphere is given in Lorenz (1967). Chapter 6 of Hartmann (1994) presents a thorough discussion of the energy and angular budgets of the general circulation; a more advanced treatment can be found in Chapter 10 of Holton (2004).

8.7. PROBLEMS

1. Consider a zonally symmetric circulation (i.e., one with no longitudinal variations) in the atmosphere. In the inviscid upper troposphere, one expects such a flow to conserve absolute angular momentum, so that $DA/Dt = 0$, where $A = \Omega a^2 \cos^2\varphi + ua \cos\varphi$ is the absolute angular momentum per unit mass (see Eq. 8-1) where Ω is the Earth rotation rate, u the eastward wind component, a the Earth's radius, and φ latitude.

    (a) Show for inviscid zonally symmetric flow that the relation $DA/Dt = 0$ is consistent with the zonal component of the equation of motion (using our standard notation, with $\mathcal{F}_x$ the x-component of the friction force per unit mass)

    $$\frac{Du}{Dt} - fv = -\frac{1}{\rho}\frac{\partial p}{\partial x} + \mathcal{F}_x ,$$

    in (x, y, z) coordinates, where $y = a\varphi$ (see Fig. 6.19).

    (b) Use angular momentum conservation to describe how the existence of the Hadley circulation explains the existence of both the subtropical jet stream in the upper troposphere and the near-surface trade winds.

    (c) If the Hadley circulation is symmetric about the equator, and its edge is at 20° latitude, determine the strength of the subtropical jet stream.

2. Consider the tropical Hadley circulation in northern winter, as shown in Fig. 8.16. The circulation rises at 10° S, moves northward across the equator in the upper troposphere, and sinks at 20° N. Assuming that the circulation outside the near-surface

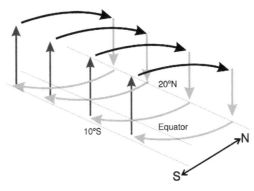

FIGURE 8.16. A schematic diagram of the Hadley circulation rising at 10° S, crossing the equator, and sinking at 20° N.

boundary layer is zonally symmetric (independent of x) and inviscid (and thus conserves absolute angular momentum about the Earth's rotation axis), and that it leaves the boundary layer at 10° S with zonal velocity $u = 0$, calculate the zonal wind in the upper troposphere at (a) the equator, (b) 10° N, and (c) 20° N.

3. Consider what would happen if a force toward the pole was applied to the ring of air considered in Problem 1 and it conserved its absolute angular momentum, A. Calculate the implied relationship between a small displacement $\delta\varphi$ and the change in the speed of the ring δu. How many kilometers northwards does the ring have to be displaced to change its relative velocity $10\,\mathrm{m\,s^{-1}}$? How does your answer depend on the equilibrium latitude? Comment on your result.

4. An open dish of water is rotating about a vertical axis at 1 revolution per minute. Given that the water is 1°C warmer at the edges than at the center at all depths, estimate, under stated assumptions and using the following data, typical azimuthal flow speeds at the free surface relative to the dish. Comment on, and give a physical explanation for, the sign of the flow.

How much does the free surface deviate from its solid body rotation form?

Briefly discuss ways in which this rotating dish experiment is a useful analogue of the general circulation of the Earth's atmosphere.

Assume the equation of state given by Eq. 4-4 with $\rho_{ref} = 1000 \, \mathrm{kg \, m^{-3}}$, $\alpha = 2 \times 10^{-4} \, \mathrm{K^{-1}}$, and $T_{ref} = 15° \, \mathrm{C}$, the mean temperature of the water in the dish. The dish has a radius of $10 \, \mathrm{cm}$ and is filled to a depth of $5 \, \mathrm{cm}$.

5. Consider the incompressible, baroclinic fluid $[\rho = \rho(T)]$ sketched in Fig. 8.17 in which temperature surfaces slope upward toward the pole at an angle s_1. Describe the attendant zonal wind field assuming it is in thermal wind balance.

   By computing the potential energy before and after interchange of two rings of fluid (coincident with latitude circles y at height z) along a surface of slope s, show that the change in potential energy $\Delta PE = PE_{final} - PE_{initial}$ is given by

   $$\Delta PE = \rho_{ref} N^2 \left(y_2 - y_1\right)^2 s\left(s - s_1\right),$$

   where $N^2 = -g / \rho_{ref} \partial \rho / \partial z$ is the buoyancy frequency (see Section 4.4),

ρ_{ref} is the reference density of the fluid, and y_1, y_2 are the latitudes of the interchanged rings. You will find it useful to review Section 4.2.3.

Hence show that for a given meridional exchange distance $(y_2 - y_1)$:

(a) energy is released if $s < s_1$;

(b) the energy released is a maximum when the exchange occurs along surfaces inclined at *half* the slope of the temperature surfaces.

This is the wedge of instability discussed in Section 8.3.3 and illustrated in Fig. 8.10.

6. Discuss qualitatively, but from basic principles, why most of the Earth's desert regions are found at latitudes of $\pm (20° - 30°)$.

7. Given that the heat content of an elementary mass dm of air at temperature T is $c_p T \, dm$ (where c_p is the specific heat of air at constant pressure), and that its northward velocity is v:

(a) show that the northward flux of heat crossing unit area (in the $x - z$ plane) per unit time is $\rho c_p v T$;

(b) hence, using the hydrostatic relationship, show that the net northward heat flux $\mathcal{H}$ in the atmosphere, at any given latitude, can be written

$$\mathcal{H} = c_p \int_{x_1}^{x_2} \int_0^\infty \rho \, vT \, dx \, dz$$
$$= \frac{c_p}{g} \int_{x_1}^{x_2} \int_0^{p_s} vT \, dx \, dp \,,$$

where the first integral (in x) is completely around a latitude circle and p_s is surface pressure.

(c) Fig. 8.18 shows the contribution of eddies to the atmospheric heat flux.

What is actually shown is the contribution of eddies to the

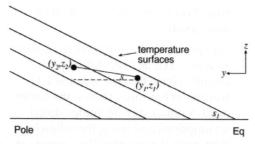

FIGURE 8.17. Schematic for energetic analysis of the thermal wind considered in Problem 5. Temperature surfaces have slope s_1; parcels of fluid are exchanged along surfaces which have a slope s.

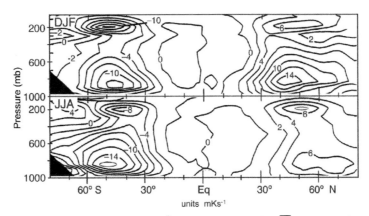

FIGURE 8.18. The contribution by eddies (mKs^{-1}) to the meridional flux $\overline{vT}$ (Eq. 8-17), plotted as a function of latitude and pressure for (top) DJF and (bottom) JJA.

quantity $\overline{vT}$, which in the above notation is

$$\overline{vT} = \frac{1}{x_2 - x_1} \int_{x_1}^{x_2} vT \, dx \qquad (8\text{-}17)$$

(i.e., the zonal average of vT), where $x_2 - x_1 = 2\pi a \cos \varphi$, where a is the Earth's radius and φ latitude.

Use the figure to *estimate* the net northward heat flux by eddies across 45° N. Compare this with the requirement (from the Earth's radiation budget and Fig. 8.13 that the net (atmosphere and ocean) heat transport must be about 5×10^{15} W.

FIGURE 8.16. The contribution by eddies (mean) to the northward flux of (Eq. 8.12), plotted as a function of latitude and pressure for (top) DJF and (bottom) JJA.

quantity $\overline{v'}$, which in the above notation is

$$\overline{v'} = \frac{1}{g} \int_{p_T}^{p_s} [\overline{v'}] \, dp$$

is the zonal average of $[\overline{v'}]$,
where $x_T = r_e \, 2\pi \cos\varphi$, where r_e
is the Earth's radius and φ latitude.

Use the figure to estimate the net
northward heat flux by eddies
across 45° N. Compare this with
the result expected (from the Earth's
radiation budget and Fig. 8.15) that
the net atmosphere and ocean
heat transport must be about
$\times 10^{15}$ W.

The ocean and its circulation

9.1. Physical characteristics of the ocean
    9.1.1. The ocean basins
    9.1.2. The cryosphere
    9.1.3. Properties of seawater; equation of state
    9.1.4. Temperature, salinity, and temperature structure
    9.1.5. The mixed layer and thermocline
9.2. The observed mean circulation
9.3. Inferences from geostrophic and hydrostatic balance
    9.3.1. Ocean surface structure and geostrophic flow
    9.3.2. Geostrophic flow at depth
    9.3.3. Steric effects
    9.3.4. The dynamic method
9.4. Ocean eddies
    9.4.1. Observations of ocean eddies
9.5. Further reading
9.6. Problems

We now begin our discussion of the circulation of the ocean. In this introductory chapter we describe the physical characteristics of the ocean, the properties of sea water, its global-scale temperature and salinity distributions, and the geography of the basins in which it is contained. We go on to describe, and interpret in terms of the balanced—geostrophic and hydrostatic—dynamics of Chapter 7, the observed pattern of ocean currents. Those readers who have a primary interest in ocean circulation and have not read all the way through to here, will find essential dynamical background reviewed in Chapters 6 and 7, to which frequent reference is made.

The ocean, like the atmosphere, is a stratified fluid on the rotating Earth. The two fluids have many similarities in their behavior and, especially, in their fluid dynamics. However, there are some important differences:

- The fluids are physically different. Water is (almost) incompressible and ocean thermodynamics has no counterpart to atmospheric moisture (as a source of latent heat).

- Unlike the atmosphere, all oceans are laterally bounded by continents (see Fig. 9.1) except in the Southern Ocean where the ocean extends all the way around the globe and fluid can pass

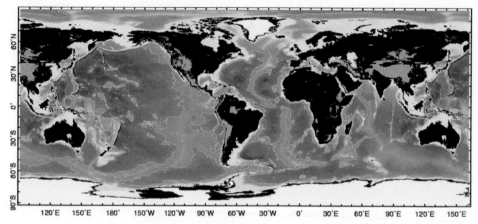

FIGURE 9.1.　World relief showing elevations over land and ocean depth. White areas over the continents mark the presence of ice at altitudes that exceed 2 km. The mean depth of the ocean is 3.7 km, but depths sometimes exceed 6 km. The thin white line meandering around the ocean basins marks a depth of 4 km.

through Drake Passage, the narrow (600 km) gap between the tip of South America and the Antarctic peninsula.

- The ocean circulation is forced in a different way than the atmosphere. We have seen that the atmosphere is largely transparent to solar radiation and is heated from below by convection. By contrast, the ocean exchanges heat and moisture with the atmosphere at its upper surface; convection in the ocean is driven by buoyancy loss *from above*.[1] In addition, there is a very important process forcing the ocean circulation that has no counterpart in the atmosphere. Winds blowing over the ocean surface exert a *stress* on it. The wind is a major driver of ocean circulation, particularly in the upper kilometer or so.

The wind-driven and buoyancy-driven circulations are intimately intertwined. Nevertheless, for pedagogical reasons here we discuss them separately, devoting Chapter 10 to the former and Chapter 11 to the latter. Finally, in Chapter 12, we discuss the role of the ocean in climate and paleo-climate.

9.1. PHYSICAL CHARACTERISTICS OF THE OCEAN

9.1.1. The ocean basins

The ocean covers about 71% of the Earth's surface and has an average total depth of 3.7 km; the distribution of land and sea and the bathymetry of the ocean basins is plotted in Fig. 9.1. A north-south section along the Greenwich meridian through Fig. 9.1 is shown in Fig. 1.2. We see that the ocean basins are highly complex and the bottom topography much more jagged than that of the land surface. See also the bathymetry shown in the hydrographic section of Fig. 9.9. This is because, as we shall see, abyssal ocean currents are, in the mean, weak, and temperature changes very slight, so the erosion of submarine relief by ocean currents occurs at a much slower rate than that of the mountains on land.

The volume of the oceans is $3.2 \times 10^{17} \, \text{m}^3$, a huge amount of water totalling roughly $1.3 \times 10^{21} \, \text{kg}$ (see Table 9.1) with a correspondingly enormous heat capacity, 1000 times that of the atmosphere. As discussed

---

[1]There are sources of geothermal heating at the bottom of the ocean but, in an average sense, this accounts for only a few milliwatts/m^2 of heat input, compared to air-sea heat fluxes of ± 10 to $100 \, \text{W/m}^2$ (see Chapter 11).

TABLE 9.1. Some key ocean numbers.

Surface area	$3.61 \times 10^{14}\,\text{m}^2$
Mean depth	$3.7\,\text{km}$
Volume	$3.2 \times 10^{17}\,\text{m}^3$
Mean density	$1.035 \times 10^3\,\text{kg}\,\text{m}^{-3}$
Ocean mass	$1.3 \times 10^{21}\,\text{kg}$

TABLE 9.2. Major constituents of seawater. The *relative proportion* of the dissolved salts tabulated here does not vary much from place to place. This is because salts largely come from weathering of continents, which is a very slow input compared to the mixing rate of the whole ocean (100,000 years compared to 1000 years).

salt	$^o/_{oo}\ \left(\text{g}\,\text{kg}^{-1}\right)$
Chloride	18.98
Sodium	10.56
Sulphate	2.65
Magnesium	1.27
Calcium	0.40
Potassium	0.38
Bicarbonate	0.14
Others	0.11
Overall salinity	34.48

in Chapter 12.1, this is one of the reasons that the ocean plays such an important role in climate.

9.1.2. The cryosphere

About 2% of the water on the planet is frozen and is known as the "cryosphere" (from the Greek *kryos*, meaning "frost" or "cold"). It includes ice sheets, sea ice, snow, glaciers, and frozen ground (permafrost). Most of the ice is contained in the ice sheets over the landmasses of Antarctica (89%) and Greenland (8%). These sheets store about 80% of the freshwater on the planet. The Antarctic ice sheet has an average depth of about 2 km (Fig. 1.2); the Greenland ice sheet is some 1.5 km thick. Of great importance for climate is not so much the mass of ice but rather the surface area covered by it. The albedo of ice can be very high, reaching 70% (see Table 2.2 and Fig. 2.5), and so reflects much of the radiation incident on it. Perennial (year-round) ice covers 11% of the land area and 7% of the ocean. Moreover, sea ice is also important because it regulates exchange of heat, moisture, and salinity in the polar oceans, and insulates the relatively warm ocean water from the cold polar atmosphere.

9.1.3. Properties of seawater; equation of state

Some key properties of water are set out in Table 9.3. The density of pure water at 4°C is 0.999×10^3 kg m^{-3}. The mean density of seawater is only slightly greater, 1.035×10^3 kg m^{-3}. Unlike air, the density

of seawater varies rather little, by only a few %, but, as we shall see, these variations turn out to be central to the dynamics. Cold water is denser than warm water; salty water is denser than fresh water; pressure increases the density of water. Density depends on temperature T, salinity S, and pressure p in a rather complicated, nonlinear way (deduced by very careful laboratory measurements), which we represent symbolically as:

$$\rho = \rho(T, S, p). \qquad (9\text{-}1)$$

Salinity is a measure of the amount of salt dissolved in the water, about 85% of which is sodium and chloride (see Table 9.2). Modern salinity measurements are dimensionless; the Practical Salinity Scale defines salinity in terms of a conductivity ratio very nearly equal to g kg^{-1} or, equivalently, $^o/_{oo}$ (parts per thousand). Typically, seawater has a salinity of 34.5 psu (practical salinity units). Thus one kg of seawater typically has 34.5 g of salt dissolved in it. Seawater is almost incompressible; not quite, since at enormous pressures in the interior ocean compressibility effects are not always negligible.[2]

---

[2]It is interesting to note that if seawater were really incompressible, sea level would be about 50 m higher than it is.

TABLE 9.3.   **Physical properties of liquid water.**

Specific heat	c_w	4.18×10^3	$J\ kg^{-1}\ K^{-1}$
Latent heat of fusion	L_f	3.33×10^5	$J\ kg^{-1}$
Latent heat of evaporation	L_e	2.25×10^6	$J\ kg^{-1}$
Density of fresh water	ρ_{fresh}	0.999×10^3	$kg\ m^{-3}$
Viscosity	μ_{water}	10^{-3}	$kg\ m^{-1}\ s^{-1}$
Kinematic viscosity	$\nu = \frac{\mu_{water}}{\rho}$	10^{-6}	$m^2\ s^{-1}$
Thermal diffusivity	k	1.4×10^{-7}	$m^2\ s^{-1}$

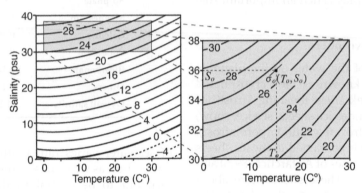

FIGURE 9.2.   Contours of seawater density anomalies ($\sigma = \rho - \rho_{ref}$ in $kg\,m^{-3}$) plotted against salinity (in psu $= g\,kg^{-1}$) and temperature (°C) at the sea surface. Note that seawater in the open ocean has σ in the range 20–29 $kg\,m^{-3}$, T in the range 0–30°C and S in the range 33–36 psu. The panel on the right zooms in on the region of oceanographic relevance. An approximation to the equation of state in the vicinity of the point $\sigma_o(T_o, S_o)$ is given by Eq. 9-5.

However for many purposes it suffices to assume that density is independent of pressure.

The dependence of the density of seawater on S and T at the surface of the ocean is shown in Fig. 9.2. What is actually plotted is the density *anomaly* σ (see Section 7.3),

$$\sigma = \rho - \rho_{ref}, \qquad (9\text{-}2)$$

the difference between the actual density and a reference value $\rho_{ref} = 1000\,kg\,m^{-3}$. From Fig. 9.2 we see that:

1. Salty water is more dense than fresh water; warm water is (almost always) less dense than cold water.

2. Fresh water ($S = 0$) has a maximum density at about 4°C: fresh water that is colder than this is less dense. This is why ice forms on the top of freshwater

lakes. Cooling from the surface in winter forms ice rather than denser water.

3. In the (rather narrow) range of temperatures and salinities found in the open ocean (0–30°C and 33–36 psu, see Figs. 9.3 and 9.4), temperature typically influences density more than salinity. At the sea surface, σ is typically 26 kg m^{-3} and varies monotonically with temperature.

The thermal expansion coefficient of water , α_T, defined by:

$$\alpha_T = -\frac{1}{\rho_{ref}}\frac{\partial \rho}{\partial T}, \qquad (9\text{-}3)$$

(S and p kept constant) has a typical value of 1×10^{-4}°C^{-1} and is larger at higher temperatures (note how the isopleths of σ in Fig. 9.2

Sea Surface Temperature (°C)

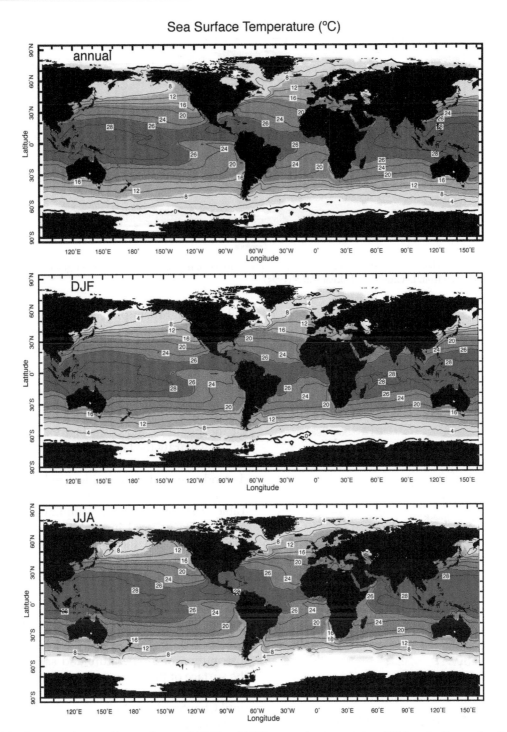

FIGURE 9.3. Average (a) annual-mean, (b) DJF, and (c) JJA sea-surface temperature (°C) from the Comprehensive Ocean-Atmosphere Data Set (COADS). Note that there are data void regions in the proximity of ice-covered areas, particularly around Antarctica in the southern hemisphere winter. The Antarctic ice edge is farthest north in September and occasionally crosses 60° S.

Surface Salinity (psu)

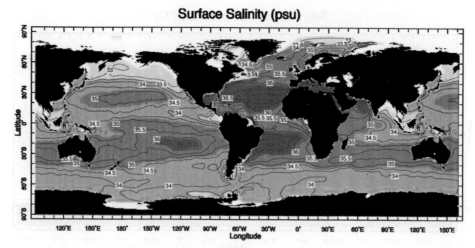

FIGURE 9.4.   The annual mean salinity distribution at the surface of the ocean (in psu). Darker green represents salty fluid. Data from Levitus World Ocean Atlas (1994).

TABLE 9.4.   The dependence of σ, α_T, and β_S on T and S at two levels in the ocean: at the surface and at a depth of 1 km.

Surface			
T_o (°C)	−1.5	5	15
α_T $\left(\times 10^{-4}\,\mathrm{K}^{-1}\right)$	0.3	1	2
S_o (psu)	34	36	38
β_S $\left(\times 10^{-4}\,\mathrm{psu}^{-1}\right)$	7.8	7.8	7.6
σ_o $\left(\mathrm{kg\,m}^{-3}\right)$	28	29	28
Depth of 1 km			
T_o (°C)	−1.5	3	13
α_T $\left(\times 10^{-4}\,\mathrm{K}^{-1}\right)$	0.65	1.1	2.2
S_o (psu)	34	35	38
β_S $\left(\times 10^{-4}\,\mathrm{psu}^{-1}\right)$	7.1	7.7	7.4
σ_o $\left(\mathrm{kg\,m}^{-3}\right)$	−3	0.6	6.9

slope increasingly upward with increasing T) and at greater pressures (see Table 9.4).

The dependence of density on salinity is defined by:

$$\beta_S = \frac{1}{\rho_{ref}} \frac{\partial \rho}{\partial S} \qquad (9\text{-}4)$$

(where T and p are kept constant). As can be seen in Fig. 9.2 and Table 9.4, β_S varies very little and has a value close to $7.6 \times 10^{-4}\,\mathrm{psu}^{-1}$.

It is sometimes useful to approximate the dependence of σ on T, S, and p using the

following highly simplified equation of state obtained by drawing short lines tangent to the isopleths of σ in Fig. 9.2 and writing:

$$\sigma = \sigma_o + \rho_{ref} \left(-\alpha_T \left[T - T_o \right] + \beta_S \left[S - S_o \right] \right);$$
$$(9\text{-}5)$$

where $\sigma_o(T_o, S_o)$ is the density anomaly (at the point T_o and S_o in Fig. 9.2), about which we draw the tangent line to compute (small) variations in σ. Note that pressure dependence in the above idealized expression is captured by making $\alpha_T = \alpha_T(p)$ (see Table 9.4). Recall that a simplified form of Eq. 9-5 was used in our discussions of convection of an incompressible fluid in Section 4.2.2.

9.1.4. Temperature, salinity, and temperature structure

Here we briefly describe the large-scale distribution of temperature and salinity at the surface of the ocean and its interior. We postpone an attempt at an "explanation" until Chapters 10 and 11.

Water has an albedo of around 10% (see Table 2.2) depending on surface conditions, and so absorbs solar radiation very efficiently. Not surprisingly sea surface temperatures, plotted in Fig. 9.3, are warmest in the tropics (up to almost 30°C) and coldest

(0°C) in high latitudes. The warmest water, that with a temperature greater than about 29°C, follows the Sun north and south. But there are also large east-west variations, particularly in the tropics, with relatively cool temperatures in the east, and the warmest temperatures west of the International Date Line. This latter region coincides with the location of greatest atmospheric convection (see Fig. 4.26), which can become very vigorous when ocean temperatures exceed about 27°C, an indication of the climatic importance of the tropical Pacific Ocean (to be discussed in Section 12.2). In middle latitudes there are large east-west differences in Sea Surface Temperature (SST) related, as we shall see, to the juxtaposition of cold boundary currents flowing toward the equation and warm boundary currents flowing toward the poles (e.g., in the region of the Gulf Stream in the Atlantic and the Kuroshio in the Pacific). There are also cold regions off California and Africa, which are not so obviously related to the advective influence of ocean currents. One also observes strong meridional gradients in SST in the Southern Ocean associated with the eastward flowing Antarctic Circumpolar Current. The coldest surface waters are found in the northern North Atlantic Ocean and around Antarctica. As will be discussed in Chapter 11, these are the regions of deepest ocean convection, where the surface ocean communicates with the abyss.

The annual mean salinity distribution at the surface of the ocean is shown in Fig. 9.4. Evaporation from the surface in the subtropics is vigorous and exceeds precipitation; since evaporation removes water, but not salt, the near-surface salinity is elevated here. In high latitudes and near the equator, there is an excess of precipitation over evaporation, so the surface waters are relatively fresh here. High values of salinity (> 38 psu) are found in the Mediterranean and the Persian Gulf. Low values of salinity (< 31 psu) are found near melting ice edges and at river outflows. Note that the surface salinity in the Atlantic is higher than that in the Pacific, making Atlantic surface waters more susceptible to convection.

Observed zonal-average mean distributions of T, S, and σ in the interior of the ocean are shown[3] in Figs. 9.5, 9.6, and 9.7. Note that the top panel in each of these figures shows a blow-up of the upper 1000 m of the ocean; the bottom panel plots properties over the full ocean depth. The reason for this is that the largest vertical gradients of properties are found near the surface. In the abyss, vertical gradients are weak and horizontal gradients are almost nonexistent; for example, the deep ocean is everywhere very cold (0–2°C) and no more that 1°C warmer in the tropics than in high latitudes (see Fig. 9.5). In the upper kilometer of the ocean, however, shown in the top panels of Figs. 9.5, 9.6, and 9.7, there are strong vertical gradients (especially of temperature and density); this is the *thermocline* of the world's oceans, having a depth of about 600 m in middle latitudes but shoaling to 100–200 m in low latitudes. The temperature contrast between high and low latitudes is not surprising; the salinity contrast, as noted above and discussed in Chapter 11, reflects the pattern of evaporation and precipitation. Notice how, particularly around Antarctica, cold surface water can be less dense than warmer fluid beneath because it is so much fresher. Note also that the contours of temperature, salinity, and density anomaly "outcrop" (rise to the surface) in high latitudes, suggesting an important linkage between the high latitude surface waters and the deep ocean.

---

[3]In fact, what is shown in Figs. 9.5 and 9.7 are *potential* temperature and *potential* density, allowing for compressibility effects. Potential temperature, in direct analogy to its definition in a compressible fluid like the atmosphere (Section 4.3.2), is the temperature that a water parcel would have if moved adiabatically to a reference pressure, commonly chosen to be the sea surface. We define potential density as the density a parcel would have when moved adiabatically to a reference pressure. If the reference pressure is the sea surface, we compute the potential temperature of the parcel, and evaluate the density at zero pressure. The measured salinity is used because it has very little pressure dependence.

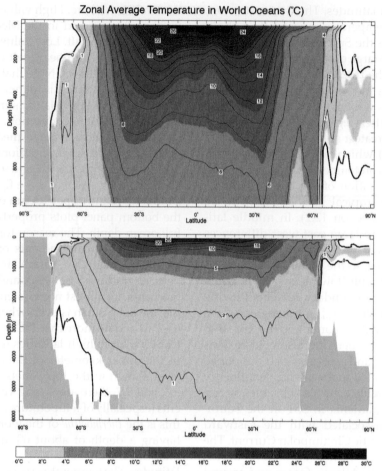

FIGURE 9.5. Annual-mean cross-section of zonal average potential temperature (in °C) in the world's oceans. The top shows the upper 1 km. The bottom shows the whole water column. Dark shading represents warm water. Note the variable contour interval in the bottom plot. Data from the Levitus World Ocean Atlas (1994).

In summary, the zonal average picture reveals a warm, salty, light lens of fluid in the subtropics of each hemisphere, shoaling on the equatorial and polar flanks, "floating" on a cold, somewhat fresher abyss. These lenses exist in each ocean basin but exhibit considerable regional characteristics and variability. For example, Fig. 9.8 maps the depth of the $\sigma = 26.5$ kg m^{-3} surface over the global ocean; we chose the depth of this surface because, as can be seen in the

zonal-average plot, Fig. 9.7, it lies roughly in the middle of the thermocline. One can readily observe the geography of the subtropical lenses of light fluid in both the northern and southern hemispheres. Note that the lens is deeper in the Pacific than the Atlantic; and outcrops in the Atlantic near 40° N and around the Southern Ocean.

Figure 9.9 shows a detailed hydrographic section[4] of T and S along (nominally) 30° W

---

[4]A hydrographic section is a section of T and S as a function of depth, obtained by lowering a CTD (a device that measures Conductivity—hence salinity—Temperature, and Depth) from a ship to the bottom of the ocean. It takes about 1 month to complete a section such as Fig. 9.9 and so gives us a snap-shot of the interior T, S structure of the ocean. The average sections shown in, for example, Figs. 9.5 and 9.6 are obtained by zonally averaging hundreds of such sections crisscrossing the ocean. Such zonal-average sections give us a coarse, blurred, but nevertheless instructive view of the meridional structure.

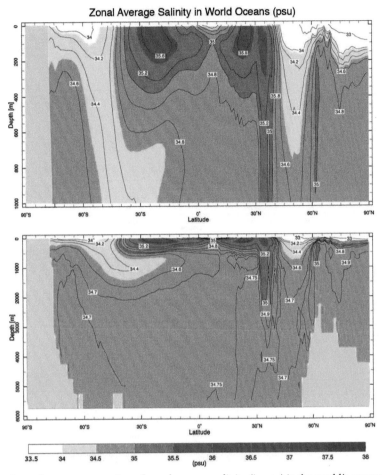

FIGURE 9.6. Annual-mean cross-section of zonal average salinity (in psu) in the world's oceans. The top shows the upper 1 km. The bottom shows the whole water column. Dark shading represents salty water. Data from the Levitus World Ocean Atlas (1994).

in the Atlantic. The thermocline is clearly visible, deep in the subtropics (±30°), shallow in equatorial regions and has much more detailed structure than the zonally averaged view of the thermocline evident in Fig. 9.5. The salinity field reveals an interesting layering of subsurface flow, which we discuss in more detail in Section 11.2.

9.1.5. The mixed layer and thermocline

At the surface of the ocean there is a well-defined *mixed layer* in direct contact with the overlying atmosphere, stirred by winds and convection, in which properties are relatively uniform in the vertical. The mixed layer depth varies with latitude and season, but is typically 50–100 m deep (see Fig. 9.10). Over the bulk of the ocean the mixed layer communicates with the underlying thermocline, except in high latitudes (particularly in the northern North Atlantic and around Antarctica) where it can get very deep (> 1 km) and thus comes into direct contact with the abyss.

The processes forming the mixed layer are illustrated schematically in Fig. 9.11. Radiation entering the ocean surface is absorbed mostly in the top few meters (depending on wavelength; IR is absorbed within a few mm, blue/green light may penetrate to almost 100 m in especially clear water, but is usually attenuated much more

Zonal-Average, Annual-Mean, Potential Density (kg/m³)

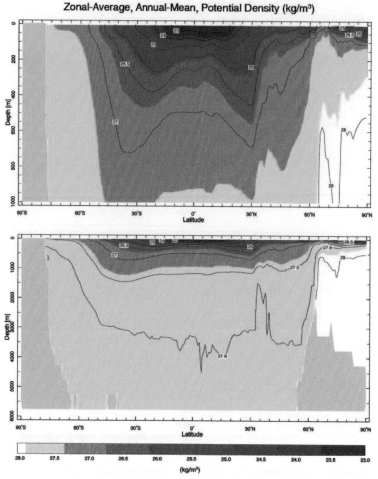

FIGURE 9.7. Annual-mean cross-section of zonal average potential density anomaly $\sigma = \rho - \rho_{ref}$ (in kg m^{-3}) for the world's oceans (referenced to the surface). The top shows the upper 1 km . The bottom shows the whole water column. Note that the contour interval is not uniform. Data from the Levitus World Ocean Atlas (1994).

rapidly). Heat loss, through IR radiation, sensible heat loss to the atmosphere, and latent heat loss through evaporation (see Chapter 11 for a more detailed discussion), occur at or within a few mm of the surface. The cooling and salinization of the surface water increases its density. Since, in an incompressible fluid, our criterion, Eq. 4-5, for the onset of convection is $\partial \rho / \partial z > 0$, surface buoyancy loss drives convective motions which stir the mixed layer and tend to homogenize its temperature and other properties, just as studied in GFD Lab II in Section 4.2.4. These turbulent

motions within the mixed layer may entrain cold water upward across the mixed layer base. In addition, wind stress at the surface drives turbulent motions within the mixed layer, which mix vertically and entrain fluid from below. Furthermore, because the base of the mixed layer slopes (Fig. 9.10), horizontal currents can carry properties to and from the mixed layer in a process known as "subduction."

Beneath the mixed layer, T and S rapidly change over the depth of the thermocline to match the properties of the relatively homogeneous abyss. Fig. 9.12(a) shows the

Depth of 26.5 Density Surface (m)

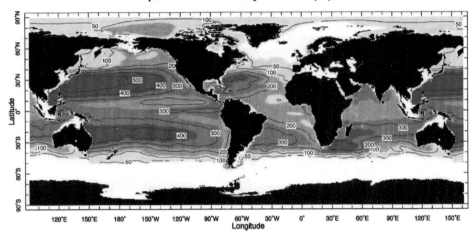

FIGURE 9.8. Depth in m of the annual-mean $\sigma = 26.5 \ \mathrm{kg\,m^{-3}}$ surface over the global ocean. Note the position of the outcrop—the line along which the σ surface cuts the sea surface—in the North Atlantic and in the southern ocean. Data from the Levitus World Ocean Atlas (1994).

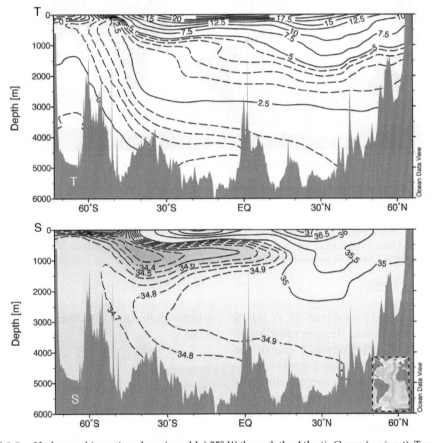

FIGURE 9.9. Hydrographic section along (roughly) 25° W through the Atlantic Ocean (see inset). Top: Potential temperature contoured every 2.5°C (solid) and every 0.5°C (dotted). Bottom: Salinity (in psu). Values greater than 35 are contoured every 0.5 psu (solid); values less than 35 are plotted every 0.1 psu (dotted). Figure produced using Ocean Data View.

Mixed Layer Depth (m)

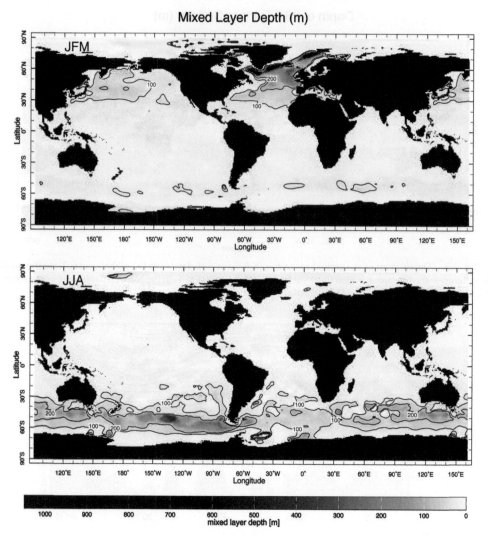

FIGURE 9.10. Mixed layer depth (in m) in (top) JFM (January, February, March, northern hemisphere winter) and (bottom) JJA (June, July, August, Southern hemisphere winter). Black contours mark the 100 and 200 m mixed layer depth isopleths. Data from the Levitus World Ocean Atlas 1994.

annual mean T and S profile at 50° W, 30° N in the Atlantic Ocean. The thermocline is clearly evident in the upper 1 km of the water column. Since density increases sharply downward across the thermocline, it is very stable, rather like the stratosphere.

We can estimate the buoyancy frequency of the thermocline as follows. Starting from Eq. 4-19 of Section 4.4 and noting that $\frac{g(\rho_P - \rho_E)}{\rho_P} \simeq \frac{-g}{\rho_E} \frac{d\rho_E}{dz} \Delta = N^2 \Delta$, the appro-

priate definition for our incompressible fluid is:

$$N^2 = -\frac{g}{\rho_{ref}} \frac{d\sigma}{dz} \simeq g\alpha_T \frac{dT}{dz} \qquad (9\text{-}6)$$

if thermocline density gradients are dominated by temperature gradients and Eq. 9-5 is used. From Fig. 9.12(a), if there is a 15°C temperature drop across the top 1000 m of the ocean, then (using $\alpha_T = 2 \times 10^{-4}\,\mathrm{K}^{-1}$ from Table 9.4) we obtain an N of, in round

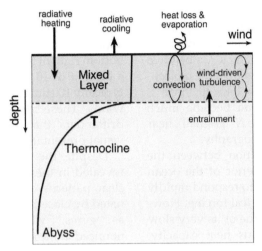

FIGURE 9.11.   A schematic diagram showing processes at work in the mixed layer of the ocean. Note that the vertical scale of the mixed layer relative to the thermocline is greatly exaggerated.

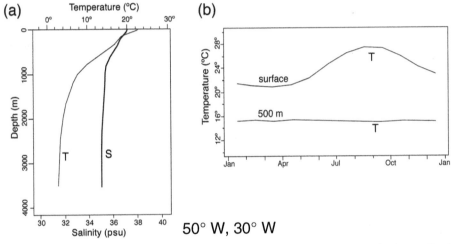

FIGURE 9.12.   (a) Annual mean T and S profile at 50° W, 30° N in the Atlantic Ocean. The thermocline is clearly evident. The T scale is at the top, the S scale at the bottom. (b) The cycle of T at 50° W, 30° N in the Atlantic Ocean in the Levitus monthly mean climatology. The strong seasonal cycle at the surface has vanished at a depth of 500 m. Data from the Levitus World Ocean Atlas (1994).

numbers, 5×10^{-3} s^{-1}. This implies a period for internal gravity waves in the ocean of $2\pi/N \simeq 20$ min[5], roughly twice that of our estimate in Section 4.4 for atmospheric gravity waves. Indeed T, S surfaces in the interior of the ocean undulate constantly with just this kind of period, excited by winds, flow over topography, tides, and myriad other processes. These internal gravity waves are not merely oceanographic noise, in the sense that averaged over many oscillation cycles they disappear. If

[5]More detailed study shows that $2\pi/N$ is the minimum period for internal waves. More energy is at frequencies closer to f rather than N.

the waves break, mixing occurs. However, the mixing due to internal waves appears to be rather weak in the thermocline (typically eddy diffusivities are 10^{-5} m^2 s^{-1}, albeit two orders of magnitude larger than the molecular diffusivity of water $k = 1.4 \times 10^{-7}$ m^2 s^{-1}; see Table 9.3), but it can become much larger in the abyss (where N^2 is small), near boundaries, and over topography.

An important distinction between the mixed layer and the interior of the ocean is that the former is able to respond rapidly to changes in meteorological forcing. However, the interior, by virtue of its very slow circulation and enormous heat capacity, can only respond very slowly to changing boundary conditions. Thus, for example, the surface mixed layer exhibits diurnal, seasonal and inter-annual variations, whereas the interior ocean evolves on interannual, decadal to centennial (and longer) time scales, the time scale typically increasing with depth. For example, Fig. 9.12(b) shows the seasonal cycle of SST and T at 500 m at 50° W, 30° N in the Atlantic Ocean. There is no seasonal cycle detectable at 500 m, although trends on decadal timescales are observed (not shown).

9.2. THE OBSERVED MEAN CIRCULATION

The global pattern of mean flow at the surface of the ocean is plotted in Fig. 9.13, where the names of the major current systems are also given. The colors separate the circulation patterns into tropical (pink), subtropical (yellow), and subpolar (blue) regimes (inspired by the dynamical discussion to be developed in Chapter 10). Because the detailed patterns are difficult to discern on this global map, regional circulations in the Pacific, Atlantic, and Indian

Oceans are also shown in Figs. 9.14, 9.15, and 9.16, respectively, with the annual-mean SST superimposed. Note the tiny vector on the bottom right of these plots, which represents a current of 10 cm s^{-1}; ocean currents are typically 100 times smaller than atmospheric winds. These maps are based on surface drifter data[6] that have been smoothed to aid visual presentation in vector form.

Despite the richness of the structures revealed in these data, on close inspection clear patterns emerge. The flow is dominated by closed circulation patterns known as "gyres," which are particularly pronounced in the northern hemisphere, where all zonal flow is blocked by coasts. In the subtropics of the northern hemisphere there are anticyclonic gyres, known as subtropical gyres (yellow shading in Fig. 9.13, with eastward flow in middle latitudes—the North Pacific and North Atlantic Currents—and westward flow in the tropics—the North Equatorial Currents of the Pacific and Atlantic. Typical current speeds in the interior of the gyres are $\lesssim 10$ cm s^{-1} (see Fig. 9.22, top). At the western edge of these subtropical gyres, there are strong, poleward currents reaching speeds $\gtrsim 100$ cm s^{-1}, the Kuroshio in the North Pacific Ocean and the Gulf Stream in the North Atlantic Ocean, as is evident in Figs. 9.14 and 9.15. These swift currents flow northward along the coasts from the tropics, turn into the interior at about 40° latitude, and then spread eastward across the ocean basin. This east-west asymmetry is perhaps the most striking aspect of the observed current system; all intense boundary currents are on the western, rather than the eastern, boundaries of ocean gyres. The interior extensions of these boundary currents are evident in the thermal structure at the ocean surface, most obviously in the strong temperature gradients near the western boundaries of the

---

[6]Most surface drifters providing these data consist of a spherical surface float tethered to a cylindrical cloth tube hanging below with holes in it, centered at a depth of 15 m. Within the float is a radio transmitter allowing it to be located through Doppler ranging by satellite. The drifters can survive for 450 days or so. For a more complete description, see Niiler (2001).

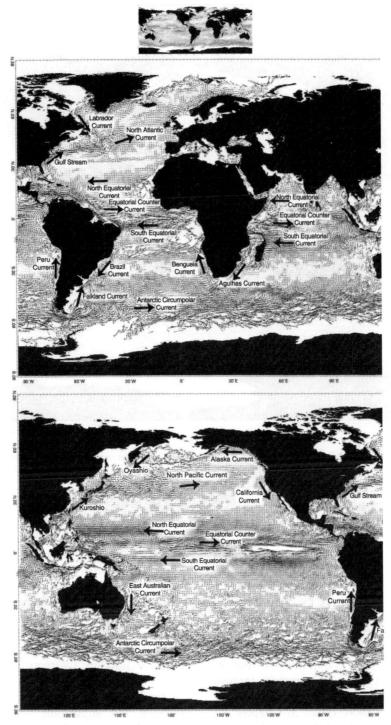

FIGURE 9.13. Major surface currents observed in the global ocean, with the names of key current systems marked. The colors separate the circulation patterns into tropical (pink), subtropical (yellow), and subpolar (blue) regimes based on the pattern of zero wind-curl lines shown in Fig. 10.11 (see Section 10.1.3). Data courtesy of Maximenko and Niiler (personal communication, 2003).

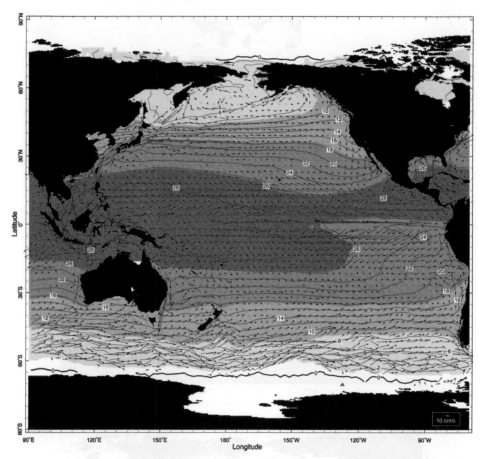

FIGURE 9.14.   Major surface currents observed in the Pacific Ocean from surface drifters with annual-mean SST superimposed. The vector at the bottom right represents a current of 10 cm s^{-1}. Data courtesy of Maximenko and Niiler (personal communication, 2003).

middle-latitude oceans and in the Antarctic Circumpolar Current of the southern hemisphere.

In the polar regions of the north Pacific and Atlantic basins there are cyclonic gyres, known as subpolar gyres (blue shading in Fig. 9.13), with southward flowing western boundary currents: the Oyashio Current in the Pacific, the Labrador Current in the Atlantic. Again we observe no intense eastern boundary currents. As discussed earlier, and in more detail in Chapter 11, the northern margin of the subpolar gyres in the Atlantic Ocean—in the Labrador and Greenland Seas—are favoured sites of deep-reaching convection.

In the tropics (pink shading in Fig. 9.13) we observe strong zonal flows in each basin, running eastward just north of the equator and counter to the prevailing winds! (See, for example, Fig. 7.28, middle panel.) The Equatorial Counter Currents very evident in Fig. 9.14 have, on either side of them, strong North and South Equatorial Currents flowing westwards. Somewhat weaker meridional temperature gradients are found in the tropics than in higher latitudes, but significant zonal temperature gradients exist along the equator. The interaction between the atmosphere and ocean in the tropical Pacific, which leads to variability of the coupled system known as

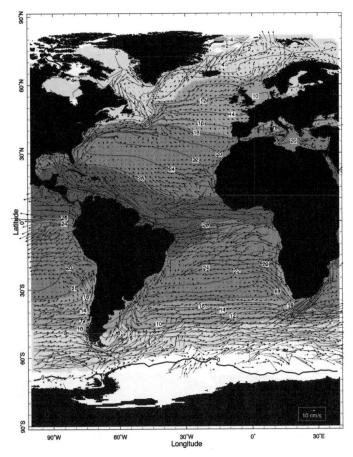

FIGURE 9.15. Mean surface currents in the Atlantic Ocean from surface drifters with annual-mean SST superimposed. The vector at the bottom right represents a current of 10 cm s^{-1}. Data courtesy of Maximenko and Niiler (personal communication, 2003).

El Niño–Southern Oscillation (ENSO) phenomenon, will be addressed in Section 12.2.

In the southern hemisphere, subtropical gyres are also evident.[7] However, on the poleward flanks of the subtropical gyres of the Southern Ocean, the strong zonal flow of the Antarctic Circumpolar Current (ACC) predominates. Fluid in the ACC can circumnavigate the Southern Ocean unimpeded by coasts. This is perhaps the ocean current that is most closely analogous to the atmospheric jet stream.

The pattern of mean currents discussed above extend downward in the water column, but the currents become weaker. For example, Fig. 9.17 (top) shows currents at a

[7] George Deacon, (1906–1984), doyen of British oceanography and father of the *Institute of Oceanographic Sciences*, elucidated the main water masses of the Southern Ocean and their circulation, using classical water mass analysis. Deacon's *Hydrology of the Southern Ocean*, published in 1937, summarized information on the circulation of the southern Atlantic and Southern Ocean and set the standard for future physical oceanographic work.

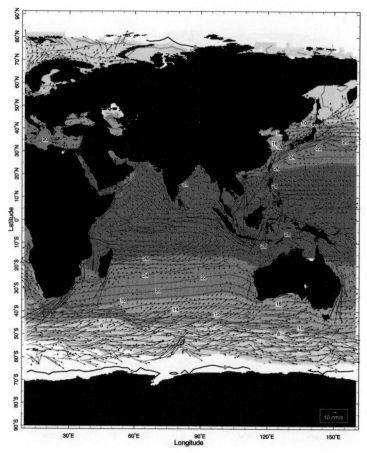

FIGURE 9.16. Mean surface currents observed in the Indian Ocean from surface drifters with annual-mean SST superimposed. The vector at the bottom right represents a current of 10 cm s^{-1}. Data courtesy of Maximenko and Niiler (personal communication, 2003).

depth of 700 m observed by neutrally buoyant floats[8] in the North Atlantic. We observe a general pattern of currents similar to that at the surface (cf. Fig. 9.15) but of reduced magnitude. Mean currents in the abyssal ocean are very weak, except in regions of western boundary currents and where flow is channeled by topography. There is significant stirring of the abyss by ocean eddies, however, because the ocean's eddy field typically decays less rapidly with depth than the mean flow.

Before begining our dynamical discussion, we estimate typical timescales associated with the horizontal circulation described above. Tropical surface waters move rather swiftly, reaching mean zonal speeds of 20 cms^{-1} or more. The Pacific basin is some $14,000 \text{ km}$ wide at the equator, yielding a transit time of 2 years or so.

[8] The subsurface currents shown in Fig. 9.17 were obtained by Profiling Autonomous Lagrangian Current Explorer (PALACE) floats. These are designed to drift at a pre-determined depth (700 m in this case) for 10 days or so and then rise to the surface, measuring water properties to obtain a temperature and salinity profile. While on the surface, data is transmitted to satellite and the geographic position of the float determined before it returns to depth to repeat the cycle. Each float has battery power for 100 cycles or so. For more details see Davis et al (2001).

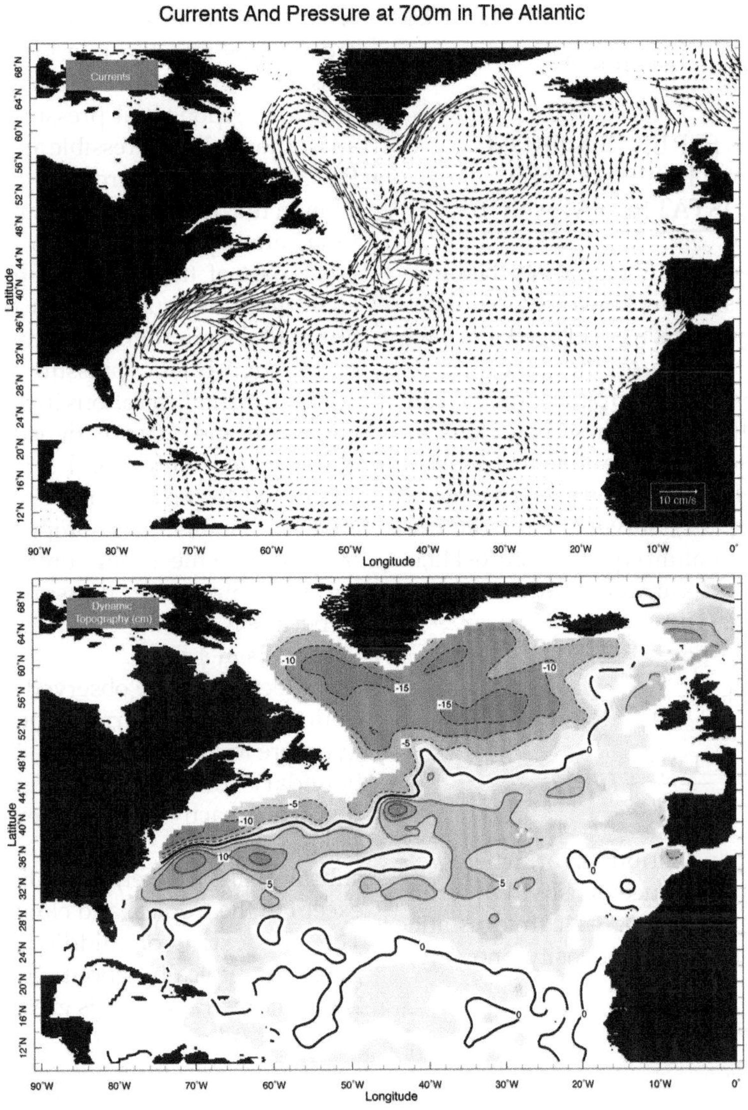

Currents And Pressure at 700m in The Atlantic

FIGURE 9.17.   Top: Currents at a depth of 700 m in the Atlantic: the arrow at bottom right corresponds to a current of 10 cm s^{-1}. Bottom: The associated geostrophic pressure field, expressed as a head of water in cm. Data courtesy of Steve Jayne, WHOI.

Typical zonal surface currents in the ACC are about 30 cm s^{-1}, and parcels thus circumnavigate the globe, a distance of some 21,000 km at 55° S, in about 2 years (see Problem 6 at the end of this chapter). It takes perhaps 5 years or more for a parcel of water to circulate around the subtropical gyre of the Atlantic Ocean. Timescales of the horizontal circulation increase with depth as the flow decreases in amplitude.

Finally it is important to emphasize that Figs. 9.14 to 9.17 are time-mean circulation patterns. But the ocean is full of time-dependent motions (due to hydrodynamical instabilities and flows driven by variable forcing) so that at any instant the circulation often looks quite different from these mean patterns, as will be discussed in Sections 9.4 and 10.5.

9.3. INFERENCES FROM GEOSTROPHIC AND HYDROSTATIC BALANCE

On the large-scale, water obeys the same fluid dynamics as air; so we have already derived the equations we will need: Eq. 6-44 applies just as well to the ocean as to the atmosphere. One simplification we can make in application of these equations to the ocean is to recognize that the density varies rather little in the ocean (by only a few %; see Fig. 9.2), so we can rewrite the horizontal momentum equations Eq. 6-44a,b thus (using our local Cartesian coordinate system; see Fig. 6.19)

$$\frac{Du}{Dt} + \frac{1}{\rho_{ref}}\frac{\partial p}{\partial x} - fv = \mathcal{F}_x;$$
$$\frac{Dv}{Dt} + \frac{1}{\rho_{ref}}\frac{\partial p}{\partial y} + fu = \mathcal{F}_y; \tag{9-7}$$

without incurring serious error, where ρ_{ref} is our constant reference density. We write the vertical equation, Eq. 6-44c (hydrostatic balance) in terms of the density anomaly, Eq. 9-2:

$$\frac{\partial p}{\partial z} = -g(\rho_{ref} + \sigma). \tag{9-8}$$

If we neglect the contribution from σ, for a moment, Eq. 9-8 implies that the pressure increases linearly downward from its surface value ($p_s = 10^5$ Pa, or one atmosphere) thus,

$$p(z) = p_s - g\rho_{ref}(z - \eta), \tag{9-9}$$

where the ocean's surface is at $z = \eta$, and z decreases into the interior of the ocean. This linear variation should be contrasted with the exponential pressure variation of an isothermal compressible atmosphere (see Eq. 3-7). The pressure at a depth of, say, 1 km in the ocean is thus about 10^7 Pa or 100 times atmospheric pressure. However, the $g\rho_{ref}z$ part of the pressure field is dynamically inert, because it does not have any horizontal variations. The dynamical part of the hydrostatic pressure is associated with horizontal variations in the free surface height, η, and interior density anomalies, σ, associated with T and S variations and connected to the flow field by geostrophic balance. Finally, it is important to realize that time-mean horizontal variations in surface atmospheric pressure, p_s, turn out to be much less important in Eq. 9-9 than variations in η and σ.[9]

In Section 9.2 we observed that the maximum horizontal current speeds in ocean gyres are found at the surface in the western boundary currents where, instantaneously, they can reach $1\,\mathrm{m\,s^{-1}}$. Elsewhere, in the interior of ocean gyres, the currents are substantially weaker, typically 5–$10\,\mathrm{cm\,s^{-1}}$, except in the tropical and circumpolar belts. The N-S extent of middle-latitude ocean gyres is typically about $20°$ latitude $\approx$ 2000 km (the E-W scale is greater). Thus setting $\mathcal{U} = 0.1\,\mathrm{m\,s^{-1}}$ and $L = 2 \times 10^6$ m with $f = 10^{-4}\,\mathrm{s^{-1}}$, we estimate a typical Rossby number, Eq. 7-1, of $R_o = \frac{\mathcal{U}}{fL} \sim 10^{-3}$. This is very small, much smaller, for example, than we found to be typical in the atmosphere where $R_o \sim 10^{-1}$ (see estimates in Section 7.1 and Fig. 7.5). Thus the geostrophic and thermal wind approximation is generally excellent for the interior of the ocean away from the equator[10] and away

---

[9]Suppose, for example, that p_s varies by 10 mbar in 1000 km (corresponding to a stiff surface wind of some $10\,\mathrm{m\,s^{-1}}$ in middle latitudes); (see, e.g., Fig. 7.25) then the surface geostrophic ocean current required to balance this surface pressure gradient is a factor $\rho_{atoms}/\rho_{ocean} \sim 1/1000$ times smaller, or $1\,\mathrm{cm\,s^{-1}}$, small relative to observed surface ocean currents. Mean surface atmospheric pressure gradients are typically considerably smaller than assumed in this estimate (see Fig. 7.27).

[10]Our estimate of R_o is applicable to the gyre as a whole. Within the western boundary currents, a more relevant estimate is $U_{bdy}/f\Delta$, where Δ is the width of the boundary current and U_{bdy} is its speed. If $\Delta \sim 100$ km and U_{bdy} reaches $2\,\mathrm{m\,s^{-1}}$, then the Rossby number within these boundary currents can approach, and indeed exceed, unity.

from surface, bottom, and side boundary layers. As discussed later in Section 9.3.4, this fact can be exploited to infer ocean currents from hydrographic observations of T and S.

Thus away from (surface, bottom, and side) boundary layers, the geostrophic equations derived in Chapter 7 will be valid: Eqs. 7-3 and 7-4 with, as in Eq. 9-8, ρ_{ref} substituted for ρ. The associated thermal wind equation, Eqs. 7-16 and 7-17, will also apply.

Using Eq. 7-16 one can immediately infer the sense of the thermal wind shear from the σ field shown in, for example, Fig. 9.7: $\partial u / \partial z > 0$ where σ increases moving northward in the northern hemisphere ($f > 0$), implying that u at the surface is directed eastward in these regions if abyssal currents are weak. Inspection of Fig. 9.7 suggests that $\partial u / \partial z$ and the surface u are positive (negative) poleward (equatorward) of 25° N, more or less as observed in Figs. 9.14 and 9.15. Moreover, Eq. 7-16 suggests a mean surface geostrophic flow of magnitude:

$$u_{surface} \sim \frac{g}{f \rho_{ref}} \frac{H \Delta \sigma}{L} \sim 8 \, \mathrm{cm \, s^{-1}},$$

if $\Delta \sigma \sim 1.5 \, \mathrm{kg \, m^{-3}}$ is the change in σ between, say, 20° N and 40° N, a distance $L \sim 2000$ km, in the top $H \sim 1000$ m of the ocean seen in Fig. 9.7. This is in good accord with direct observations shown, for example, from surface drifters in Fig. 9.14.

9.3.1. Ocean surface structure and geostrophic flow

Near-surface geostrophic flow

To the extent that the surface currents shown in Figs. 9.14 to 9.16 are in geostrophic balance, there must be a pressure gradient force balancing Coriolis forces acting on them. Consider, for example, the eastward flowing Gulf Stream of the Atlantic evident in Fig. 9.15. To balance the southward-directed Coriolis force acting on it, there must be a pressure gradient force directed northward. This is provided by a tilt in the free surface of the ocean: we shall see that the sea surface is higher[11] to the south of the Gulf Stream than to the north of it.

Consider Fig. 9.18. Integrating the hydrostatic relation, Eq. 3-3, from some horizontal surface at constant z up to the free surface at $z = \eta$ (where $p = p_s$, atmospheric pressure) we obtain:

$$p(z) = p_s + \int_z^{\eta} g \rho \, dz = p_s + g \langle \rho \rangle (\eta - z),$$

$$(9\text{-}10)$$

where $\langle \rho \rangle = 1 / (\eta - z) \int_z^{\eta} g \rho \, dz$ is the mean density in the water column of depth $\eta - z$. If we are interested in the near-surface region ($z = z_0$, say, in Fig. 9.18), fractional variations in column depth are much greater

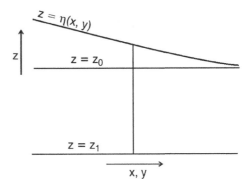

FIGURE 9.18. The height of the free surface of the ocean is $z = \eta(x, y)$. The depths of two reference horizontal surfaces, one near the surface and one at depth, are given by $z = z_0$ and $z = z_1$, respectively.

---

[11]Here "high" means measured relative to the position that the ocean surface would take up if there were no surface geostrophic flow. This equipotential reference surface is the "geoid," which is everywhere perpendicular to the local plumb line. Recall that the "geoid" of our rotating tank of water considered in Section 6.6 is the parabolic surface shown in Fig. 6.11. We saw in Fig. 6.18 that the geoid surface of the rotating Earth approximates a reference ellipsoid. Geoid height variations about this reference ellipsoid are of the order ± 100 m, 100 times larger than height variations associated with geostrophic flow.

than those of density, so we can neglect the latter, setting $\rho = \rho_{ref}$ in Eq. 9-10 and leaving

$$p(z_0) = p_s + g\rho_{ref}(\eta - z_0).$$

We see that horizontal variations in pressure in the near-surface region thus depend on variations in atmospheric pressure and in free-surface height. Since here we are interested in the mean ocean circulation, we neglect day-to-day variations of atmospheric pressure and equate horizontal components of the near-surface pressure gradient with gradients in surface elevation: $(\partial p/\partial x, \partial p/\partial y) = g\rho_{ref}(\partial \eta/\partial x, \partial \eta/\partial y)$. Thus the geostrophic flow just beneath the surface is, from Eq. 7-4, $(u_{g\text{surface}}, v_{g\text{surface}}) = \frac{g}{f}(-\partial \eta/\partial y, \partial \eta/\partial x)$ or, in vector form

$$\mathbf{u}_{g\text{surface}} = \frac{g}{f}\hat{\mathbf{z}} \times \nabla \eta \qquad (9\text{-}11)$$

Note how Eq. 9-11 exactly parallels the equivalent relationship Eq. 7-7 for geostrophic flow on an atmospheric pressure surface.

In Section 7.1.1 we saw that geostrophic winds of $15\,\mathrm{m\,s^{-1}}$ were associated with tilts of pressure surfaces by about 800 m over a distance of 5000 km. But because oceanic flow is weaker than atmospheric flow, we expect to see much gentler tilts of pressure surfaces in the ocean. We can estimate the size of η variations by using Eq. 9-11 along with observations of surface currents. If U is the eastward speed of the surface current, then η must drop by an amount $\Delta\eta$ in a distance L given by:

$$\Delta\eta = \frac{fLU}{g},$$

or 1 m in 1000 km if $U = 10^{-1}\,\mathrm{m\,s^{-1}}$ and $f = 10^{-4}\,\mathrm{s^{-1}}$. Can we see evidence of this in the observations?

Observations of surface elevation

Maps of the height of the sea surface, η, give us the same information as do maps of

the height of atmospheric pressure surfaces (and, of course, because p_s variations are so slight, the ocean surface is, to a very good approximation, a surface of constant pressure). If we could observe the η field of the ocean then, just as in the use of geopotential height maps in synoptic meteorology, we could deduce the surface geostrophic flow in the ocean. Amazingly, variations in ocean topography, even though only a few centimeters to a meter in magnitude, can indeed be measured from satellite altimeters and are mapped routinely over the globe every week or so. Orbiting at a height of about 1000 km above the Earth's surface, altimeters measure their height above the sea surface to a precision of 1–2 cm. And, tracked by lasers, their distance from the center of the Earth can also be determined to high accuracy, permitting η to be found by subtraction.

The 10 y-mean surface elevation (*relative to the mean geoid*) is shown in Fig. 9.19. Consistent with Figs. 9.14 and 9.15, the highest elevations are in the anticyclonic subtropical gyres (where the surface is about 40 cm higher than near the eastern boundary at the same latitude), and there are strong gradients of height at the western boundary currents and near the circumpolar current of the Southern Ocean, where surface height changes by about 1 m across these currents.

9.3.2. Geostrophic flow at depth

At depths much greater than variations of η (at $z = z_1$, say, in Fig. 9.18), we can no longer neglect variations of density in Eq. 9-10 compared with those of column depth. Again neglecting atmospheric pressure variations, horizontal pressure variations at depth are therefore given by, using Eq. 9-10,

$$\hat{\mathbf{z}} \times \nabla p = g\langle\rho\rangle\,\hat{\mathbf{z}} \times \nabla\eta + g(\eta - z)\,\hat{\mathbf{z}} \times \nabla\langle\rho\rangle.$$

Thus the deep water geostrophic flow is given by

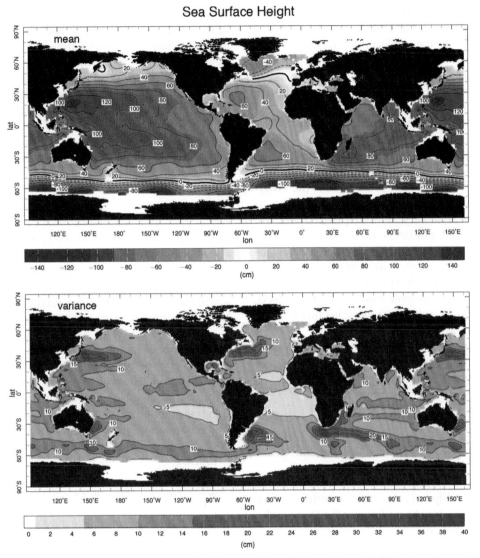

FIGURE 9.19. Top: The 10-year mean height of the sea surface relative to the geoid, $\overline{\eta}$, (contoured every 20 cm) as measured by satellite altimeter. The pressure gradient force associated with the tilted free surface is balanced by Coriolis forces acting on the geostrophic flow of the ocean at the surface. Note that the equatorial current systems very evident in the drifter data, Figs. 9.14 and 9.22, are only hinted at in the sea surface height. Near the equator, where f is small, geostrophic balance no longer holds. Bottom: The variance of the sea surface height, $\sigma\eta = \sqrt{\overline{\eta'^2}}$, Eq. 9-17, contoured every 5 cm.

$$\mathbf{u} = \frac{1}{f\rho_{ref}}\widehat{\mathbf{z}} \times \nabla p$$

$$= \frac{g}{f\rho_{ref}}\left[\langle\rho\rangle\,\widehat{\mathbf{z}} \times \nabla\eta + (\eta - z)\,\widehat{\mathbf{z}} \times \nabla\langle\rho\rangle\right]$$

$$\simeq \frac{g}{f}\widehat{\mathbf{z}} \times \nabla\eta + \frac{g(\eta - z)}{f\rho_{ref}}\widehat{\mathbf{z}} \times \nabla\langle\rho\rangle, \quad (9\text{-}12)$$

since we can approximate $\langle\rho\rangle \simeq \rho_{ref}$ in the first term, because we are not taking its gradient. We see that $\mathbf{u}$ has two contributions: that associated with free-surface height variations, and that associated with interior ocean density gradients. Note that if the ocean density were to be uniform, the second term would vanish and the deep-water geostrophic flow would be the same as that at the surface: *geostrophic flow in an ocean of uniform density is independent of depth.* This, of course, is a manifestation of the Taylor-Proudman theorem, discussed at length in Section 7.2. However, observed currents and pressure gradients at depth are smaller than at the surface (cf. Figs. 9.15 and 9.17) suggesting that the two terms on the rhs of Eq. 9-12 tend to balance one another.

The second term in Eq. 9-12 is the "thermal wind" term (cf. Section 7.3), telling us that **u** will vary with depth if there are horizontal gradients of density. Thus the presence of horizontal variations in surface height, manifested in surface geostrophic currents given by Eq. 9-11, does not guarantee a geostrophic flow at depth. Indeed, as already noted, the flow becomes much weaker at depth.

How much would interior density surfaces need to tilt to cancel out deep pressure gradients and hence geostrophic flow? If **u** $\longrightarrow$ 0 in Eq. 9-12 then, at depth $z = -H$ (assuming $H \gg \eta$),

$$\frac{H}{\rho_{ref}} |\nabla \langle \rho \rangle| \approx |\nabla \eta|, \qquad (9\text{-}13)$$

or, dividing through by N^2, Eq. 9-6, we find that the slope of *isopycnals*—interior density surfaces—must be related to the slope of the free surface by:

$$|\text{isopycnal slope}| \approx \frac{g}{N^2 H}$$
$$\times |\text{free surface slope}|.$$

Using our estimate of N^2 in Section 9.1.5 and rearranging, we find that $g/N^2 H \approx 400$ if we assume that $H = 1\,\text{km}$. Thus we see that for every meter the free surface tilts up, density surfaces must tilt down by around 400 m if deep pressure gradients, and hence deep geostrophic flows, are to be cancelled out. But this is just what we observe in, for example, Fig. 9.5. In those regions where the sea surface is high—over the subtropical gyres (see Fig. 9.19)—density surfaces

bow down into the interior (see Fig. 9.8). In the subpolar gyre (where the sea surface is depressed) we observe density surfaces bowing up toward the surface. It is these horizontal density gradients interior to the ocean that "buffer out" horizontal gradients in pressure associated with the tilt of the free surface; in the time-mean, the second term in Eq. 9-12 does indeed tend to cancel out the first. This is illustrated schematically in Fig. 9.20.

9.3.3. Steric effects

A major contribution to the spatial variations in the height of the ocean surface shown in Fig. 9.19 is simply the expansion (contraction) of water columns that are warm (cold) relative to their surroundings. Note that the sea surface is high over the subtropical gyres, which are warm, and low over the subpolar gyres and around Antarctica, which are relatively cold. Similarly, salty columns of water are shorter than fresh columns, all else being equal. This expansion/contraction of water columns due to T and S anomalies is known as the *steric effect.* We can estimate its magnitude from Eq. 9-13 as follows:

$$\frac{\Delta \eta}{H} \simeq \left(\alpha_T \langle T - T_o \rangle - \beta_S \langle S - S_o \rangle \right), \qquad (9\text{-}14)$$

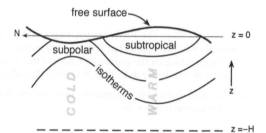

FIGURE 9.20. Warm subtropical columns of fluid expand relative to colder polar columns. Thus the sea surface (measured relative to the geoid) is higher, by about 1 m, in the subtropics than the pole, and is thus greatly exaggerated in this schematic. Pressure gradients associated with the sea-surface tilt are largely compensated by vertical thermocline undulations, of about 400 m, ensuring that abyssal pressure gradients are much weaker than those at the surface.

where, as before, the angle brackets denote $\langle () \rangle = \frac{1}{(H+\eta)} \int_{-H}^{\eta} () \, dz$ and Eq. 9-5 has been used. From Fig. 9.5 we estimate that $\langle T - T_o \rangle \simeq 10°C$ over the top kilometer of the warm water lens of the subtropical gyres and, from Fig. 9.6, $\langle S - S_o \rangle \simeq 0.5$psu. Thus if $\alpha_T = 2 \times 10^{-4}°C^{-1}$ and $\beta_S = 7.6 \times 10^{-4}$psu^{-1} we find that:

$$\frac{\Delta \eta}{H} \simeq \left(\underbrace{2}_{temp} + \underbrace{-0.38}_{salt} \right) \times 10^{-3}.$$

We see that over the top kilometer or so of the column, height variations due to salt are more than offset by the 2 m of expansion due to the warmth of the surface lens. These estimates of gradients in surface elevation are broadly consistent with Fig. 9.19 (top).

9.3.4. The dynamic method

The thermal wind relation, Eq. 7-17, is the key theoretical relationship of observational oceanography, providing a method by which observations of T and S as a function of depth can be used to infer ocean currents.[12] Let us vertically integrate, for example, Eq. 7-16 a from level z to z_1 (see Fig. 9.18) to obtain:

$$u_g(z) - u_g(z_1) = \frac{g}{f} \int_{z_1}^{z} \frac{1}{\rho_{ref}} \frac{\partial \sigma}{\partial y} dz = \frac{g}{f} \frac{\partial D}{\partial y},$$
$$(9\text{-}15)$$

where

$$D = \int_{z_1}^{z} \frac{\sigma}{\rho_{ref}} dz \qquad (9\text{-}16)$$

is known as the *dynamic height*.

Given observations of T and S with depth (for convenience, pressure is typically used as the vertical coordinate) from a hydrographic section, such as Fig. 9.21, we compute σ from our equation of state of seawater, Eq. 9-1, and then vertically integrate to obtain D from Eq. 9-16. The geostrophic flow u_g at any height relative to the geostrophic current at level $u_g(z_1)$ is then obtained from Eq. 9-15 by taking horizontal derivatives. This is called the *dynamic method*. Note, however, that it only enables one to determine geostrophic velocities relative to some reference level. If we assume, or choose, z_1 to be sufficiently deep that, given the general decrease of flow with depth, $u_g(z_1)$ is far smaller than $u_g(z)$, then we can proceed. Indeed, upper ocean velocities are insensitive to the assumption of a deep "level of no motion." However, deep water *transport* calculations, $\int u_g dz$, are extremely sensitive to this assumption that is rarely true . Alternatively we can use altimetric measurements or surface drifters, such as those shown in Figs. 9.13 and 9.19, choose the sea surface as our level of known geostrophic flow and integrate downwards.

As an example of the dynamic method, we show D in Fig. 9.21 (bottom) cutting across the Gulf Stream computed relative to a depth of 2 km. It increases from zero at 2 km to an order of 1 m at the surface. Taking

[12] Georg Wust (1890–1975), a Berliner who dedicated his life to marine research, was the first to systematically map the vertical property distribution and circulation of the ocean. In 1924, using the Florida Current as an example, Wust apparently confirmed that the geostrophic shear measured by current meters were broadly consistent with those calculated from the pressure field using geostrophic balance and the dynamic method.

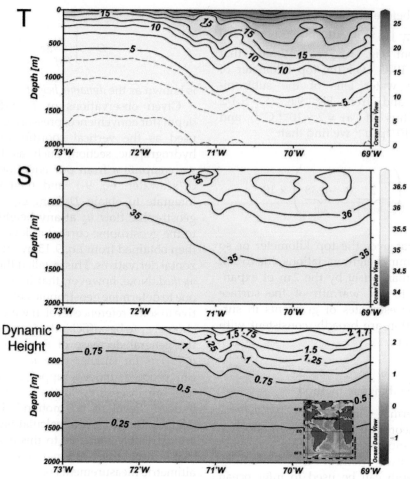

FIGURE 9.21. Top: Temperature section (in °C) over the top 2 km of the water column crossing the Gulf Stream along 38° N, between 69° W and 73° W (as marked on the inset). Middle: Salinity (in psu) across the same section. Bottom: Dynamic height, D (in m computed from Eq. 9-16) relative to 2 km. Produced using Ocean Data View.

horizontal derivatives and using Eq. 9-15, we deduce that $u_g(z)$ relative to 2 km is $\frac{9.81\ \mathrm{m\,s^{-2}}}{2\Omega \sin 52°\ \mathrm{s^{-1}}} \times \frac{1\ \mathrm{m}}{100 \times 10^3\ \mathrm{m}} = 0.85\ \mathrm{m\,s^{-1}}$. This is a swift current but typical of Gulf Stream speeds which, instantaneously, can reach up to 1–2 m s^{-1}.

9.4. OCEAN EDDIES

9.4.1. Observations of ocean eddies

Figures 9.14 to 9.16 show currents averaged over many (about 20) years of observations. However, just as in the atmosphere

(in fact, even more so) the picture we have described of the general circulation of the ocean, while appropriate to the time-averaged flow, is inadequate for describing the instantaneous flow. There are large variations of currents and of surface height that, instantaneously, can mask the time-averaged picture.

The altimetric and drifter data can be analyzed to yield statistics of the time variability. For example, if the sea surface height is $\eta(x, y, t)$, then we can write

$$\eta(x, y, t) = \bar{\eta}(x, y) + \eta'(x, y, t),$$

where the overbar denotes a long time average and η' is the instantaneous departure of η from the average. An impression of the magnitude of these variations and their geographical distribution can be obtained by mapping the variance of surface height about the time mean defined as

$$\sigma_\eta = \sqrt{\overline{\eta'^2}}. \qquad (9\text{-}17)$$

The global map is shown in Fig. 9.19 (bottom). There are distinct maxima of σ_η in regions of strong flow, such as in the western boundary currents of the Gulf Stream and Kuroshio and in the Antarctic Circumpolar Current (especially at the southern tip of Africa and south of Australia and New Zealand). In these regions, $\sigma_h > 20$ cm, so the temporal variance is not very much less than mean spatial variations seen in the top frame of Figure 9.19. Figure 9.22 shows the mean speed of the surface currents together with eddy speeds based on surface drifter data. Mean surface currents in the middle of the gyres are less than $10 \, \text{cm s}^{-1}$, but note how eddy speeds considerably exceed the mean almost everywhere. This implies that the instantaneous flow can be directed opposite to the time-mean flow. Note also how the tropical variability that is very evident in drifter observations, Fig. 9.22, is less evident in the surface elevation field measured by altimetry, Fig. 9.19. This is not surprising. Geostrophic balance, as expressed by Eq. 9-11, tells us that, for a given current amplitude, the height gradient vanishes as $f \to 0$ near the equator. Of course this is true for both mean and eddy components of the flow.

These observations show us that the ocean is not steady and laminar; rather, just like the atmosphere, it is highly turbulent. Indeed, ocean eddies are dynamically analogous to the baroclinic eddies studied in GFD Lab XI, Section 8.2.2, and can be called *ocean weather systems*. However, as will be discussed in Section 10.5, because of the weaker vertical stratification in the ocean compared to the atmosphere, these eddies are typically only 100 km in lateral scale, much smaller than their meteorological counterparts. Ocean eddies typically have a lifetime of many months and intense eddies shed by boundary currents, such as the Gulf Stream, can survive for a year or so.

The SST distribution over the North Atlantic obtained from satellite observations, shown in Fig. 9.23, presents an instantaneous picture of ocean variability in the region of the Gulf Stream. There is a strong gradient of SST across the Gulf Stream, which exhibits large meanders and undulations, some of which break off to form closed eddies ("warm core rings" to the north of the stream and "cold core rings" to the south). The eddies and rings have a scale of 100 km or so. In Section 10.5 their formation by the baroclinic instability of the thermal wind shear associated with the substantial temperature gradient across the Gulf Stream will be discussed. Fig. 9.21, a hydrographic section across the Gulf Stream, shows that there is a strong and systematic gradient of temperature, more or less coincident with the main current, extending all the way down to about 1 km depth. Recall the map of the mean currents at 700 m in the North Atlantic in Fig. 9.17. The mean spatial gradient of surface height seen in Fig. 9.19 and the surface currents observed by drifters, Fig. 9.15, is coincident with this temperature gradient.

High-resolution numerical simulations also give a vivid impression of what the real ocean is probably like; near-surface current speeds in a numerical model are shown in Fig. 9.24. Again, we see large-amplitude spatial variations characteristic of eddy motions in the western boundary and circumpolar currents. Just as in the observations, Fig. 9.22, there is also much eddy activity straddling the equator.

9.5. FURTHER READING

Reviews of the physical properties of seawater and a description of the large-scale circulation of the ocean can be found in

Surface Current Speeds (cm/s)

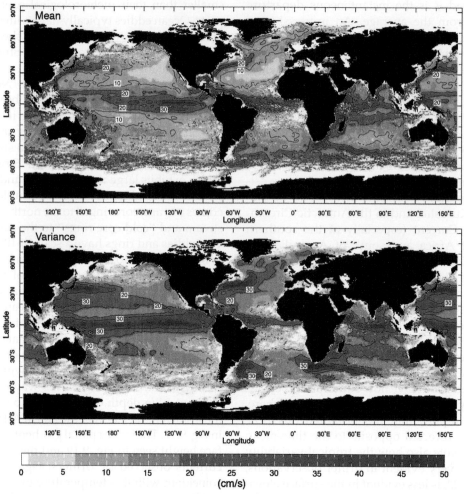

FIGURE 9.22. Top: Mean surface drifter speeds, $\sigma_{\overline{u}} = \sqrt{\left(\overline{u}^2 + \overline{v}^2\right)^{\frac{1}{2}}}$, and (bottom) eddy drifter speeds, $\sigma_{u'} = \sqrt{\left(\overline{u'^2} + \overline{v'^2}\right)^{\frac{1}{2}}}$, computed from 20 y of drifter observations. Regions of the ocean in which observations are sparse, particularly in the southern oceans around Antarctica, appear as white gaps. Data courtesy of Maximenko and Niiler (personal communication, 2003).

Pickard and Emery (1990). A more advanced text which reviews the underlying dynamical principles is Pond and Pickard (1983). Another useful introductory resource is the Open University text on Ocean Circulation. Stommel's (1965) book on the Gulf Stream is an excellent and very accessible account of ocean dynamics in the context of the observations. Wunsch (1996) discusses how one quantitatively infers ocean circulation from observations.

9.6. PROBLEMS

1. Consider an ocean of uniform density $\rho_{ref} = 1000$ kg m^{-3}.

   (a) From the hydrostatic relationship, determine the pressure at a depth of 1 km and at 5 km. Express your answer in units of atmospheric surface pressure, $p_s = 1000$ mbar $= 10^5$ Pa.

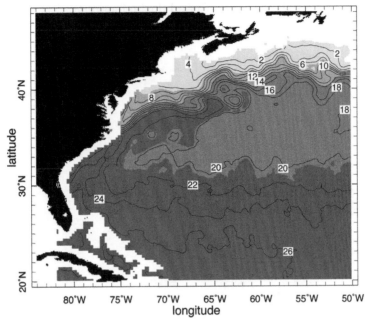

FIGURE 9.23. A satellite-derived Sea Surface Temperature map (in °C) over the Gulf Stream, from June 20, 2003. Note the advection of warm tropical waters northwards by the strong western boundary current and the presence of strong meanders and eddies in the seaward extension of the Gulf Stream.

(b) Given that the heat content of an elementary mass dm of dry air at temperature T is $c_p T\, dm$ (where c_p is the specific heat of air at constant pressure), find a relationship for, and evaluate, the (vertically integrated) heat capacity (heat content per degree Kelvin) of the atmosphere per unit horizontal area. Find how deep an ocean would have to be to have the same heat capacity per unit horizontal area. (You will need information in Tables 1.4 and 9.3.)

2. Simple models of mixed layers.

   (a) Assume that in the surface mixed layer of the ocean, mixing maintains a vertically uniform temperature. A heat flux of $25\ \mathrm{W\,m^{-2}}$ is lost at the ocean surface. If the mixed layer depth does not change and there is no entrainment from its base, determine how long it takes for the mixed layer to cool down by 1°C. [Assume the mixed layer has a depth of 100 m, and use data in Table 9.3.]

   (b) Consider the development of a simplified, convective, oceanic mixed layer in winter. Initially, at the start of winter, the temperature profile is given by

   $$T(z) = T_s + \Lambda z$$

   where z is depth (which is zero at the sea surface and increases upwards) is depth, and the gradient $\Lambda > 0$.

   During the winter heat is lost from the surface at a rate $Q\ \mathrm{W\,m^{-2}}$.

Surface Current Speed (Instantaneous)

0   0.5

m/s

FIGURE 9.24.   Instantaneous map of surface current speed from a global "eddy-resolving" numerical model of ocean circulation. The scale is in units of m s^{-1}. Modified from Menemenlis et al (2005).

As the surface cools, convection sets in and mixes the developing, cold, mixed layer of depth $h(t)$, which has uniform temperature $T_m(t)$. (Recall GFD Lab II, Section 4.2.4.) Assume that temperature is continuous across the base. By matching the heat lost through the surface to the changing heat content of the water column, determine how $h(t)$ and $T_m(t)$ evolve in time over the winter period. Salinity effects should be assumed negligible, so density is related to temperature through Eq. 4-4.

(c) If $Q = 25\text{W m}^{-2}$ and $\Lambda = 10°C$ per kilometer, how long will it take the mixed layer to reach a depth of 100 m?

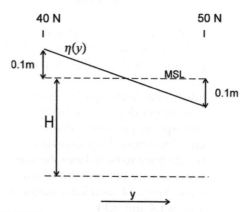

FIGURE 9.25.   Schematic of ocean surface for Problem 3.

3. Consider an ocean of uniform density $\rho_{ref} = 1000$ kg m^{-3}, as sketched in Fig. 9.25. The ocean surface, which is flat in the longitudinal direction (x), slopes linearly with latitude (y) from

$\eta = 0.1$ m above mean sea level (MSL) at 40° N to $\eta = 0.1$ m below MSL at 50° N. Using hydrostatic balance, find the pressure at depth H below MSL. Hence show that the latitudinal pressure gradient $\partial p / \partial y$ and the geostrophic flow are independent of depth. Determine the magnitude and direction of the geostrophic flow at 45° N.

4. Consider a straight, parallel, oceanic current at 45° N. For convenience, we define the x- and y-directions to be along and across the current, respectively. In the region $-L < y < L$, the flow velocity is

$$u = U_0 \cos\left(\frac{\pi y}{2L}\right) \exp(\frac{z}{d})$$

where z is height (note that $z = 0$ at mean sea level and decreases downward), $L = 100$ km, $d = 400$ m, and $U_0 = 1.5$ m s^{-1}. In the region $|y| > L$, $u = 0$.

The surface current is plotted in Fig. 9.26:

Using the geostrophic, hydrostatic, and thermal wind relations:

(a) Determine and sketch the profile of surface elevation as a function of y across the current.

(b) Determine and sketch the density difference, $\rho(y,z) - \rho(0,z)$.

(c) Assuming the density is related to temperature by Eq. 4-4, determine the temperature difference, $T(L,z) - T(-L,z)$, as a function of z. Evaluate this difference at a depth of 500 m. Compare with Fig. 9.21.

5. Figure 9.21 is a section across the Gulf Stream at 38° N (in a plane normal to the flow), showing the distribution of temperature as a function of depth and of horizontal distance across the flow. Assume for the purposes of this question (all parts) that the flow is geostrophic.

(a) Using hydrostatic balance, and assuming that atmospheric pressure is uniform and that horizontal pressure gradients vanish in the deep ocean, *estimate* the differences in surface elevation across the Gulf Stream (i.e., between 70° W and 72° W). Neglect the effect of salinity on density, and assume that the dependence of density ρ on temperature T is adequately described by Eq. 4-4.

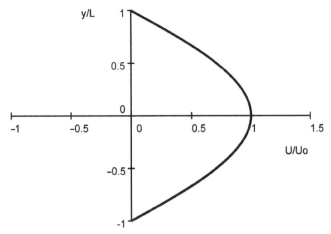

FIGURE 9.26. A plot of u/U_0 at the surface, $z = 0$, against y/L.

(b) The near-surface geostrophic flow **u** is related to surface elevation η by

$$\mathbf{u} = \frac{g}{f}\hat{\mathbf{z}} \times \nabla\eta,$$

where g is gravity and f the Coriolis parameter. Explain how this equation is consistent with the geostrophic relationship between Coriolis force and pressure gradient.

(c) Assuming (for simplicity) that the flow is uniform to a depth $D = 500$ m, and that the flow is zero below this depth, show that the net water transport (volume flux) along the Gulf Stream at this latitude is

$$\frac{gD}{f}\,\delta\eta,$$

where $\delta\eta$ is the elevation difference you estimated in part (a). Evaluate this transport.

6. Figure 9.27 shows the trajectory of a "champion" surface drifter, which made one and a half loops around Antarctica between March, 1995, and March, 2000 (courtesy of Nikolai Maximenko). Red dots mark the position of the float at 30 day interval.

(a) Compute the mean speed of the drifter over the 5 years.

(b) Assuming that the mean zonal current at the bottom of the ocean is zero, use the thermal wind relation (neglecting salinity effects) to compute the depth-averaged temperature gradient across the Antarctic Circumpolar Current.

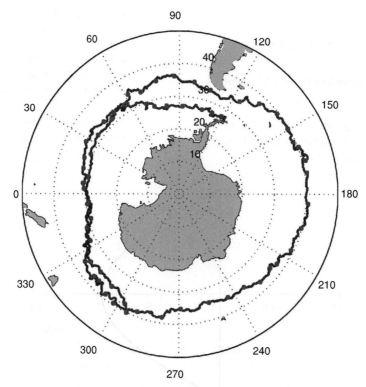

FIGURE 9.27.   The trajectory of a surface drifter which made one and a half loops around Antarctica between March, 1995, and March, 2000 (courtesy of Nikolai Maximenko): Red dots mark the position of the float at 30-day intervals. The float is moving in a clockwise direction.

Hence estimate the mean temperature drop across the 600 km-wide Drake Passage.

(c) If the zonal current of the ACC increases linearly from zero at the bottom of the ocean to a maximum at the surface (as measured by the drifter), estimate the zonal transport of the ACC through Drake Passage, assuming a meridional velocity profile as in Fig. 9.26 and that the depth of the ocean is 4 km. The observed transport through Drake Passage is 130 Sv. Is your estimate roughly in accord? If not, why not?

10

The wind-driven circulation

10.1. The wind stress and Ekman layers
    10.1.1. Balance of forces and transport in the Ekman layer
    10.1.2. Ekman pumping and suction and GFD Lab XII
    10.1.3. Ekman pumping and suction induced by large-scale wind patterns
10.2. Response of the interior ocean to Ekman pumping
    10.2.1. Interior balances
    10.2.2. Wind-driven gyres and western boundary currents
    10.2.3. Taylor-Proudman on the sphere
    10.2.4. GFD Lab XIII: Wind-driven ocean gyres
10.3. The depth-integrated circulation: Sverdrup theory
    10.3.1. Rationalization of position, sense of circulation, and volume transport of ocean gyres
10.4. Effects of stratification and topography
    10.4.1. Taylor-Proudman in a layered ocean
10.5. Baroclinic instability in the ocean
10.6. Further reading
10.7. Problems

In Chapter 9 we saw that the ocean comprises a warm, salty, stratified lens of fluid, the thermocline, circulating on top of a cold, fresh, relatively well mixed abyss, as sketched in Fig. 10.1. The time-mean circulation of thermocline waters is rapid relative to the rather sluggish circulation of the abyss.

There are two processes driving the circulation of the ocean:

1. tangential stresses at the ocean's surface due to the prevailing wind systems, which impart momentum to the ocean—the *wind-driven circulation*, and

2. convection, induced by loss of buoyancy in polar latitudes, due to cooling and/or salt input, causing surface waters to sink

to depth, ventilating the abyss—the *thermohaline circulation*.

This separation of the circulation into wind-driven and thermohaline components is somewhat artificial but provides a useful conceptual simplification. In this chapter we will be concerned with the circulation of the warm, salty thermocline waters sketched in Fig. 10.1 that are brought into motion by the wind. We shall see that the effects of the wind blowing over the ocean is to induce, through Ekman pumping or suction (see Section 10.2), a pattern of vertical motion indicated by the arrows on the figure. Pumping down of buoyant surface water in the subtropics and sucking up of

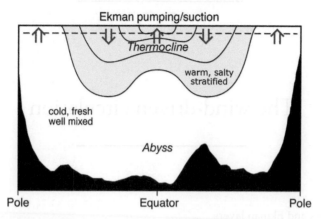

FIGURE 10.1.   The ocean comprises a warm, salty, stratified lens of fluid, the thermocline, circulating on top of a cold, fresh, relatively well mixed abyss. The surface layer, above the horizontal dotted line at a depth of about 100 m, is driven directly by the wind. The thermocline below is brought in to motion through a pattern of vertical velocity driven by the wind (Ekman pumping and suction), which induces flow in the ocean beneath.

heavier interior fluid at the pole and the equator, tilts density surfaces, as sketched in Fig. 10.1 and evident in Fig. 9.7, setting up a thermal wind shear and geostrophic motion. The presence of jagged topography acts to damp strong mean currents in the abyss. Vertical shears build up across the tilted thermocline, however, supporting strong surface currents. The tilting of the thermocline induced by the collusion of horizontal surface density gradients and vertical motion induced by the wind, leads to a vast store of available potential energy (recall our discussion in Section 8.3.2). We shall see in Section 10.5 that this potential energy is released by baroclinic instability, leading to an energetic eddy field, the ocean's analogue of atmospheric weather systems. Ocean eddies have a horizontal scale of typically 100 km, and as discussed in Chapter 9, are often much stronger than the mean flow, leading to a highly turbulent, chaotic flow (see Fig. 9.24). The mean pattern of currents mapped in Figs. 9.14 to 9.16 only emerges after averaging over many years.

10.1. THE WIND STRESS AND EKMAN LAYERS

One cannot escape noticing the similarity between the pattern of surface currents in the ocean and that of the low-level winds in the atmosphere. Compare, for example, the pattern of surface elevation of the ocean in Fig. 9.19 with the pattern of surface atmospheric pressure, Fig. 7.27. Winds, through turbulent transfer of momentum across the atmospheric boundary layer, exert a *stress* on the ocean's surface that drives ocean currents.

The surface wind stress can, to a useful degree of accuracy, be related to the wind velocity through the following relationship (known as a 'bulk formula'):

$$\left(\tau_{wind_x}, \tau_{wind_y} \right) = \rho_{air} c_D u_{10} \left(u_a, v_a \right), \quad (10\text{-}1)$$

where $\tau_{wind_x}, \tau_{wind_y}$ are, respectively, the zonal and meridional stress components, c_D is a bulk transfer coefficient for momentum (typically $c_D = 1.5 \times 10^{-3}$), ρ_{air} is the density of air at the surface, and u_{10} is the speed of the wind at a height of 10 m. The observed annual average of the stress, computed from Eq. 10-1, is shown in Fig. 10.2.

Although there is some similarity between the pattern of surface ocean currents and the pattern of surface wind stress—compare Fig. 10.2 with Fig. 9.19 (top)—the way in which the ocean responds to this wind stress is fascinating and rather subtle, as we are about to see.

Surface Wind Stress (N/m²)

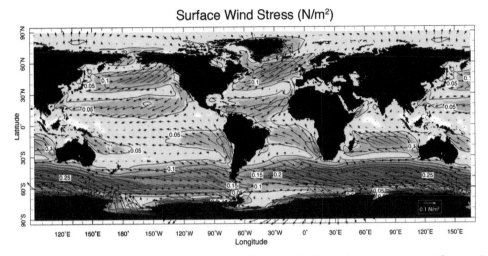

FIGURE 10.2. Annual mean wind stress on the ocean. The green shading and contours represent the magnitude of the stress. Stresses reach values of 0.1 to 0.2 $\mathrm{N\,m^{-2}}$ under the middle-latitude westerlies, and are particularly strong in the southern hemisphere. The arrow is a vector of length 0.1 $\mathrm{N\,m^{-2}}$. Note that the stress vectors circulate around the high and low pressure centers shown in Fig. 7.27, as one would expect if the surface wind, on which the stress depends, has a strong geostrophic component.

10.1.1. Balance of forces and transport in the Ekman layer

If the Rossby number is small, we can neglect the D/Dt terms in the horizontal momentum Eq. 9-7, reducing them to a three way balance between the Coriolis force, the horizontal pressure gradient, and the applied wind stress forcing. This is just Eq. 7-25 in which $\mathcal{F}$ is interpreted as an applied body force due to the action of the wind on the ocean. First we need to express $\mathcal{F}$ in terms of the wind stress, τ_{wind}.

Consider Fig. 10.3 showing a stress that varies with depth, acting on a body of ocean. The stress component of interest here, $\tau_x(z)$, is the x-component of force acting at depth z, per unit horizontal area on the layer beneath. Note that the units of τ are $\mathrm{N\,m^{-2}}$. The slab of thickness δz at level z is subjected to a force per unit horizontal area $\tau_x(z + \delta z)$ at its upper surface, but also subjects the layers beneath it to a force $\tau_x(z)$ per unit horizontal area. Therefore the net force per unit horizontal area felt by the layer is $\tau_x(z + \delta z) - \tau_x(z)$. Since the slab has thickness δz, it has volume δz per unit horizontal area; and if the slab has uniform density ρ_{ref}, it has

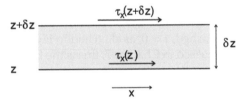

FIGURE 10.3. The stress applied to an elemental slab of fluid of depth δz is imagined to diminish with depth.

mass $\rho_{ref}\delta z$ per unit horizontal area. Therefore the force per unit mass, $\mathcal{F}_x$, felt by the slab is

$$\mathcal{F}_x = \frac{\text{force per unit area}}{\text{mass per unit area}}$$
$$= \frac{\tau_x(z + \delta z) - \tau_x(z)}{\rho_{ref}\delta z} = \frac{1}{\rho_{ref}}\frac{\partial \tau_x}{\partial z} \, ,$$

for small slab thickness. We can obtain a similar relationship for $\mathcal{F}_y$ and hence write:

$$\mathcal{F} = \frac{1}{\rho_{ref}}\frac{\partial \tau}{\partial z} \qquad (10\text{-}2)$$

for the horizontal stress vector $\tau = (\tau_x, \tau_y)$. Hence our momentum equation for the steady circulation becomes Eq. 7-25 with

$\mathcal{F}$ defined above which, for convenience, we write out in component form here:

$$-fv + \frac{1}{\rho_{ref}}\frac{\partial p}{\partial x} = \frac{1}{\rho_{ref}}\frac{\partial \tau_x}{\partial z};$$

$$fu + \frac{1}{\rho_{ref}}\frac{\partial p}{\partial y} = \frac{1}{\rho_{ref}}\frac{\partial \tau_y}{\partial z}. \qquad (10\text{-}3)$$

Equation 10-3 describes the balance of forces in the directly wind-driven circulation, but it does not yet tell us what that circulation is. The stress at the surface is known—it is the wind stress, τ_{wind}, plotted in Fig. 10.2—but we do not know the vertical distribution of stress beneath the surface. The wind stress will be communicated downward by turbulent, wind-stirred motions confined to the near-surface layers of the ocean. The direct influence of wind-forcing decays with depth (rather rapidly, in a few tens of meters or so, depending on wind strength) so that by the time a depth $z = -\delta$ has been reached, the stress has vanished, $\tau = 0$, as sketched in Fig. 10.4. As discussed in Chapter 7, this is the Ekman layer.

Conveniently, we can bypass the need to know the detailed vertical distribution of τ by focusing on the transport properties of the layer by integrating vertically across it. As in Section 7.4, we split the flow into geostrophic and ageostrophic parts. With $\mathcal{F}$ given by Eq. 10-2, the ageostrophic component of Eq. 7-25 is

$$f\hat{\mathbf{z}} \times \mathbf{u}_{ag} = \frac{1}{\rho_{ref}}\frac{\partial \tau}{\partial z}. \qquad (10\text{-}4)$$

Multiplying Eq. 10-4 by ρ_{ref} and integrating across the layer from the surface where $\tau = \tau_{wind}$, to a depth $z = -\delta$, where $\tau = 0$ (see Fig. 10.4) we obtain:

$$f\hat{\mathbf{z}} \times \mathbf{M}_{Ek} = \tau(z = 0) = \tau_{wind},$$

where

$$\mathbf{M}_{Ek} = \int_{-\delta}^{0} \rho_{ref}\mathbf{u}_{ag}\,dz$$

is the lateral mass transport over the layer. Noting that $\hat{\mathbf{z}} \times (\hat{\mathbf{z}} \times \mathbf{M}_{Ek}) = -\mathbf{M}_{Ek}$, we may rearrange the above to give:

$$\mathbf{M}_{Ek} = \frac{\tau_{wind} \times \hat{\mathbf{z}}}{f}. \qquad (10\text{-}5)$$

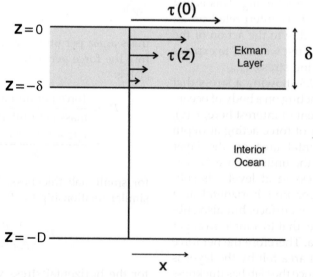

FIGURE 10.4. The stress at the sea surface, $\tau(0) = \tau_{wind}$, the wind stress, diminishes to zero at a depth $z = -\delta$. The layer directly affected by the stress is known as the Ekman layer.

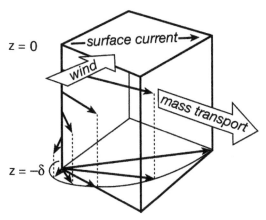

FIGURE 10.5. The mass transport of the Ekman layer is directed to the right of the wind in the northern hemisphere (see Eq. 10-5). Theory suggests that horizontal currents, u_{ag}, within the Ekman layer spiral with depth as shown.

Since $\hat{\mathbf{z}}$ is a unit vector pointing vertically upwards, we see that the mass transport of the Ekman layer is exactly to the right of the surface wind (in the northern hemisphere). Eq. 10-5 determines $\mathbf{M}_{Ek}$, which depends *only* on τ_{wind} and f. But Eq. 10-5 does not predict typical velocities or boundary layer depths that depend on the details of the turbulent boundary layer. A more complete analysis (carried out by Ekman, 1905[1]) shows that the horizontal velocity vectors within the layer trace out a spiral, as shown in Fig. 10.5. Typically $\delta \simeq 10 - 100$ m. So the direct effects of the wind are confined to the very surface of the ocean.

10.1.2. Ekman pumping and suction and GFD Lab XII

Imagine a wind stress blowing over the northern hemisphere ocean with the general anticyclonic pattern sketched in Fig. 10.6

(left). We have just seen that the Ekman transport is directed to the right of the wind and hence there will be a mass flux directed inward (as marked by the broad open arrows in the figure), leading to convergence into the center. Since the mass flux across the sea surface is zero (neglecting evaporation and precipitation, which, as discussed in Chapter 11, are typically $\pm 1\,\mathrm{m\,y^{-1}}$), water cannot accumulate in the steady state, and so it must be driven down into the interior. Conversely, the cyclonic pattern of wind stress sketched in Fig. 10.6 (right) will result in a mass flux directed outward from the center, and therefore water will be drawn up from below.

If the wind stress pattern varies in space (or, more precisely, as we shall see, if the wind stress has some "curl") it will therefore result in vertical motion across the base of the Ekman layer. The flow within the Ekman layer is convergent in *anticyclonic* flow and divergent in *cyclonic* flow, as sketched in Fig. 10.6. The convergent flow drives downward vertical motion (called Ekman pumping); the divergent flow drives upward vertical motion from beneath (called Ekman suction). We will see that it is this pattern of vertical motion from the base of surface Ekman layers that brings the interior of the ocean into motion. But first let us study Ekman pumping and suction in isolation, in a simple laboratory experiment.

GFD Lab XII: Ekman pumping and suction

The mechanism by which the wind drives ocean circulation through the action of Ekman layers can be studied in a simple laboratory experiment in which the cyclonic

[1] Vagn Walfrid Ekman (1874–1954), a Swedish physical oceanographer, is remembered for his studies of wind-driven ocean currents. The role of Coriolis forces in the wind-driven layers of the ocean was first suggested by the great Norwegian explorer Fridtjof Nansen, who observed that sea ice generally drifted to the right of the wind and proposed that this was a consequence of the Coriolis force. He suggested the problem to Ekman—at the time a student of Vilhelm Bjerknes—who, remarkably, worked out the mathematics behind what are now known as Ekman spirals in one evening of intense activity.

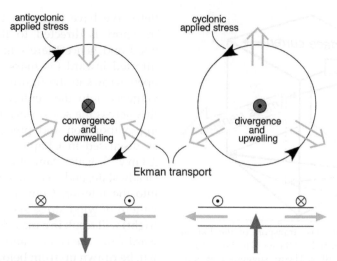

FIGURE 10.6. The Ekman transport is directed perpendicular to the applied stress (to the right if $\Omega > 0$, to the left if $\Omega < 0$) driving (left) convergent flow if the stress is anticyclonic and (right) divergent flow if the stress is cyclonic. (The case $\Omega > 0$—the northern hemisphere—is shown.)

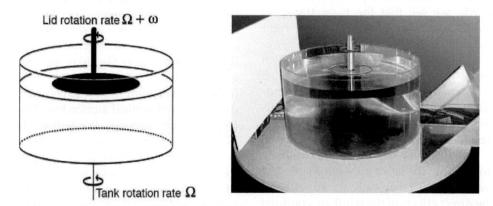

FIGURE 10.7. We rotate a disc at rate ω on the surface of a cylindrical tank of water (the disc is just submerged beneath the surface). The tank of water and the disc driving it are then rotated at rate Ω using a turntable; 10 rpm works well. We experiment with disc rotations of both signs, $\omega = \pm 5$ rpm. Low values of $|\omega|$ are used to minimize the generation of shearing instabilities at the edge of the disc. The apparatus is left for about 20 minutes to come to equilibrium. Once equilibrium is reached, dye crystals are dropped into the water to trace the motions. The whole system is viewed from above in the rotating frame; a mirror can be used to capture a side view, as shown in the photograph on the right.

and anticyclonic stress patterns sketched in Fig. 10.6 are created by driving a disc around on the surface of a rotating tank of water (see Fig. 10.7 and legend). We apply a stress by rotating a disc at the surface of a tank of water that is itself rotating, as depicted in Fig. 10.7. If $\Omega > 0$, the stress imparted to the fluid below by the rotating disc will induce an ageostrophic flow to the right of

the stress (Eq. 10-5): outward if the disc is rotating cyclonically relative to the rotating table ($\omega/\Omega > 0$), inward if the disc is rotating anticyclonically ($\omega/\Omega < 0$), as illustrated in Fig. 10.6. The Ekman layers and patterns of upwelling and downwelling can be made visible with dye crystals.

With Ω, $\omega > 0$ (i.e., the disk is rotating cyclonically and faster than the table) the

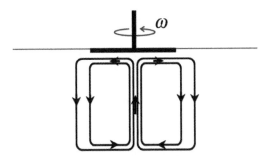

FIGURE 10.8. Schematic of the ageostrophic flow driven by the cyclonic rotation of a surface disc relative to a homogeneous fluid that is itself rotating cyclonically below. Note that the flow in the bottom Ekman layer is in the same sense as Fig. 7.23, top panel, from GFD Lab X. The flow in the top Ekman layer is divergent. If the disc is rotated anticyclonically, the sense of circulation is reversed.

column of fluid is brought into cyclonic circulation and rubs against the bottom of the tank. The flow in the bottom Ekman layer is then just as it was for the cyclonic vortex studied in GFD Lab X (Fig. 7.23, top panel) and is directed inward at the bottom as sketched in Fig. 10.8. Convergence in the bottom Ekman layer thus induces upwelling, drawing fluid up toward the rotating disc at the top, where it diverges in the Ekman layer just under the disk. This is clearly evident in the photographs of upwelling fluid shown in Fig. 10.9. The Ekman layer directly under the disc drives fluid outward to the periphery, drawing fluid up from below. This process is known as *Ekman suction*.

If the disc is rotated anticyclonically, convergence of fluid in the Ekman layer underneath the disc drives fluid downward into the interior of the fluid in a process known as *Ekman pumping* (in this case the sign of all the arrows in Fig. 10.8 is reversed).

Before going on to discuss how Ekman pumping and suction manifest themselves on the large scale in the ocean, it should be noted that in this experiment the Ekman layers are laminar (nonturbulent) and controlled by the viscosity of water.[2] In the ocean the momentum of the wind is carried down into the interior by turbulent motions rather than by molecular processes. Nevertheless, the key result of Ekman theory, Eq. 10-5, still applies. We now go on to estimate typical Ekman pumping rates in the ocean.

10.1.3. Ekman pumping and suction induced by large-scale wind patterns

Figure 10.10 shows a schematic of the midlatitude westerlies (eastward wind stress) and the tropical easterlies (westward stress) blowing over the ocean, as suggested by Fig. 10.2. Because the Ekman transport is to the right of the wind in the northern hemisphere, there is convergence and downward Ekman pumping in the subtropics. This Ekman layer convergence also explains why the sea surface is higher in the subtropics than in subpolar regions (cf. Fig. 9.19): the water is "piled up" by the wind through the action of counter-posed Ekman layers, as sketched in Fig. 10.10. So the interior of the subtropical ocean "feels" the wind stress indirectly, through Ekman-induced downwelling. It is this downwelling that, for example, causes the $\sigma = 26.5$ surface plotted in Fig. 9.8 to bow down in the subtropics. Over the subpolar oceans, by contrast, where the westward stress acting on the ocean diminishes in strength moving northward (see Fig. 10.2), Ekman suction is induced, drawing fluid from the interior

---

[2]Theory tells us (not derived here, but see Hide and Titman, 1967) that:

$$w_{Ek} = \frac{\omega}{2} \left(\frac{\nu}{2\Omega + \omega} \right)^{\frac{1}{2}},$$

where $\nu = 10^{-6} \, \mathrm{m^2 \, s^{-1}}$ is the kinematic viscosity of water (see Table 9.3). In the experiment of Fig. 10.7, typically $\Omega = 1 \, \mathrm{rad \, s^{-1}}$ (that is 10 rpm) and $\omega = 0.5 \, \mathrm{rad \, s^{-1}}$, and so $w_{Ek} \simeq 1.5 \times 10^{-4} \, \mathrm{m \, s^{-1}}$. Thus in Fig. 10.9 , fluid is sucked up at a rate on the order of 9 cm every 10 minutes.

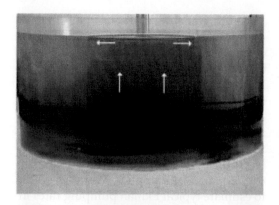

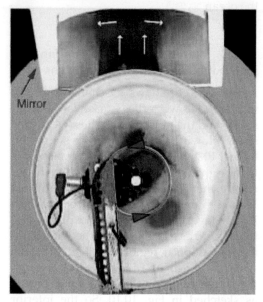

FIGURE 10.9. (Top) The anticlockwise rotation of the disc at the surface induces upwelling in the fluid beneath (Ekman suction), as can be clearly seen from the "dome" of dyed fluid being drawn up from below. (Bottom) We see the experiment from the top and, via the mirror, from the side (and slightly below). Now, sometime later, the dye has been drawn up in to a column reaching right up to the disk and is being expelled outward at the top. The white arrows indicate the general direction of flow. The yellow line marks the rotating disk.

upward into the Ekman layer (as in GFD Lab XII, just described). Hence isopycnals are drawn up to the surface around latitudes of 60° N, S. This general pattern of Ekman pumping/suction imposed on the interior ocean is represented by the vertical arrows in the schematic Fig. 10.1.

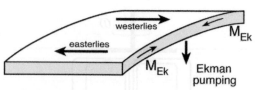

FIGURE 10.10. A schematic showing midlatitude westerlies (eastward wind stress) and tropical easterlies (westward stress) blowing over the ocean. Because the Ekman transport is to the right of the wind in the northern hemisphere, there is convergence and downward Ekman pumping into the interior of the ocean. Note that the sea surface is high in regions of convergence.

We can obtain a simple expression for the pattern and magnitude of the Ekman pumping/suction field in terms of the applied wind stress as follows. Integrating the continuity equation, Eq. 6-11, across the Ekman layer, assuming that geostrophic flow is nondivergent (but see footnote in Section 10.2.1):

$$\nabla_h \cdot \mathbf{u}_{ag} + \frac{\partial w}{\partial z} = 0, \qquad (10\text{-}6)$$

and noting $w = 0$ at the sea surface, then the divergence of the Ekman layer transport results in a vertical velocity at the bottom of the Ekman layer that has magnitude, using Eq. 10-5

$$w_{Ek} = \frac{1}{\rho_{ref}} \nabla_h \cdot \mathbf{M}_{Ek} \qquad (10\text{-}7)$$

$$= \frac{1}{\rho_{ref}} \hat{\mathbf{z}} \cdot \nabla \times \left(\frac{\tau_{wind}}{f} \right)$$

$$= \frac{1}{\rho_{ref}} \left(\frac{\partial}{\partial x} \frac{\tau_{wind_y}}{f} - \frac{\partial}{\partial y} \frac{\tau_{wind_x}}{f} \right) \qquad (10\text{-}8)$$

(see Appendix A.2.2, III). The Ekman pumping velocity defined in Eq. 10-8 depends on the curl of (τ_{wind}/f). Note, however, that typically τ_{wind} varies much more than f, and so the pattern of w_{Ek} is largely set by variations in τ_{wind}. We can estimate the magnitude of w_{Ek} as follows. Figure 10.2 shows that τ_{wind} changes from $+0.1\,\mathrm{N\,m^{-2}}$ to $-0.1\,\mathrm{N\,m^{-2}}$ over 20° of latitude, or 2000 km. Thus Eq. 10-8 suggests

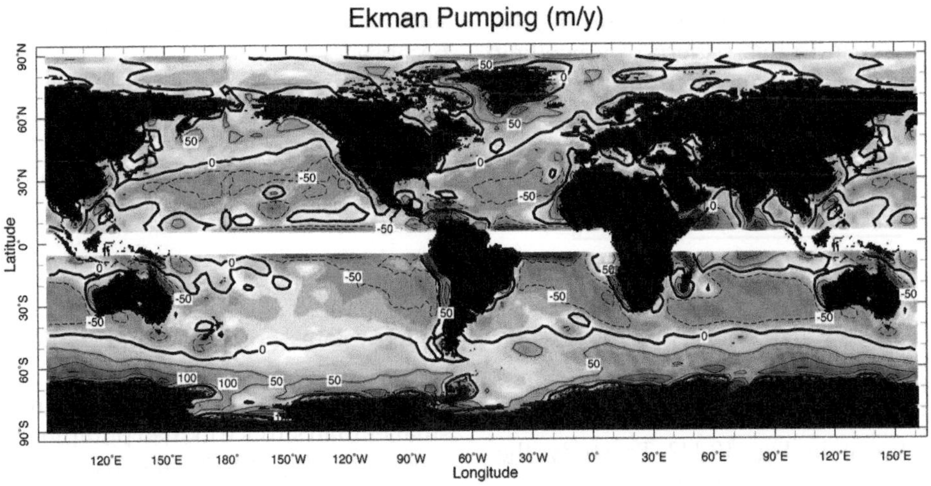

Ekman Pumping (m/y)

FIGURE 10.11. The global pattern of Ekman vertical velocity ($\mathrm{m\,y^{-1}}$) computed using Eq. 10-8 from the annual mean wind stress pattern shown in Fig. 10.2. Motion is upward in the green areas, downward in the brown areas. w_{Ek} is not computed over the white strip along the equator because $f \longrightarrow 0$ there. The thick line is the zero contour. Computed from Trenberth et al (1989) data. The broad regions of upwelling and downwelling delineated here are used to separate the ocean into different dynamical regimes, as indicated by the colors in Fig. 9.13.

$w_{Ek} \approx \frac{1}{10^3\,\mathrm{kg\,m^{-3}}} \times \frac{0.2\,\mathrm{N\,m^{-2}}}{10^{-4}\,\mathrm{s^{-1} \times 2 \times 10^6\,m}} = 32\,\mathrm{m\,y^{-1}}$. It is interesting to compare this with the annual-mean precipitation rate over the globe of about $1\,\mathrm{m\,y^{-1}}$ (this quantity is plotted in Fig. 11.6). We see that the wind, through the action of Ekman layers, achieves a vertical volume flux that is some 30–50 times larger than a typical annual-mean precipitation rate!

Figure 10.11 shows the global pattern of w_{Ek} computed from the surface stress distribution shown in Fig. 10.2, using Eq. 10-7. First of all, note the white band along the equator, where the calculation was not attempted because $f \longrightarrow 0$ there. In fact, however, the equatorial strip is a region of upwelling, because the trade winds on either side of the equator drive fluid away from the equator in the surface Ekman layer, and so demand a supply of fluid from below (as will be seen in Section 12.2).

Away from the equator we observe downwelling in the subtropics and upwelling in subpolar regions with typical magnitudes of $50\,\mathrm{m\,y^{-1}}$, pushing down and pulling up the isopycnals in Fig. 9.7. It takes about $8\,\mathrm{y}$ to pump water down from the surface to $400\,\mathrm{m}$, a typical thermocline depth (cf. Fig. 9.8) indicative of the timescale operating in the thermocline. Note that the zero Ekman pumping contours in Fig. 10.11 separate the ocean into geographical domains that will be central to our understanding of the pattern of ocean gyres seen in Figs. 9.14 to 9.16 and large-scale property distributions.[3] It is these broad domains, demarcated by zero wind stress curl lines, that are color-coded in Fig. 9.13: subpolar regions (blue) are generally subjected to upwelling, subtropical regions (yellow) to downwelling, and tropical regions (red) to upwelling.

---

[3]The pattern of Ekman pumping imposed on the ocean by the wind, Fig. 10.11, has a profound influence on the distribution of dynamically important tracers (such as T, S) the focus of attention here, but also on biologically important properties, such as nutrients. Subpolar gyres, for example, are replete in nutrients because of upwelling of nutrient-rich waters from below and so are regions of high biological productivity. Conversely, subtropical gyres are relative deserts, biologically speaking, because downwelling driven by the wind pushes the nutrients away from the sunlit upper layer where photosynthesis can take place. Thus Ekman pumping has both physical and biogeochemical consequences for the ocean.

What, then, is the response of the interior ocean to this pattern of upwelling and downwelling imposed from above?

10.2. RESPONSE OF THE INTERIOR OCEAN TO EKMAN PUMPING

10.2.1. Interior balances

Beneath the Ekman layer the flow is in geostrophic balance. How does this geostrophic flow respond to the imposed pattern of vertical velocity from the Ekman layer shown in Fig. 10.11? To study the effect of w on the interior ocean, we make use of the continuity equation, Eq. 6-11, applied to the interior flow, assumed to be geostrophic. Taking the horizontal divergence of the geostrophic flow we find:

$$\nabla_h \cdot \mathbf{u}_g = \frac{\partial}{\partial x}\left(-\frac{1}{\rho_{ref} f}\frac{\partial p}{\partial y}\right)$$

$$+\frac{\partial}{\partial y}\left(\frac{1}{\rho_{ref} f}\frac{\partial p}{\partial x}\right) = -\frac{\beta}{f}v_g, \quad (10\text{-}9)$$

where we have remembered that f varies with y and, noting that $dy = a d\varphi$,

$$\beta = \frac{df}{dy} = \frac{1}{a}\frac{df}{d\varphi} = \frac{2\Omega}{a}\cos\varphi \quad (10\text{-}10)$$

is the meridional gradient in the Coriolis parameter, Eq. 6-42. The variation of f with latitude is known as the *β- effect*.

We see, then, that because f varies with latitude, the geostrophic flow is horizontally divergent. Hitherto (in Eq. 10-6, for example) we have assumed that $\nabla_h \cdot \mathbf{u}_g = 0$. However, on the planetary scales being considered here, we can no longer ignore variations in f and the resulting horizontal divergence of geostrophic flow is associated

with vertical stretching of water columns, because:[4]

$$\nabla_h \cdot \mathbf{u}_g + \frac{\partial w}{\partial z} = 0. \quad (10\text{-}11)$$

Combining Eqs. 10-9 and 10-11, we obtain the very useful expression:

$$\beta v_g = f\frac{\partial w}{\partial z}, \quad (10\text{-}12)$$

which relates horizontal and vertical currents.

If vertical velocities in the abyss are much smaller than surface Ekman pumping velocities, then Eq. 10-12 tells us that ocean currents will have a southward component in regions where $w_{Ek} < 0$, and northward where $w_{Ek} > 0$. This is indeed observed in the interior regions of ocean gyres. Consider Figs. 9.14 and 9.15, for example, in the light of Eq. 10-12 and Fig. 10.11. Does Eq. 10-12 make any quantitative sense? Putting in numbers, $f = 10^{-4}\,\text{s}^{-1}$, $w_{Ek} = 30\,\text{my}^{-1}$, H the depth of the thermocline ~ 1 km, we find that $v = 1$ cm s^{-1}, typical of the gentle currents observed in the interior of the ocean on the large scales. Note that Fig. 9.22 (top) shows *surface* currents, whereas our estimate here is an average current over the depth of the thermocline. This is the basic mechanical drive of the ocean circulation: the pattern of Ekman pumping imposed on the ocean from above induces meridional motion through Eq. 10-12.

10.2.2. Wind-driven gyres and western boundary currents

We have seen that Ekman pumping, downward into the interior of the ocean, must drive an equatorward flow in the subtropical latitude belt between the mid-latitude westerlies and tropical easterlies. Although this flow is indeed observed (see the equatorward interior flow over the subtropics in Figs. 9.14 and 9.15), we have

---

[4]Eq. 10-11 tells us that $\nabla_h \cdot \mathbf{u}_g \sim w_{Ek}/H$ where H is the vertical scale of the thermocline, and so is δ/H smaller than $\nabla_h \cdot \mathbf{u}_{ag} = w_{Ek}/\delta$ in Eq. 10-6, where δ is the Ekman layer depth. Because $\delta/H \lesssim 0.1$, we are thus justified in neglecting $\nabla_h \cdot \mathbf{u}_g$ in comparison with $\nabla_h \cdot \mathbf{u}_{ag}$ in the computation of w_{Ek}, as was assumed in deriving Eq. 10-7.

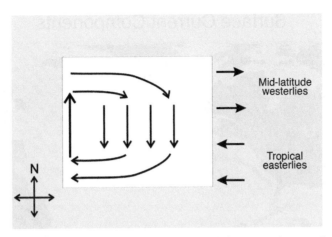

FIGURE 10.12.   Schematic diagram showing the sense of the wind-driven circulation in the interior and western boundary regions of subtropical gyres.

not yet explained the remainder of the circulation. If we take a very simple view of the geometry of the midlatitude ocean basins (Fig. 10.12), then the equatorward flow induced by Ekman pumping could be "fed" at its poleward side by eastward or westward flow, and in turn feed westward or eastward flow at the equatorward edge. Either would be consistent with Eq. 10-12, which only dictates the N-S component of the current. However, the general sense of circulation must mirror the anticyclonic sense of the driving wind stress in Fig. 10.12. Hence the required poleward return flow must occur on the western margin, as sketched in Fig. 10.12.[5] In the resulting intense *western boundary current*, our assumptions of geostrophy and/or of negligible friction break down (as will be discussed in more detail in Section 11.3.3), and therefore Eq. 10-12 is not applicable. This current is the counterpart in our simple model of the Gulf Stream or the Kuroshio, and other western boundary currents.

The preference for western as opposed to eastern boundary currents is strikingly evident in the surface drifter observations shown in Fig. 10.13, which displays the same data as in Fig. 9.15, but zooms in at high resolution on the surface circulation in the North Atlantic. We clearly observe the northward flowing Gulf Stream and the southward flowing Labrador Current hugging the western boundary. Here mean currents can exceed $40\,\mathrm{cm\,s^{-1}}$. Note how the path of the Gulf Stream and its interior extension, the North Atlantic Current, tends to follow the zero Ekman pumping line in Fig. 10.11, which marks the boundary between the subtropical and subpolar gyres and the region of enhanced temperature gradients.

10.2.3. Taylor-Proudman on the sphere

Before going on in Section 10.3 to a fuller discussion of the implications of Eq. 10-12 on ocean circulation, we now discuss what it means physically. Equation 10-12 can be simply understood in terms of the attendant rotational and geometrical constraints on the fluid motion, i.e., the Taylor-Proudman

---

[5]Indeed, to sustain the circulation against frictional dissipation at the bottom and side boundaries, the wind stress must *do work* on the ocean. Since the rate of doing work is proportional to the product of wind stress and current velocity ($\mathbf{u} \cdot \mathcal{F} > 0$) these two quantities must, on average, be in phase for the work done to be positive. This is the case if the sense of circulation is as depicted in Fig. 10.12, whereas the work done would be negative if the circulation were returned to the east.

Surface Current Components

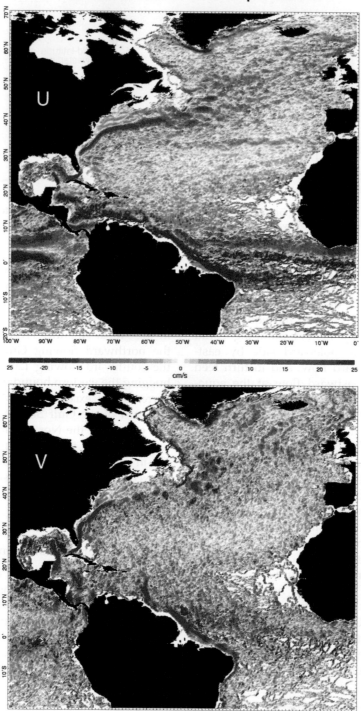

FIGURE 10.13.   (Top) Time-mean zonal velocity and (Bottom) the meridional velocity in cm s^{-1} computed from surface drifters averaged over a 0.25° × 0.25° grid over the North Atlantic. Green colors denote positive values, brown colors negative. Data courtesy of Maximenko and Niiler (2003).

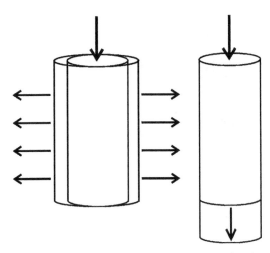

FIGURE 10.14. In Section 10.2.3 we consider the possibility that a Taylor column subjected to Ekman pumping at its top might conserve mass by expanding its girth, as sketched on the left. We argue that such a scenario is not physically plausible. Instead the column maintains its cross-sectional area and increases its length, as sketched on the right.

theorem, on the sphere. This can be seen as follows.

Let us first consider rotational constraints on the possible motion. If the ocean were homogeneous then, as described in Section 7.2, the steady, inviscid, low-Rossby-number flow of such a fluid must obey the Taylor-Proudman theorem, Eq. 7-14. Thus the velocity vector cannot vary in the direction parallel to the rotation vector, and flow must be organized into columns parallel to Ω, an expression of gyroscopic rigidity, as illustrated in GFD Lab VII, Section 7.2.1, and sketched in Fig. 7.8.

Now, consider what happens to such a column of fluid subjected to Ekman pumping at its top. According to Eq. 7-14, we might think that mass continuity would be satisfied by uniform flow sideways out of the column, as sketched in Fig. 10.14 left. This satisfies the constraint that the flow be independent of height, but cannot be sustained in a steady flow. Why not? If the flow is axisymmetric about the circular column, it will conserve its angular momentum, $\Omega r^2 + v_\theta r$, where v_θ is the azimuthal

component of flow around the column, and r is the column radius. The column must continuously expand as fluid is being pumped into it at its top; thus r must increase and so v_θ must change by an amount $\delta v_\theta \simeq -2\Omega \delta r$ (this, of course, is just Eq. 7-22) as the column increases its girth by an amount δr. Thus v_θ must become increasingly negative as r increases. This is obviously inconsistent with our assumption of steady state and not physically reasonable.

So, what else can happen? Let's now introduce the geometrical constraint that our Taylor columns must move in a spherical shell, as sketched in Fig. 10.15a. (We have obviously exaggerated the depth h of the fluid layer in the figure—recall Fig. 1.1!) We see that the columns have greatest length, in the direction parallel to Ω, near the equator. Therefore, if supplied by fluid from above by systematic Ekman pumping, a fluid column can expand in volume without expanding its girth (which we have seen is not allowed in steady state), by *moving systematically equatorward* in the spherical shell and hence increasing its length in the manner illustrated in Fig. 10.15b. The column will move equatorward at just the rate required to ensure that the "gap" created between it and the spherical shell is at all times filled by the pumping down of water from the surface. This, in essence, is how the wind, through the Ekman layers, drives the circulation in the interior of the ocean. The rate at which fluid is pumped down from the Ekman layer must be equal to the rate of change of the volume of the Taylor columns beneath.

Let's think about this process in more detail following Fig. 10.15. The Taylor columns are aligned parallel to the rotation axis; hence, since $\cos\theta = \sin\varphi = h/d$ where θ is co-latitude and φ latitude (see Fig. 10.15a), their length is given by:

$$d = \frac{h}{\sin\varphi} \qquad (10\text{-}13)$$

if the shell is thin (this is inaccurate within less than 1% of the equator). If the change in

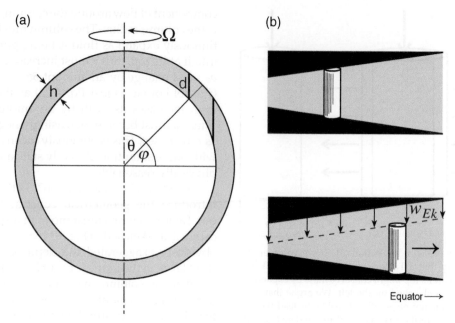

FIGURE 10.15. (a) An illustration of Taylor-Proudman on a rotating sphere. We consider a spherical shell of homogeneous fluid of constant thickness h. Taylor columns line up parallel to Ω with length d. The latitude is φ and the co-latitude θ. (b) A Taylor column in a wedge. If the wedge narrows, or fluid is pumped down from the top at rate w_{Ek}, the Taylor column moves sideways to the thicker end of the wedge. This is just how one flicks a lemon seed. The downward motion between finger and thumb generates lateral (shooting) motion as the seed slips sideways. Modified from a discussion by Rhines (1993).

the volume of the Taylor column is ADd/Dt, where (see Fig. 10.16) A is its cross-sectional

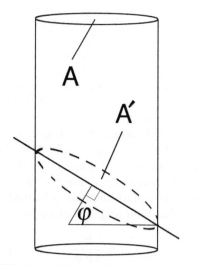

FIGURE 10.16. The area $A' = A/\sin\varphi$ is the cross-sectional area of a Taylor column, A, projected on to the surface of the sphere, where φ is the latitude.

area measured perpendicular to Ω (which, as discussed above, does not change in time), and d is its length parallel to Ω, then

$$A'w_{Ek} = \frac{A}{\sin\varphi}w_{Ek} = \frac{A}{\sin\varphi}\left(-\frac{Dh}{Dt}\right) = -A\frac{Dd}{Dt}$$

since $A' = A/\sin\varphi$ is the area of the Taylor column projected onto the surface of the sphere over which fluid is being pumped down from the Ekman layer at rate w_{Ek}. Note that the minus sign ensures that if $w_{Ek} < 0$ (pumping down into the ocean), then $Dd/Dt > 0$: so that *following the Taylor column along its length increases.*

Rewriting,

$$\frac{w_{Ek}}{\sin\varphi} = -\frac{Dd}{Dt} = -\frac{D\varphi}{Dt}\frac{dd}{d\varphi} = \frac{D\varphi}{Dt}\left(\frac{h\cos\varphi}{\sin^2\varphi}\right)$$

$$= \frac{v}{a}\left(\frac{h\cos\varphi}{\sin^2\varphi}\right) \qquad (10\text{-}14)$$

where Eq. 10-13 has been used and $v = aD\varphi/Dt$ is the meridional velocity of the column. Multiplying both sides by 2Ω and rearranging, Eq. 10-14 may be written in the form:

$$\beta v = f \frac{w_{Ek}}{h}, \qquad (10\text{-}15)$$

where f is given by Eq. 6-42 and β by Eq. 10-10. Note that, setting $\partial w/\partial z = w_{Ek}/h$, we have arrived at a version of Eq. 10-12.

The simple mechanism sketched in Fig. 10.15b is the basic drive of the wind-driven circulation; the gentle vertical motion induced by the prevailing winds, w_{Ek}, is amplified by a large geometrical factor, $\frac{f}{\beta h} = \frac{a}{h} \tan \varphi \simeq$ radius of Earth/depth of ocean ~ 1000, to create horizontal currents with speeds that are 1000 times that of Ekman pumping rates, which is $1\,\text{cm s}^{-1}$ compared to $30\,\text{m y}^{-1}$. Thus we see that the stiffness imparted to the fluid by rotation results in strong lateral motion as the Taylor columns are squashed and stretched in the spherical shell.

There are two useful mechanical analogies:

1. 'pip' flicking: a lemon seed shoots out sideways on being squashed between finger and thumb (see Fig. 10.15b).

2. a child's spinning top: the "pitch" of the thread on the spin axis results in horizontal motion when the axis is pushed down (see Fig. 10.17). The tighter the pitch the more horizontal motion one creates (note, however, that in practice friction prevents use of a very tight pitch).

10.2.4. GFD Lab XIII: Wind-driven ocean gyres

The previous discussion motivates a laboratory demonstration of the wind-driven circulation. We need the following three essential ingredients: (i) geometrical and (ii) rotational constraints, and (iii) a representation of Ekman pumping. The apparatus, shown in Fig. 10.18, consists of a rotating

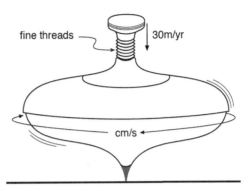

FIGURE 10.17. The mechanism of wind-driven ocean circulation can be likened to that of a child's spinning top. The tight pitch of the screw thread (analogous to rotational rigidity) translates weak vertical motion (Ekman pumping of order 30 m y^{-1}) into rapid horizontal swirling motion (ocean gyres circulating at speeds of cm s^{-1}).

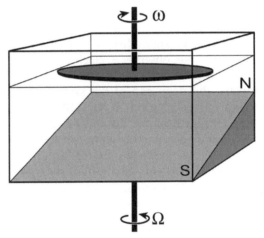

FIGURE 10.18. A tank with a false sloping bottom is filled with water so that the water depth varies between about 5 cm at the shallow end and 15 cm at the deep end. The slope of the bottom represents the spherical "beta-effect": the shallow end of the tank corresponds to high latitudes. (The labels N/S correspond to $\Omega > 0$, appropriate to the northern hemisphere.) A disc is rotated clockwise very slowly at the surface of the water; a rate of 1 rpm works well. To minimize irregularities at the surface, the disc can be submerged so that its upper surface is a millimeter or so underneath the surface. The whole apparatus is then rotated anticlockwise on a turntable at a speed of $\Omega = 10$ rpm. It is left to settle down for 20 minutes or so. Holes bored in the rotating disc can be used to inject dye and visualize the circulation beneath, as was done in Fig. 10.19.

Plexiglas disc (to represent the action of the wind as in GFD Lab XII) on the surface of a anticlockwise-rotating square tank of water with a sloping bottom (to represent, as we shall see, spherical geometry). The stress applied by the rotating lid to the underlying water is analogous to the wind stress at the ocean surface. With clockwise differential rotation of the disc (Fig. 10.6, left), fluid is drawn inwards in the Ekman layer just under the lid and pumped downward in to the interior, mimicking the pumping down of water in subtropical gyres by the action of the winds, as sketched in Fig. 10.10. The varying depth of the tank mimics the variation in the depth of the spherical shell measured in the direction parallel to the rotation vector on the sphere (Fig. 10.15). The shallow end of the tank is thus analogous to the poleward flank of the ocean basin (the 'N' in Figs. 10.18 and 10.19) and the deep end to the tropics.

On introduction of dye (through bore holes in the Plexiglas disc) to help visualize the flow, we observe a clockwise (anticyclonic) gyre with interior flow moving toward the deep end of the tank ("equatorward") as charted in Fig. 10.19. This flow (except near the lid and the bottom) will be independent of depth, because the interior flow obeys the Taylor-Proudman theorem. Consistent with the discussion (see Section 10.2.2), a strong "poleward" return flow forms at the "western" boundary; this is the tank's equivalent of the Gulf Stream in the Atlantic or the Kuroshio in the Pacific.

In a manner directly analogous to that described in Section 10.2.3, we can relate the strength of the north-south flow to the Ekman pumping from under the disc and the slope of the bottom, as follows (compare with Eq. 10-14):

$$w_{Ek} = -\frac{Dd}{Dt} = -\frac{Dy}{Dt}\frac{\partial d}{\partial y} = -v\frac{\partial d}{\partial y}, \quad (10\text{-}16)$$

where d is the depth of the water in the tank, y represents the upslope ("poleward") coordinate, and v is the velocity in that direction. In our experiment, $w_{Ek} \simeq 5 \times 10^{-5}\,\mathrm{m\,s^{-1}}$

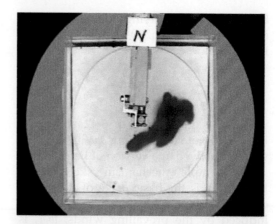

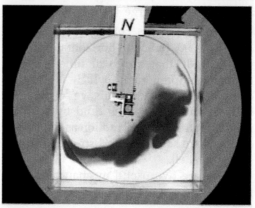

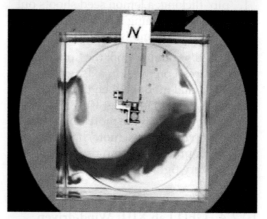

FIGURE 10.19. A time sequence (every 7 min) showing the evolution of red dye injected through a hole in the rotating disc. The label 'N' marks the shallow end of the tank. The plume of dye drifts "equatorward" in the "Sverdrup" interior where Eq. 10-15 holds. In the bottom picture we see the dye being returned poleward in a western boundary current, the laboratory model's analogue of the Gulf Stream or the Kuroshio. Equatorward flow is broad and gentle, poleward flow much swifter and confined to a western boundary current.

and the bottom has a slope $\partial d/\partial y = -0.2$. Thus v should reach speeds of $2.5 \times 10^{-4}\,\mathrm{m\,s^{-1}}$, or 15 cm in 10 minutes, directed equatorward (toward the deep end). This is broadly in accord with observed flow speeds.

The above relation cannot hold, however, over the whole domain, because it implies that the flow is southward everywhere, draining fluid from the northern end of the tank. As can be seen in Fig. 10.19, the water returns in a poleward flowing western boundary current, just as sketched in Fig. 10.12.

10.3. THE DEPTH-INTEGRATED CIRCULATION: SVERDRUP THEORY

One might wonder about the relevance of the previously discussed homogeneous model of ocean circulation to the real ocean. The ocean is not homogeneous and it circulates in basins with complicated geometry. It is far from the homogeneous spherical shell of fluid sketched in Fig. 10.15. And yet, as we will now see, the depth-integrated circulation of the ocean is indeed governed by essentially the same dynamics as a homogeneous fluid in a shallow spherical shell.

We begin by, just as in Section 10.2.1, eliminating (by cross-differentiation) the pressure gradient terms between the horizontal momentum balances, Eq. 10-3, to obtain, using continuity, Eq. 6-11:

$$\beta v = f\frac{\partial w}{\partial z} + \frac{1}{\rho_{ref}}\frac{\partial}{\partial z}\left(\frac{\partial \tau_y}{\partial x} - \frac{\partial \tau_x}{\partial y}\right),$$

which is Eq. 10-12 modified by wind stress curl terms. Now, integrating up from the bottom of the ocean at $z = -D$ (where we imagine $w = 0$ and $\tau = 0$) to the very top, where w is again zero but $\tau = \tau_{wind}$, we obtain:

$$\beta V = \frac{1}{\rho_{ref}}\left(\frac{\partial \tau_{wind_y}}{\partial x} - \frac{\partial \tau_{wind_x}}{\partial y}\right)$$

$$= \frac{1}{\rho_{ref}}\hat{\mathbf{z}}\cdot\nabla\times\tau_{wind}, \qquad (10\text{-}17)$$

where

$$V = \int_{-D}^{0} v\,dz \qquad (10\text{-}18)$$

is the depth-integrated meridional transport. Equation 10-17 is the result we seek. It is known as the *Sverdrup relation*[6] and relates the vertically-integrated meridional flow to the curl of the wind stress. The key assumption that must be satisfied for its validity—in addition to $R_o \ll 1$—is that mean flow in the deep ocean must be sufficiently weak (well supported by observation), so that both frictional stress on the ocean bottom and vertical motion are negligibly small.

Note the close connection between Eq. 10-17 and Eq. 10-15, (see Problem 3 at the end of the chapter, for a detailed look), when we realize that $hv = V$ is the transport and $f\,w_{EK}$ is (if f were constant) the wind-curl divided by ρ_{ref} (cf. Eq. 10-8). Thus, despite the apparent restrictions in the assumptions made in deriving Eq. 10-15, it turns out to have wide applicability. Moreover, when combined with the preference

---

[6] Harald Ulrik Sverdrup (1888–1957), a Norwegian who began his career studying meteorology with Vilhelm Bjerknes in Oslo, was appointed director of Scripps Oceanographic Institution in 1936. His 1947 paper "Wind-driven currents in a baroclinic ocean," which showed the link between meridional currents and the curl of the wind stress, began the modern era of dynamical oceanography and initiated large-scale modeling of ocean circulation.

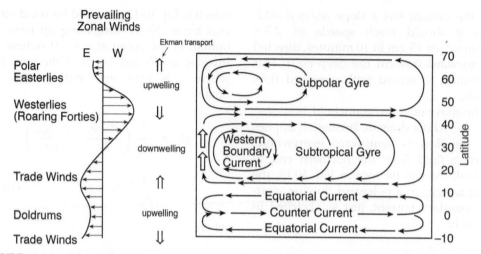

FIGURE 10.20. Schematic diagram showing the classification of ocean gyres and major ocean current systems and their relation to the prevailing zonal winds. The pattern of Ekman transport and regions of upwelling and downwelling are also marked.

for western as opposed to eastern boundary currents, it gives us deep physical insights into the mechanism underlying the wind-driven circulation.

10.3.1. Rationalization of position, sense of circulation, and volume transport of ocean gyres

The Sverdrup relation, Eq. 10-17, is one of the cornerstones of dynamical oceanography. We can use it to derive a simple expression for the transport of ocean gyres as follows. The depth integrated flow must be horizontally nondivergent, and so we can introduce a streamfunction, Ψ, to map it out, [just as a streamfunction was used to represent the geostrophic flow in Section 7.1] where

$$U = -\frac{\partial \Psi}{\partial y}; \quad V = \frac{\partial \Psi}{\partial x}. \qquad (10\text{-}19)$$

Combining Eqs. 10-17 and 10-19 and integrating westward from the eastern

boundary, where we set $\Psi = 0$ (no transport through the eastern boundary[7]) we obtain:

$$\Psi(x, y) = \frac{1}{\rho_{ref}\beta} \int_{\text{eastern bdy}}^{x} \hat{\mathbf{z}} \cdot \nabla \times \tau_{wind}\, dx.$$

$$(10\text{-}20)$$

This simple formula is remarkably successful—it predicts the sense of circulation and volume transport of all the major ocean gyres, rationalizing the patterns of ocean currents shown in Figs. 9.14, 9.15 and 9.16 in terms of the pattern of imposed winds. Note, in particular, that Eq. 10-17 tells us that $V = 0$ along the lines of zero wind stress curl, the thick black lines in Fig. 10.11. Where $\hat{\mathbf{z}} \cdot \nabla \times \tau_{wind} < 0$, $V < 0$ and visa-versa, allowing us to define subpolar and subtropical gyres, etc., and rationalizing the pattern of zonal jets observed in the tropical oceans, as sketched in the schematic diagram in Fig. 10.20.

---

[7]Since $U = 0$ at the eastern boundary, Eq. 10-19 tells us that Ψ is a constant there. Equation 10-19 allows us to add any arbitrary constant to Ψ, so we are free to set $\Psi = 0$ at the boundary. Note, however, that we cannot simultaneously satisfy a vanishing normal flow condition through the western *and* eastern boundary using Eq. 10-17, because it involves only one derivative in x. As discussed in Section 11.3.3, dissipative processes must be invoked at the western boundary to obtain a complete solution. In Fig. 10.21, the western boundary is sketched in as a feature required by mass continuity.

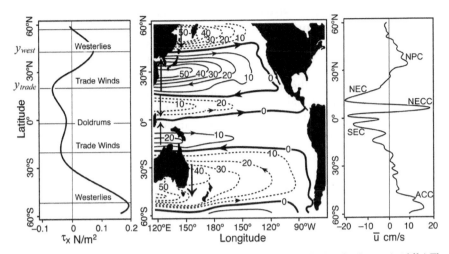

FIGURE 10.21. (left) The zonal-average of the zonal wind stress over the Pacific Ocean. (middle) The Sverdrup transport stream function (in $Sv = 10^6 \, m^3 \, s^{-1}$) obtained by evaluation of Eq. 10-20 using climatological wind stresses, Fig. 10.2. Note that no account has been made of islands; we have just integrated right through them. The transport of the western boundary currents (marked by the $N \leftrightarrow S$ arrows) can be read off from Ψ_{west_bdy}. (right) The zonal-average zonal current over the Pacific obtained from surface drifter data shown in Fig. 9.14. Key features corresponding to Fig. 9.13 are indicated.

Figure 10.21 shows Ψ in the Pacific sector and should be compared with Fig. 10.20. The curl of the annual-mean surface stress shown in Fig. 10.2 was computed and integrated westward from the eastern boundary to yield Ψ from Eq. 10-20 as a function of horizontal position. The units are in $10^6 \, m^3 \, s^{-1}$. It is convenient to use 1 million cubic meters per second as a unit of volume transport, which is known as the "Sverdrup" (or Sv, after Harald Sverdrup). To put things in perspective, the flow of the Amazon river as it runs into the sea is about 0.2 Sv. Thus transport of the subtropical gyre of the North Pacific is about 50 Sv, or 250 times that of the Amazon, the biggest river on Earth!

One interesting rationalization provided by Sverdrup theory is that it accounts for the countercurrents observed in the tropical oceans discussed in Chapter 9, which are currents that flow in a direction *opposite* to the prevailing winds. As can be seen in Figs. 10.20 and 10.21, Sverdrup balance implies meridional flow away from the Doldrums in the interior with a return flow in a western boundary current.

Convergence of these boundary currents drives an eastward flow just north of the equator, even though the winds are blowing toward the west here. This is just as observed in surface drifter data (see Fig. 9.14 and the zonal-average surface currents across the Pacific plotted in the right-most panel in Fig. 10.21).

The transport of the western boundary current must be equal and opposite to that in the interior. The total interior transport is

$$\text{Int. Trans.} = \int_{eastern \, bdy}^{western \, bdy} V \, dx$$

$$= \frac{1}{\rho_{ref} \beta} \int_{eastern \, bdy}^{western \, bdy} \hat{\mathbf{z}} \cdot \nabla \times \tau_{wind} \, dx$$

$$= \Psi_{west\_bdy}, \qquad (10\text{-}21)$$

and so the balancing boundary current transport, Ψ_{west_bdy}, can be read off Fig. 10.21. It is the value of Ψ at the western margin of the Pacific. However, it is instructive to obtain an explicit expression for the transport as follows. Let us assume that the wind stress only has an east-west component (a useful approximation to reality, cf. Fig. 10.2) so $\tau_{wind} = \left(\tau_{wind_x}, 0 \right)$ with $\tau_{wind_x} = -\tau_x \cos \left(\frac{\pi(y - y_{trade})}{L_y} \right)$, where $L_y = y_{west} - y_{trade}$ is a measure of the meridional scale over which the zonal wind changes from the westerly wind belt to the trades, as marked in Fig. 10.21 (left most panel). Then, since $\hat{\mathbf{z}} \cdot \nabla \times \tau_{wind} = -\frac{\partial \tau_{wind_x}}{\partial y}$, Eq. 10-21 yields:

$$\text{Transport} = \frac{\pi \tau_x}{\rho_{ref} \beta} \frac{L_x}{L_y} \sin \left(\frac{\pi \left(y - y_{trade} \right)}{L_y} \right),$$

where L_x is the east-west extent of the basin, assumed constant. Inserting numbers into the above expression typical of the Pacific—$\tau_x = 0.1 \, \mathrm{N\,m^{-2}}$, $\beta = 1.98 \times 10^{-11} \, \mathrm{m^{-1}\,s^{-1}}$, $L_y = 3000 \, \mathrm{km}$; $L_x = 8000 \, \mathrm{km}$— we find a maximum transport of the subtropical gyre of 44 Sv, roughly in accord with the detailed calculation given in Fig. 10.21. The transport of the subtropical gyre of the Atlantic Ocean is somewhat smaller, about 30 Sv, largely on account of the much reduced east-west scale of the basin.

The interior Sverdrup transport of the gyre is thus returned meridionally in narrow western boundary currents as marked in Fig. 10.21 in the Pacific. In the Atlantic, it is clear from Fig. 10.13 that the horizontal extent of the Gulf Stream is about 100 km. The hydrographic section of Fig. 9.21 (top) shows that the vertical extent of the region of strong lateral temperature gradients in the Gulf Stream is about 1 km. If a current of these dimensions is to have a transport of 30 Sv, then it must have a mean speed of some 30 $\mathrm{cm\,s^{-1}}$, roughly in accord with, but somewhat smaller than, direct measurement.[8]

Before going on, it should be mentioned that there is one major current system in Fig. 9.13 that cannot be addressed in the context of Sverdrup theory—the Antarctic Circumpolar Current. The ACC is not in Sverdrup balance because there are no meridional barriers that allow water to be "propped up" between them, hence supporting a zonal pressure gradient and meridional motion. Dynamically, the ACC is thought to have much in common with the atmospheric jet stream discussed in Chapter 8; eddy processes are central to its dynamics.

10.4. EFFECTS OF STRATIFICATION AND TOPOGRAPHY

Our analysis of the wind-driven circulation in Section 10.2 assumed the ocean to have constant density, whereas (see, e.g., Fig. 9.7) the density of the ocean varies horizontally and with depth. In fact, the variation of density with depth helps us out of a conceptual difficulty with our physical interpretation in terms of the Taylor-Proudman theorem on the sphere presented in Section 10.2.3. We described how the Taylor columns of fluid must lengthen in the subtropical gyres to accommodate Ekman pumping from above; where there is no Ekman pumping, they must maintain constant length. But the bottom of the ocean is far from flat. In the Atlantic Ocean, for example, there is a mid-ocean ridge that runs almost the whole length of the ocean rising about 2 km above the ocean bottom (see Fig. 9.1). If the ocean were really homogeneous, water columns simply could not cross this ridge: there would be an enormous, elongated, stagnant Taylor column above it. And yet we have seen that homogeneous theory accounts qualitatively for the observed circulation, including the observation that, for example, the Atlantic subtropical gyre does indeed involve water flowing over the ridge. How does this happen?

---

[8]The discrepancy is due to the fact that the transport of the Gulf Stream can considerably exceed the prediction based on Sverdrup theory, because a portion of the fluid that flows in it recirculates in closed loops that do not extend far into the interior of the ocean.

In most regions the mean circulation, in fact, does not extend all the way to the bottom because, as discussed in Section 9.3.2, the interior stratification of the ocean largely cancels out surface pressure gradients. The thermal wind relation tells us that there can be no vertical shear in the flow of a homogeneous fluid since there are no horizontal density gradients. In the presence of density gradients the constraint of vertical coherence is weakened. Consider Fig. 10.22. We suppose for simplicity that the ocean has two layers of different density ρ_1, ρ_2 (with $\rho_1 > \rho_2$, of course). The density difference produces a stable interface (somewhat like an atmospheric inversion described in Section 4.4) which effectively decouples the two layers. Thus Taylor columns in the upper layer, driven by Ekman pumping/suction from the surface, "feel" the interface rather than the ocean bottom. As long as the interface is above the topography, they will be uninfluenced by its presence. Thus the density stratification "buffers" the flow from control by bottom topography. If we look at the density stratification in the real ocean (Fig. 9.7), we see that most of the density stratification is found in the main thermocline, within a few hundred meters of the surface. Thus the mean wind-driven circulation is largely confined to these upper layers.

10.4.1. Taylor-Proudman in a layered ocean

If we suppose that the ocean is made up of many layers of fluid with slightly differing densities, $\Delta\rho = \rho_1 - \rho_2$, and so on then we can imagine miniature Taylor columns within each homogeneous layer, and because each layer is homogeneous, T-P applies. Let us suppose that each Taylor column has a length d, measured parallel to the axis of rotation within each layer, as sketched in Fig. 10.23. An interior column will "try" to maintain its length. Thus d is constant, as will be the quantity

$$\frac{2\Omega}{d} = \text{constant},$$

which, using Eq. 10-13, can be written:

$$\frac{f}{h} = \text{constant}, \qquad (10\text{-}22)$$

where f is the Coriolis parameter, Eq. 6-42, and h is the thickness of the layer in the direction of gravity measured in the vertical, as sketched in Fig. 10.23.

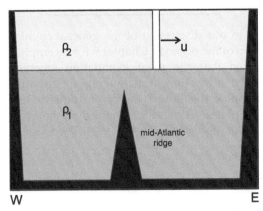

FIGURE 10.22. Upper-layer Taylor column in a two layer idealization of the ocean moving over topography, such as the mid-Atlantic ridge, confined to the lower layer.

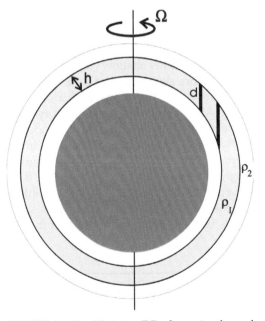

FIGURE 10.23. Miniature T-P columns in a layered fluid. The layers increase in density going downward with $\rho_1 > \rho_2$.

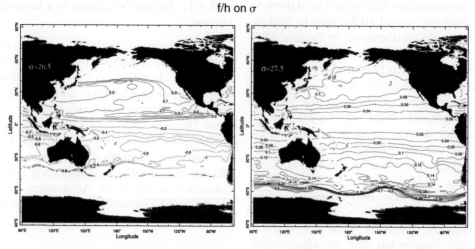

FIGURE 10.24. The quantity $(-f/\rho_{ref}\, \partial\sigma/\partial z)$ where σ is the potential density on a shallow density surface in the Pacific (left; $\sigma = 26.5$; the depth of this surface is plotted in Fig. 9.8) and on a deeper density surface (right; $\sigma = 27.5$). Note that $\partial\sigma/\partial z < 0$ because the ocean is stably stratified and so $(-f/\rho_{ref}\, \partial\sigma/\partial z)$ is positive in the northern hemisphere and negative in the southern hemisphere.

If we focus on one particular layer then, away from the direct influence of Ekman pumping, f/h will be conserved following that column of fluid around the ocean. What do the f/h contours look like in the ocean? Some are plotted in Fig. 10.24 for chosen density surfaces in the Pacific. In fact what is actually plotted is $\left(-f/\rho_{ref}\, \partial\sigma/\partial z\right)$, a "continuous" version of f/h,[9] where σ is the potential density. On the $\sigma = 26.5$ surface in the North Pacific, for example, which is everywhere rather shallow (see Fig. 9.8) the f/h contours sweep around and turn back on themselves. Variations in h dominate over f, allowing the strong circulatory flow of the gyre to persist within the thermocline without violation of the Taylor-Proudman theorem. On the $\sigma = 27.5$ surface, much deeper in the water column (see Fig. 9.7), f/h contours are more zonal and they intersect the coast. On these deeper surfaces, variations in f are much more important than

variations in h. Since interior fluid columns conserve their f/h, there can be little flow deep down, because the columns would run into the coast. So the deep ocean interior is, in the mean, largely quiescent. However, as we shall see in Section 11.3 and 11.4, a weak abyssal circulation does exist, fed by deep western boundary currents driven by convective sources at the poles. This is particularly true in the Atlantic Ocean (see, e.g., Fig. 11.24).

10.5. BAROCLINIC INSTABILITY IN THE OCEAN

In our discussion of the general circulation of the ocean in Chapter 9 it was emphasized that the mean circulation emerges only after long time-averages. Instantaneously the flow is highly turbulent (see, e.g., Figs. 9.19 and 9.22) and the numerical simulation shown in Fig. 9.24. The sloping

---

[9]If h is the thickness of the layer across which the density changes by $\Delta\sigma$, then multiplying f/h by $\Delta\sigma/\rho_{ref}$ we arrive at:

$$\frac{f}{h} \times \frac{\Delta\sigma}{\rho_{ref}} \longrightarrow \frac{f}{\rho_{ref}}\frac{\partial\sigma}{\partial z}$$

as h and $\Delta\sigma$ become small.

isopycnals evident in Fig. 9.7 suggest that there is available potential energy (*APE*) in the ocean's thermocline. Indeed, following the analysis of Section 8.3.2, the ratio of *APE* to kinetic energy in the flow is, from Eq. 8-11, of order $(L/L_\rho)^2$. In the ocean the deformation radius, Eq. 7-23, has a value of $L_\rho = NH/f \simeq 5 \times 10^{-3}\,\mathrm{s}^{-1} \times 10^3\,\mathrm{m}/10^{-4}\,\mathrm{s}^{-1} = 50\,\mathrm{km}$ and so $(L/L_\rho)^2 \sim (1000\,\mathrm{km}/50\,\mathrm{km})^2 = 400$ assuming that the mean flow changes over scales of 1000 km. Thus the potential energy stored up in the sloping density surfaces of the main thermocline represents a vast reservoir available to power the motion. This energy is tapped by baroclinic instability which fills the ocean with small-scale energetic eddies that often mask the mean flow. We can estimate expected eddy length scales and time scales using the arguments developed in Section 8.2.2 (see Eqs. 8-3 and 8-4). A deformation radius of 50 km yields eddy length scales of order 100 km. Growth rates are $\mathcal{T}_{eddy} \sim L_\rho/U = 50\,\mathrm{km}/10\,\mathrm{cm\,s}^{-1} = 5 \times 10^5\,\mathrm{s}$, a week or so, significantly slower than that of atmospheric weather patterns. Eddy lifetimes in the ocean are considerably longer than those of their atmospheric counterparts; weeks and months rather than the few days of a typical weather system.

The mechanism by which *APE* is built up in the ocean is very different from the atmosphere and, as we now describe, involves the collusion of mechanical (wind) and thermodynamic processes. In our discussion of the atmospheric general circulation in Section 8.3, we described how the net radiative imbalance led to warming in the tropics, cooling over the pole and hence the equator-to-pole tilt of θ surfaces and a store of available potential energy which can power the eddy field. In contrast, horizontal density gradients in the interior of the ocean and their associated store of *APE* are produced mechanically, by the same processes that drive the wind-driven currents, as illustrated in Fig. 10.25 . In the anticyclonic gyre, on the equatorward flank of the Gulf Stream, for example, Ekman pumping depresses isopycnals and isotherms by pumping light, warm water downward; poleward of the Gulf Stream, where the curl of the wind stress is cyclonic and there is a cyclonic gyre, Ekman suction lifts up the isopycnals and isotherms, as we saw in Fig. 10.11. Thus a horizontal density and temperature

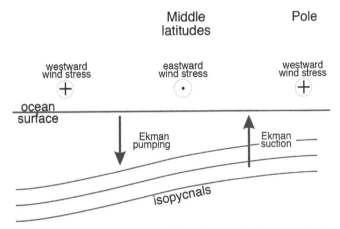

FIGURE 10.25. A schematic of the mechanism by which a large-scale sub-surface horizontal density gradient is maintained in the middle-latitude ocean. Ekman suction draws cold, dense fluid up to the surface in subpolar regions; Ekman pumping pushes warm, light fluid down in the subtropics. The resulting horizontal density gradient supports a thermal wind shear. Its baroclinic instability spawns an energetic eddy field which tends to flatten out the horizontal gradients.

contrast is established across the current. In middle latitudes, then, the oceans have a thermal structure that is somewhat similar to that of the atmosphere, albeit maintained by a very different mechanism. In particular, the isopycnals slope poleward/upward, as sketched in Fig. 10.25, just as do θ surfaces in the atmosphere. Along with horizontal density gradients there is a reservoir of available potential energy which, through the agency of baroclinic instability, can be released into kinetic energy in the form of eddies, just as in GFD Lab XI.[10] Thus oceanic eddies are analogs of midlatitude atmospheric eddies and have many similar properties. It is less clear, however, just how important oceanic eddies are in the "big picture" of the ocean circulation. They are certainly less crucial to poleward heat transport than their atmospheric counterparts. This (as discussed in Chapter 11), is because the wind-driven and thermohaline circulations are quite capable of efficient poleward heat transport even in the absence of eddies. This contrasts with the atmosphere where, in middle latitudes, the zonal flow that would exist in the absence of eddies cannot transport heat, or anything else, in the N-S direction. In this context, the crucial difference between ocean and atmosphere is the E-W confinement of the oceans by continents, permitting western boundary currents and meridional transport. Note, however, that the ACC is not blocked by coasts and the mean flow is west-east: this is one place in the ocean where eddies are known to play a central role in meridional property transport.

10.6. FURTHER READING

Hartmann (1994) briefly reviews aspects of wind-driven ocean circulation theory and describes key observations. Much more detailed theoretical discussions can be found in Gill (1982), Rhines (1993), Pedlosky (1996), and Vallis (2006).

10.7. PROBLEMS

1. Consider the Ekman layer experiment, GFD Lab XII, Section 10.1.2. Assume the lid rotates cyclonically with respect to the turntable, as sketched in Fig. 10.8. In addition to the Ekman layer at the base, there is a second layer at the lid. In this top Ekman layer, the effect of friction is to *drive* the flow, rather than to slow it down as at the base. Draw a schematic diagram showing the balance of forces in the top Ekman layer and use it to deduce the sense of the radial component of the flow. Contrast this with the bottom Ekman layer.

2. Figure 10.11 shows the pattern and magnitude of Ekman pumping acting on the ocean. Estimate how long it would take a particle of fluid to move a vertical distance of 1 km if it had a speed w_{Ek}. If properties are diffused vertically at a rate $k = 10^{-5} \, \mathrm{m^2 \, s^{-1}}$ (typical of the main thermocline), compare this to the implied diffusive timescale. Comment.

3. Use the results of Ekman theory to show that when one adds the meridional volume transport in the Ekman layer (given by Eq. 10-5) to the meridional transport in the geostrophic interior (obtained from Eq. 10-12) one obtains the Sverdrup transport, Eq. 10-17:

$$\frac{1}{\rho_{ref}} \mathbf{M}_{Ek_y} + \int_{-D}^{-\delta} v_g \, dz = \text{Sverdrup transport.}$$

4. Consider the Atlantic Ocean to be a rectangular basin, centered on 35° N,

---

[10]This statement refers to the midlatitude eddies evident in the height variance maps, Fig. 9.19 (bottom). The near-equatorial eddies evident in the surface current variance maps, Fig. 9.22 (bottom), are produced by another mechanism.

of longitudinal width $L_x = 5000\,\text{km}$ and latitudinal width $L_y = 3000\,\text{km}$.

The ocean is subjected to a zonal wind stress of the form

$$\tau_x(y) = -\tau_s \cos\left(\frac{\pi y}{L_y}\right); \quad (10\text{-}23)$$

$$\tau_y(y) = 0;$$

where $\tau_s = 0.1\,\text{N}\,\text{m}^{-2}$. Assume a constant value of $\beta = df/dy$ appropriate to 35°N, and that the ocean has uniform density $1000\,\text{kg}\,\text{m}^{-3}$.

(a) From the Sverdrup relation, Eq. 10-17, determine the magnitude and spatial distribution of the depth-integrated meridional flow velocity in the interior of the ocean.

(b) Using the depth-integrated continuity equation, and assuming no flow at the eastern boundary of the ocean, determine the magnitude and spatial distribution of the depth-integrated zonal flow in the interior.

(c) If the return flow at the western boundary is confined to a width of 100 km, determine the depth-integrated flow in this boundary current.

(d) If the flow is confined to the top 500 m of the ocean (and is uniform with depth in this layer), determine the northward components of flow velocity in the interior, and in the western boundary current.

(e) Compute and sketch the pattern of Ekman pumping, Eq. 10-7, implied by the idealized wind pattern, Eq. 10-23.

5. From your answer to Problem 4, determine the net volume flux at 35° N (the volume of water crossing this latitude in units of Sverdrups: $\text{Sv} = 10^6\,\text{m}^3\,\text{s}^{-1}$):

(a) for the entire ocean, excluding the western boundary current.

(b) for the western boundary current only.

(c) Assume again that the flow is confined to the top 500 m of the ocean. Determine the volume of the top 500 m of the ocean and, by dividing this number by the volume flux you calculated in part (a), come up with a timescale. Discuss what this timescale means.

(d) Assume now that the water in the western boundary current has a mean temperature of 20°C, whereas the rest of the ocean has a mean temperature of 5°C. Show that $\mathcal{H}_{ocean}$, the net flux of heat across 35° N, is

$$\mathcal{H}_{ocean} = \rho_{ref} c_p \mathcal{V} \, \Delta T \, ,$$

where $\mathcal{V}$ is the volume flux you calculated in part (c), and ΔT is the temperature difference between water in the ocean interior and in the western boundary current. Recall that Fig. 5.6 shows that the Earth's energy balance requires a poleward heat flux of around 5×10^{15} W. Calculate and discuss what contribution the Atlantic Ocean makes to this flux.

6. Describe how the design of the laboratory experiment sketched in Fig. 10.18 captures the essential mechanism behind the wind-driven ocean circulation. By comparing Eq. 10-16 with Eq. 10-12, show that the slope of the bottom of the laboratory tank plays the role of the β–effect: that is the bottom slope $\longleftrightarrow (1/\tan\varphi)$ (h/a), where h is the depth of the ocean and a is the radius of the Earth.

7. Imagine that the Earth was spinning in the opposite direction to the present.

(a) What would you expect the pattern of surface winds to look like, and why (read Chapter 8 again)?

(b) On what side (east or west) of the ocean basins would you expect to find boundary currents in the ocean, and why?

   If you live in the southern hemisphere perhaps you are not scratching your head.

8. Use Sverdrup theory and the idea that only western boundary currents are allowed, to sketch the pattern of ocean currents you would expect to observe in the basin sketched on the right in which there is an island. Assume a wind pattern of the form sketched in the diagram.

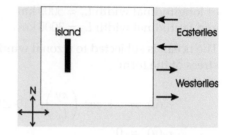

9. Fig. 5.5 shows the observed net radiation at the top of the atmosphere as a function of latitude. Taking this as a starting point, describe the chain of dynamical processes that leads to the existence of anticyclonic gyres in the upper subtropical oceans. Be sure to discuss the key physical mechanisms and constraints involved in each step.

11

The thermohaline circulation of the ocean

11.1. Air-sea fluxes and surface property distributions
    11.1.1. Heat, freshwater, and buoyancy fluxes
    11.1.2. Interpretation of surface temperature distributions
    11.1.3. Sites of deep convection
11.2. The observed thermohaline circulation
    11.2.1. Inferences from interior tracer distributions
    11.2.2. Time scales and intensity of thermohaline circulation
11.3. Dynamical models of the thermohaline circulation
    11.3.1. Abyssal circulation schematic deduced from Taylor-Proudman on the sphere
    11.3.2. GFD Lab XIV: The abyssal circulation
    11.3.3. Why western boundary currents?
    11.3.4. GFD Lab XV: Source sink flow in a rotating basin
11.4. Observations of abyssal ocean circulation
11.5. The ocean heat budget and transport
    11.5.1. Meridional heat transport
    11.5.2. Mechanisms of ocean heat transport and the partition of heat transport between the atmosphere and ocean
11.6. Freshwater transport by the ocean
11.7. Further reading
11.8. Problems

The *thermohaline circulation* is that part of the ocean circulation induced by deep-reaching convection driven by surface buoyancy loss in polar latitudes, as sketched in Fig. 11.1.[1] As we shall see, deep convection in the ocean is highly localized in space and only occurs in a few key locations; in particular, the northern North Atlantic Ocean and around Antarctica. However, the response of the ocean to this localized forcing is global in scale. Giant patterns of meridional overturning circulation are set up that cross the equator and connect the hemispheres together. Unlike the faster wind-driven circulation, which is confined to the top kilometer or so, the thermohaline circulation plays a major role in setting properties of the abyssal ocean. Both

---

[1]The phrase *thermohaline circulation* is widely used but not precisely defined (see Wunsch, 2002). It means different things to different people. Perhaps its most literal interpretation is the circulation of heat and salt in the ocean and thus involves both wind-driven and buoyancy-driven circulation. Here, however, we adopt its more common, narrow usage, to mean the circulation induced by polar convection.

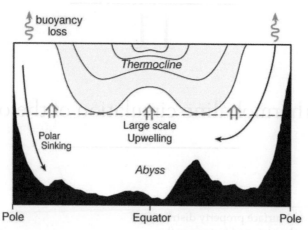

FIGURE 11.1. The deep ocean is ventilated by localized convection at polar latitudes, induced by loss of buoyancy (due to cooling and/or salt input), causing surface waters to sink to depth. Compensating upwelling is thought to occur on the large-scale, as indicated by the vertical arrows at mid-depth.

wind-driven (Chapter 10) and buoyancy-driven circulations play an important role in meridional ocean heat transport. However, because of the very long timescales and very weak currents involved, the thermohaline circulation is much less well observed or understood. We shall see that theory has played a central role in shaping our conception of the likely circulation patterns and mechanisms.

In this chapter, then, we describe the patterns of air-sea heat and fresh water fluxes that drive the thermohaline circulation and go on to discuss inferences of abyssal flow patterns and rates from observations of interior tracer distributions such as salinity and oxygen. We then develop a simple conceptual model and associated laboratory experiments of the thermohaline circulation employing the same dynamical framework, Taylor-Proudman on the sphere, used to discuss the wind-driven circulation in Chapter 10. This model predicts the existence of deep western boundary currents carrying fluid away from their source regions whose existence is confirmed by observations. Finally we discuss the role of the wind-driven and thermohaline circulations in the meridional flux of heat and freshwater.

11.1. AIR-SEA FLUXES AND SURFACE PROPERTY DISTRIBUTIONS

11.1.1. Heat, freshwater, and buoyancy fluxes

As discussed in Chapter 4, atmospheric convection is triggered by warming at the surface. Vertical mass transport is confined to a few regions of strong updrafts driven by deep convection over the warmest oceans and land masses in the tropics, with broader areas of subsidence in between (see the atmospheric mean meridional circulation plotted in Fig. 5.21). In contrast to the atmosphere, the ocean is forced from above by air-sea fluxes. We therefore expect ocean convection to be most prevalent in the *coldest* regions, where the interior stratification is small, most likely at high latitudes in winter where surface density can increase through:

1. direct cooling, reducing temperature and hence increasing density.

2. brine rejection in ice formation, thus increasing salinity (and hence density) of the water immediately below the ice.

Whether a parcel of water sinks depends on its buoyancy anomaly (as described in Section 4.2.1), defined by Eq. 4-3, which we write out again here adopting our oceanographic notation:

$$b = -g\frac{(\sigma - \sigma_o)}{\rho_{ref}}, \qquad (11\text{-}1)$$

where g is the acceleration due to gravity, and $\sigma - \sigma_o$ is the difference between the density of the parcel and its surroundings (see Eq. 9-5). As discussed in Section 9.1.3, the buoyancy of seawater at the surface depends on both the T and S distribution. To determine whether convection will occur we must therefore consider both the flux of heat and freshwater across the ocean surface, which induce T, S and hence buoyancy changes, as well as the ambient, pre-existing stratification of the water column.

The equations governing the evolution of T and S are:

$$\frac{DT}{Dt} = -\frac{1}{\rho_{ref}c_w}\frac{\partial Q}{\partial z} \qquad (11\text{-}2)$$

$$\frac{DS}{Dt} = S\frac{\partial \mathcal{E}}{\partial z}, \qquad (11\text{-}3)$$

where c_w is the heat capacity of water (see Table 9.3), and Q and $\mathcal{E}$ are, respectively, the turbulent vertical flux of heat and freshwater driven by air-sea exchange, convection, ice formation, and vertical mixing in the ocean (as sketched in Fig. 9.11). At the surface $Q = Q_{net}$, the net heat flux across the sea surface (see Eq. 11-5) and $\mathcal{E} = \mathcal{E}_{surface} = E - P$ (evaporation minus precipitation, including that due to river runoff and ice formation processes) is the net fresh water flux across the sea surface. Typically Q and $\mathcal{E}$ decay with depth over the mixed layer from these surface values. Note that the minus sign in Eq. 11-2 ensures that if heat is lost from the ocean ($Q_{net} > 0$), its temperature decreases.

The buoyancy equation can be deduced by taking D/Dt of Eq. 11-1 and using Eqs. 9-5, 11-2, and 11-3 to obtain

$$\frac{Db}{Dt} = -\frac{g}{\rho_{ref}}\left[\frac{\alpha_T}{c_w}\frac{\partial Q}{\partial z} + \rho_{ref}\beta_S S\frac{\partial \mathcal{E}}{\partial z}\right]$$

$$= -\frac{\partial B}{\partial z},$$

where B is the vertical buoyancy flux, allowing us to identify the air-sea buoyancy flux thus:

$$B_{\text{surface}} = \frac{g}{\rho_{ref}}\left(\underbrace{\frac{\alpha_T}{c_w}Q_{\text{net}}}_{\text{thermal}} + \underbrace{\rho_{ref}\beta_S S\,(E-P)}_{\text{haline}}\right)$$

$$(11\text{-}4)$$

where α_T, β_S are defined in Eqs. 9-3 and 9-4, respectively. The units of buoyancy flux are $\text{m}^2\,\text{s}^{-3}$, that of velocity × acceleration. We see that the buoyancy flux is made up of both thermal and haline components.

The net heat flux through the sea surface is itself made up of a number of components:

$$Q_{\text{net}} = \underbrace{Q_{SW}}_{\text{shortwave}} + \underbrace{Q_{LW}}_{\text{longwave}} + \underbrace{Q_S}_{\text{sensible flux}}$$

$$+ \underbrace{Q_L}_{\text{latent flux}} \qquad (11\text{-}5)$$

Estimates of the various terms from observations are shown in Figs. 11.2, 11.3, and 11.4. The units are in W m^{-2}.

The shortwave flux is the incoming solar radiation that reaches the sea surface and penetrates the ocean (the ocean has a low albedo; see Table 2.2), warming it down to a depth of 100–200 meters, depending on the transparency of the water. The longwave flux is the net flux of longwave radiation at the sea surface due to the radiation beamed out by the ocean according to the blackbody law, Eq. 2-2, less the "back radiation" from the atmospheric cloud and water vapor layer (see Chapter 2).

The sensible heat flux is the flux of heat through the sea surface due to turbulent exchange. It depends on the wind speed and

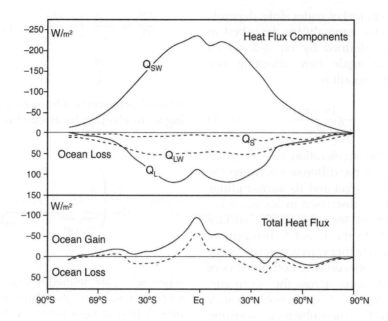

FIGURE 11.2. Upper: Zonal averages of heat transfer to the ocean by insolation Q_{SW}, and loss by long-wave radiation Q_{LW}, sensible heat flux Q_S, and latent heat flux Q_L, calculated by DaSilva, Young, and Levitus (1995) using the COADS data set. Lower: Net heat flux through the sea surface calculated from the data above (solid line) and net heat flux constrained to give heat and fresh water transports by the ocean that match independent calculations of these transports. The area under the lower curves ought to be zero, but it is 16 W m^{-2} for the unconstrained case (solid line) and -3 W m^{-2} for the constrained case (dotted line). From Stewart (2005).

the air-sea temperature difference according to the following (approximate) formula:

$$Q_S = \rho_{air} c_p c_S u_{10} \left(SST - T_{air}\right), \qquad (11\text{-}6)$$

where ρ_{air} is the density of air at the surface, c_S is a stability-dependent bulk transfer coefficient for heat (which typically has a value of about 10^{-3}), c_p is the specific heat of air, T_{air} and u_{10} are, respectively, the air temperature and wind speed at a height of 10 m and SST is the sea surface temperature. Note that if $SST > T_{air}$, $Q_S > 0$ and the sensible heat flux is out of the ocean which therefore cools. The global average temperature of the surface ocean is indeed 1 or 2 degrees warmer than the atmosphere and so, on the average, sensible heat is transferred from the ocean to the atmosphere; see the zonal-average curves in Fig. 11.2.

The latent heat flux is the flux of heat carried by evaporated water. The water vapor leaving the ocean eventually condenses into water droplets forming clouds, as described in Chapter 4 and sketched in Fig. 11.5, releasing its latent heat of vaporization to the atmosphere. The latent heat flux depends on the wind-speed and relative humidity according to Eq. 11-6,

$$Q_L = \rho_{air} L_e c_L u_{10} \left(q_*(SST) - q_{air}\right), \qquad (11\text{-}7)$$

where c_L is a stability-dependent bulk transfer coefficient for water vapor (which, like c_S in Eq. 11-6, typically has a value $\sim 10^{-3}$), L_e is the latent heat of evaporation, q_{air} is the specific humidity (in kg vapor per kg air), and q_* is the specific humidity at saturation which depends on SST (see Section 4.5.1). High winds and dry air evaporate much more water than weak winds and moist air. Evaporative energy loss rises steeply with

Terms in Net Upward Surface Heat Flux (W/m²)

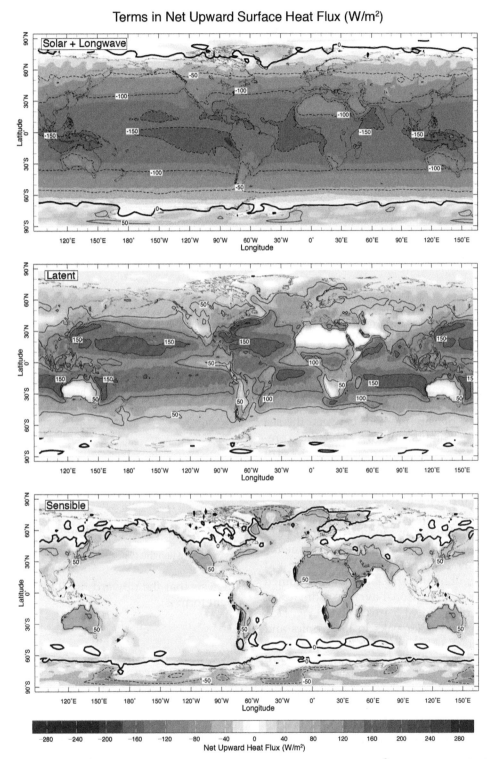

FIGURE 11.3. Global map of $Q_{SW} + Q_{LW}$, Q_L and Q_S across the sea surface in Wm^{-2}. Areas in which the fluxes are upward, into the atmosphere, are positive and shaded green; areas in which the flux is downward, into the ocean, are negative and shaded brown. Contour interval is 50 Wm^{-2}. From Kalnay et al. (1996).

Net Upward Heat Flux (W/m²)

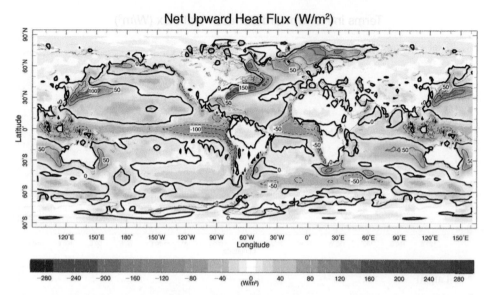

FIGURE 11.4. Global map of net annual-mean constrained heat flux, Q_{net}, across the sea surface in Wm^{-2}. Areas in which the fluxes are upward, into the atmosphere, are shaded green; areas in which the flux is downward, into the ocean, are shaded brown. Contour interval is 50 Wm^{-2}. From Kalnay et al. (1996).

water temperature due to the sensitivity of saturation vapor pressure to temperature (see Fig. 1.5) and the concomitant increase in vapor density gradient between the sea and air. At higher latitudes where these gradients are smaller, evaporative transfer is of lesser importance and sensible heat transfer, which can be of either sign, becomes more important (see Fig. 11.3).

It is very difficult to directly measure the terms that make up Eq. 11-5. Estimates can be made by combining *in situ* measurements (when available), satellite observations, and the output of numerical models constrained by observations. Zonal-average estimates of each term in Eq. 11-5 are shown in Fig. 11.2. We see that Q_{SW} peaks in the tropics and is somewhat balanced by the evaporative processes Q_L and outgoing longwave radiation, Q_{LW}, with Q_S making only a small contribution. Note that solar radiation is the only term that warms the ocean. The major source of cooling is Q_L. Its typical magnitude can be estimated by noting that the net upward transfer of H_2O in evaporation must equal precipitation (see Fig. 11.5). This

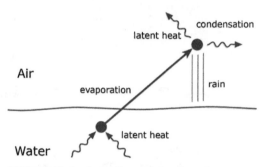

FIGURE 11.5. Latent heat is taken from the ocean to evaporate water that is subsequently released into the atmosphere when the vapor condenses to form rain.

suggests that the upward flux of energy in latent form is:

$$Q_L \sim L_e \frac{dm}{dt},$$

where L_e is the latent heat of evaporation and m is the mass of water falling per square meter (note that the precipitation rate $P = (1/\rho_{ref})(dm/dt)$ and has units of velocity, most often expressed in meters per year). Inserting numerical values from Table 9.3, the previous equation yields $71 W\,m^{-2}$ for every $m\,y^{-1}$ of rainfall. This is broadly in accord with Fig. 11.2, because the annual mean rainfall rate is around $1\,\mathrm{m\,y^{-1}}$.

Note that if the ocean is not to warm up or cool down in the long-run, the net air-sea heat flux integrated over the surface of the ocean, the area beneath the continuous black line in the lower panel in Fig. 11.2, should be zero. In fact, due to uncertainties in the data, it is $16\,\mathrm{W\,m^{-2}}$. That this is an unrealistically large net flux can easily be deduced as follows. If the global ocean were heated by an air-sea flux of magnitude Q_{net} to a depth h over time Δt, it would warm (assuming it to be well mixed) by an amount ΔT given by, see Eq. 11-2:

$$\Delta T = \frac{Q_{net}\Delta t}{h\rho_{ref}c_w},$$

or $0.75°\mathrm{C}$ for every $1\,\mathrm{W\,m^{-2}}$ of global imbalance sustained for a 100 year period, assuming $h = 1\,\mathrm{km}$, and the data in Table 9.3. This is a full $12°\mathrm{C}$ if the imbalance is $16\,\mathrm{W\,m^{-2}}$! The "observed" warming of the ocean's thermocline during the second half of the 20th century is an almost imperceptible few tenths of a $°\mathrm{C}$. Clearly Q_{net} integrated over the global ocean must, in reality, be very close to zero. Thus, in the lower panel of Fig. 11.2, a constrained (adjusted) zonal-average estimate of Q_{net} is shown by the dotted line, the area under which is close to zero. This is much the more likely distribution. We now see that the ocean gains heat in the tropics and loses it at high latitudes, as seems intuitively reasonable.

The geographical distribution of Q_{net} and $E - P$ are shown in Figs. 11.4 and 11.6. Both fields are constrained to have near-zero global integral. We generally see cooling of the oceans in the northern subtropics and high latitudes, and particularly intense regions of cooling over the Kuroshio and Gulf Stream extensions, exceeding $100\,\mathrm{W\,m^{-2}}$ in the annual mean. These latter regions are places where, in winter, very cold air blows over the ocean from the adjacent cold land-masses and where the western boundary currents of the ocean carry warm fluid from the tropics to higher latitudes. Such intense regions of heat loss are not seen in the southern hemisphere because the juxtaposition of land and sea is largely absent. Moreover, because the 'fetch' of ocean is so much larger there, the air-sea temperature difference and hence air-sea flux is much reduced. In the tropics we observe warming of the ocean due to incoming shortwave solar radiation.

The pattern of $E - P$, Fig. 11.6, shows excess of evaporation over precipitation in the subtropics, creating the region of high salinity seen in Fig. 9.4. Precipitation exceeds evaporation in the tropics (in the rising branch of the Hadley circulation), and also in high latitudes. This creates anomalously fresh surface water (cf. see Fig. 9.4) and acts to stabilize the water column.

Another perspective on the forcing of thermohaline circulation is given by Fig. 11.7 which shows the zonally averaged air-sea buoyancy flux $B_{surface}$ defined by Eq. 11-4, and the thermal and haline components that make it up. Buoyancy loss from the ocean peaks in the subtropics, yet we do not observe deep mixed layers at these latitudes (see Fig. 9.10). Evidently here buoyancy loss is not strong enough to "punch through" the strong stratification of the main thermocline (cf. Fig. 9.7). It is clear from Fig. 11.7 that the haline component stabilizes the polar oceans (excess of precipitation over evaporation at high latitudes), but is generally considerably weaker in magnitude than the thermal contribution to the buoyancy flux. Nevertheless, in the present climate, the weak stratification of the polar oceans enables the buoyancy lost there to trigger deep-reaching convection which ventilates the abyssal ocean. In past climates it is thought that the freshwater supply to the polar oceans may have been different (due to, for example, enhanced atmospheric moisture transport in a warm climate or melting of polar ice releasing fresh water) and could be an important driver of climate variability (see Section 12.3.5).

Global Precipitation and Evaporation (m/y)

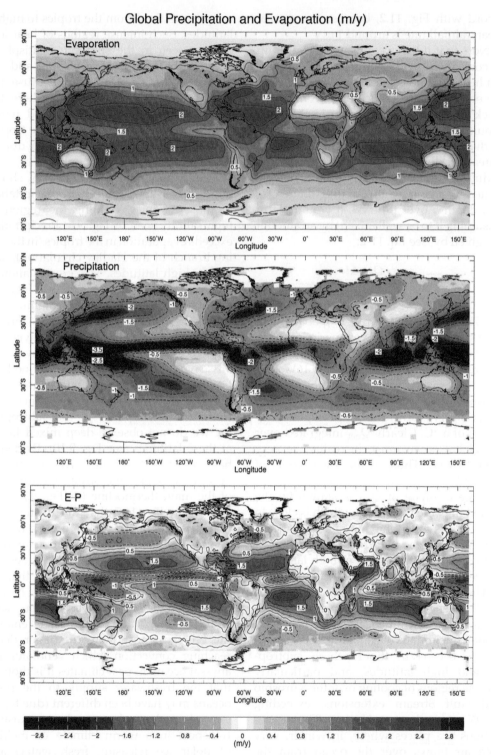

FIGURE 11.6.   A map of anual-mean evaporation (*E*), precipitation (*P*), and evaporation minus precipitation (*E* − *P*) over the globe. In the bottom map, *E* > *P* over the green areas; *P* > *E* over the brown areas. The contour interval is $0.5 \ \mathrm{m \, y^{-1}}$. From Kalnay et al. (1996).

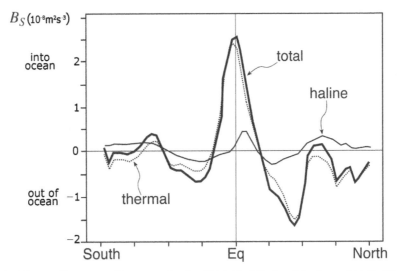

FIGURE 11.7.   The zonally-averaged buoyancy forcing (thick black line) and the thermal (dotted line) and haline (thin line) components that make it up, Eq. 11-4 in units of $m^2 s^{-3}$. Courtesy of Arnaud Czaja (Imperial College). Note that a heat flux of 50 W m^{-2} is (roughly) equal to a buoyancy flux of 2×10^{-8} m^{-8} m^2 s^{-3}. Data from Kalnay et al. (1996).

Finally we note that Figs. 11.4, 11.6, and 11.7 only show components of the buoyancy flux associated with air-sea exchange; they ignore the effects of, for example, brine rejection in ice formation which is thought to be a key mechanism in the creation of dense water around Antarctica. Here ice is blown equatorwards in the surface Ekman layers (to the left of the wind in the southern hemisphere) leaving salty and hence dense water behind, which is susceptible to convection.[2]

11.1.2. Interpretation of surface temperature distributions

The observed SST distribution shown in Figs. 9.3, 9.14, 9.15, and 9.16 is maintained by (i) heat flux through the sea surface, (ii) heat flux through the base of the mixed layer by mixing and upwelling/downwelling, and (iii) horizontal advection.

Let us first consider tropical latitudes. Fig. 11.8 illustrates important aspects of the

air-sea flux contribution there. We take a mean air temperature $T_{air} = 27°C$ which yields a specific humidity of about 15 g kg^{-1} if the relative humidity is 70%. Assuming a mean wind-speed of 3 m s^{-1}, the contributions of the Q_{SW}, Q_{LW}, Q_S, and Q_L are plotted as functions of SST using bulk formulae, Eqs. 11-6 and 11-7. The shortwave incoming radiation reaching the surface is assumed to be 341 W m^{-2} and is offset by back radiation Q_{LW}, sensible heat loss Q_S (when SST > 27°C) and evaporative loss Q_L. Because Q_L rises very steeply with temperature, there is a natural limit on the tropical SST that depends on the radiation available and the wind speed. For the parameters chosen, energy supply and loss balance at an SST of about 30°C. At temperatures only slightly above this limit, evaporative losses far exceed the possible input of radiation, limiting SST. This evaporative feedback explains why tropical temperatures are so stable. In fact some energy is carried down

---

[2]At low temperatures typical of polar oceans, α_T varies strongly with T and p and becomes smaller at lower temperatures and increases with depth, especially in the Weddell and Greenland Seas (see Table 9.4). The excess acceleration of a parcel resulting from the increase of α_T with depth (known as the thermobaric effect), can result in a destabilization of the water column if the displacement of a fluid parcel (as a result of gravity waves, turbulence, or convection) is sufficiently large.

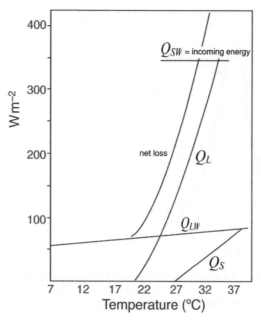

FIGURE 11.8. Contributions of the air-sea flux terms: Q_{SW}, the incoming solar radiation, and Q_{LW}, Q_S, and Q_L making up the net loss term (see Eq. 11-5), plotted as a function of SST using bulk formulae, Eqs. 11-6 and 11-7.

into the interior ocean and the solar energy is absorbed in the top few meters rather than right at the surface. Nevertheless, the limit on SST due to surface processes illustrated in Fig. 11.8 is very much at work in regulating surface temperatures in the tropics.

In the subtropics, evaporation is the principal heat loss term (see Fig. 11.3) and varies comparatively little throughout the year. Due to the seasonal cycle of insolation, there is a net deficit of energy in the winter and a net excess in the summer. Along with cooling to the atmosphere, upwelling of deep cold fluid in the subpolar gyres helps to maintain low surface temperatures in polar latitudes. At very high latitudes there is a net buoyancy loss out of the ocean in regions of very weak stratification. This, as discussed below, can trigger deep-reaching convection creating very deep mixed layers that can ventilate the abyss.

Small-scale turbulent mixing in the ocean allows temperature changes at the surface to be communicated to deeper layers,

as sketched in Fig. 9.11. Wind-generated turbulence often creates almost isothermal conditions in the top 20–50 m of the ocean with a sharp discontinuity at the base. Another major factor in determining the distribution of surface properties is the pattern of Ekman pumping shown in Fig. 10.11. For example, in the equatorial belt the trade winds drive Ekman transport away from the equator drawing cold water up from below. The surface temperature can be actually lower directly along the equator than immediately to the north or south!

Such considerations of energy balance at the surface and wind-induced Ekman pumping and turbulence, provide a first-order explanation of many of the major features of the SST distribution and its seasonal variation shown in Fig. 9.3, and reflect variations in the available solar radiation modified by air-sea fluxes of sensible and latent heat, advection (both horizontal and vertical), and mixing of properties with deeper layers. These processes are frequently interpreted in terms of vertical one-dimensional models which attempt to represent the turbulent transfer of heat and buoyancy through the mixed layer, by wind and convectively driven turbulence, and its communication with the ocean below through entrainment and vertical motion.

It is also worthy of note that the North Atlantic is warmer than both the North Pacific and the Southern Ocean at the same latitude in the same season. The warmth of the Atlantic relative to the Pacific is thought to be largely a consequence of differences in the surface wind patterns. In the North Atlantic the zero wind-stress curl line (along which the interior extension of the western boundary currents tends to flow) slants much more than in the Pacific, allowing the Gulf Stream to carry warm surface waters into far northern latitudes in the Atlantic.

11.1.3. Sites of deep convection

A comparison of Figs. 11.4 and 11.6 with Fig. 9.10 shows that there is no direct relationship between the pattern of air-sea buoyancy forcing and the pattern of mixed layer depth. This is because the strength of the underlying stratification plays an important role in "preconditioning" the ocean for convection. The deepest mixed layers are seen in the polar regions of the winter hemisphere and are particularly deep in the Labrador and Greenland Seas of the North Atlantic, where they can often reach depths well in excess of 1 km. Here the ambient stratification of the ocean is sufficiently weak and the forcing sufficiently strong to trigger deep-reaching convection and bring fluid from great depth into contact with the surface. Note that deep mixed layers are notably absent in the North Pacific Ocean. Waters at the surface of the North Pacific are relatively fresh (note, for example, how much fresher the surface of the Pacific is than the Atlantic in Fig. 9.4) and remain buoyant even when cooled. Deep mixed layers over wide areas of the southern oceans in winter are also observed, but they are considerably shallower than their counterparts in the northern North Atlantic.

Evidence from observations of mixed layer depth and interior tracer distributions reviewed below in Section 11.2.1, suggest that convection reaches down into the abyssal ocean only in the Atlantic (in the Labrador and Greenland Seas) and also in the Weddell Sea, as marked in Fig. 11.9. These sites, despite their small areal extent, have global significance in setting and maintaining the properties of the abyss. They are thought to play a major role in climate variability (see Section 12.3.5). Observations suggest that there are certain common features and conditions that predispose these regions to deep-reaching convection. First, there is strong atmospheric forcing because of thermal and/or haline surface fluxes (see Fig. 11.4). Thus open ocean regions adjacent to boundaries are favored, where cold, dry winds from land or ice surfaces blow over water inducing large sensible and latent heat and moisture fluxes. Second, the stratification beneath the surface-mixed layer is weak, made weak perhaps by previous convection. And third, the weakly stratified underlying waters are brought up toward

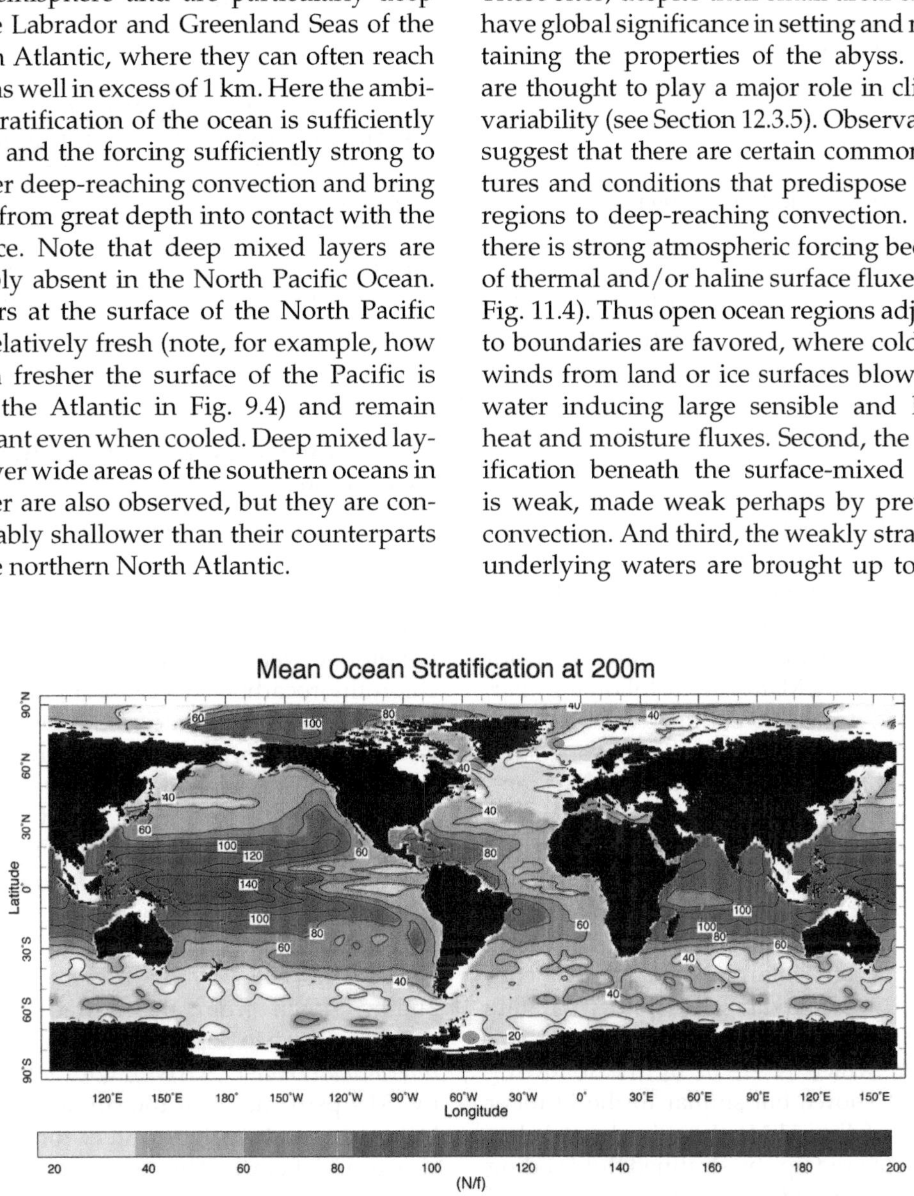

FIGURE 11.9.　The annual-mean stratification of the ocean at a depth of 200 m, as measured by N/f_{ref}: i.e., the buoyancy frequency, Eq. 9-6, normalized by a reference value of the Coriolis parameter, $f_{ref} = 10^{-4} \, \text{s}^{-1}$. Note that $N/f_{ref} \lesssim 20$ in regions where deep mixed layers are common (cf. Fig. 9.10). Sites of deep-reaching convection are marked in the Labrador Sea, the Greenland Sea, the Western Mediterranean and the Weddell Sea.

the surface so that they can be readily and directly exposed to buoyancy loss from the surface. This latter condition is favored by cyclonic circulation associated with density surfaces, which "dome up" to the surface, drawn upward by Ekman suction over sub-polar gyres (Fig. 10.11). In places where deep convection is occurring, weak vertical buoyancy gradients are observed (see Fig. 11.9, which plots N, Eq. 9-6, at a depth of 200 m over the global ocean) and isopycnals dome up toward the surface (see Fig. 9.7).

Observations of deep convection

Observations at sea during deep convection are rare because of the inhospitable conditions in which wintertime convection occurs (see the photographs taken from a research vessel in the Labrador Sea in winter, presented in Fig. 11.10). The best-observed region of deep convection is the Labrador Sea. Fig. 11.11 shows sections of σ and N through the Labrador Sea just before (fall 1996) and during/after winter-time deep convection. In October we see a near-surface stratified layer, some 500 m or so in depth, overlaying a relatively well-mixed intermediate layer, formed by prior convection. By March of the following year, however, convection triggered by cooling from the surface has broken through the stratified layer, mixing intermediate and surface fluid and leading to a well mixed patch of some 200 km in horizontal extent by, in places, 1500 m in depth. Just as in our studies of convection in water heated from below described in Section 4.2.4, cooling of the ocean from above results in convection which returns the fluid to a state of neutral stability with a well mixed column in which $N \rightarrow 0$. By the following fall (not shown but similar to the October section of Fig. 11.11) the mixed patch has been "covered up" by stratified fluid sliding over from the side. The water mass formed by convection in the previous winter, now exists as a subsurface bolus of fluid which in

the subsequent months and years is drawn into the interior of the ocean. The process has been likened to a chimney, but is perhaps better described as analogous to the way in which a snake swallows an egg, as sketched in the schematic diagram setting out the phases of deep convection in the ocean shown in Fig. 11.12.

11.2. THE OBSERVED THERMOHALINE CIRCULATION

The time-mean abyssal flow in the ocean is so weak that it cannot be measured directly. However abyssal circulation, and the convective processes forcing it, leaves its signature in the distribution of water properties, from which much can be inferred.

11.2.1. Inferences from interior tracer distributions

Water masses modified by deep convection are tagged with T and S values characteristic of their formation region, together with other tracers, such as tritium from the atomic weapon tests of the 1960s and chlorofluorocarbons (CFCs) from industrial and household use. Tracers can be tracked far from their formation region, revealing interior pathways through the ocean.

Zonal-average sections of T and S across the Atlantic Ocean are shown in Fig. 11.13; see also the hydrographic section along 25° W in the Atlantic shown in Fig. 9.9. We see three distinct layers of deep and abyssal ocean water, fed from different sources. Sliding down from the surface in the southern ocean to depths of 1 km is *Antarctic Intermediate Water* (AAIW), with low salinity (34.4 psu) and, near the surface, slightly lower temperature than water immediately above and below. This water appears to originate from about 55° S and is associated with regions of deep mixed layers in the circumpolar ocean seen in Fig. 9.10. At a

FIGURE 11.10. The Woods Hole ship KNORR cuts through harsh Labrador Sea conditions during the winter Labrador Sea Deep Convection Experiment (Feb–Mar, 1997) taking observations shown in Fig. 11.11. Waves such as those shown at the top caused continual ice build-up on the ship, as can be seen at the bottom. Courtesy of Bob Pickart, WHOI.

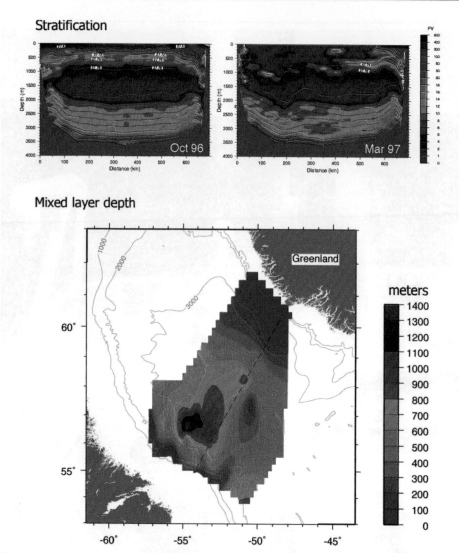

FIGURE 11.11.   Top: Sections of potential density, σ (contoured), and stratification, $\partial\sigma/\partial z$ (colored), across the Labrador Sea in October, 1996, before the onset of convection, and in March, 1997, after and during wintertime convection. Purple indicates regions of very weak stratification. Bottom: A horizontal map of mixed layer depth observed in Feb-Mar, 1997, showing convection reaching to depths in excess of 1 km. The position of the sections shown at the top is marked by the dotted line. Courtesy of Robert Pickart, WHOI.

depth of 2 km or so—indeed, filling most of the Atlantic basin—is *North Atlantic Deep Water* (NADW), with high salinity (34.9 psu) originating in high northern latitudes, but identifiable as far south as 40° S and beyond. At the very bottom of the ocean is the *Antarctic Bottom Water* (AABW), less saline but colder (and denser) than NADW. Together, these give us a picture of

a multilayered pattern of localized sinking and horizontal spreading of the dense water, which were represented schematically by the arrows in Fig. 11.1.

Another useful tracer of the circulation is dissolved oxygen. Surface waters are near saturation in oxygen content (in fact, they are slightly super-saturated). As the water leaves the surface (the source of oxygen), its

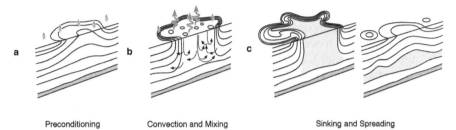

Preconditioning   Convection and Mixing   Sinking and Spreading

FIGURE 11.12. Schematic diagram of the three phases of open-ocean deep convection: (a) preconditioning, (b) deep convection and mixing, and (c) sinking and spreading. Buoyancy flux through the sea surface is represented by curly arrows, and the underlying stratification/outcrops are shown by continuous lines. The volume of fluid mixed by convection is shaded. From Marshall and Schott (1999).

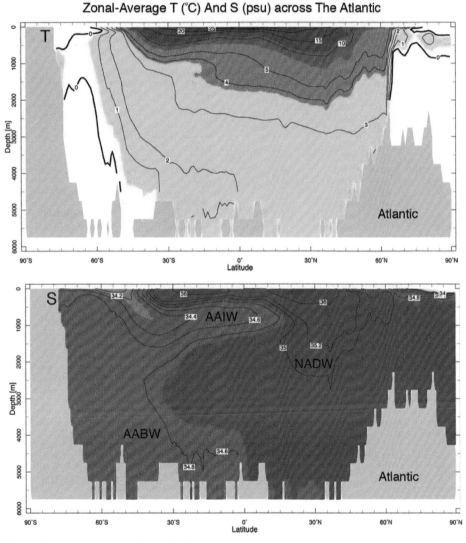

FIGURE 11.13. Zonal average (0° → 60° W) temperature (top) and salinity (bottom) distributions across the Atlantic Ocean. Antarctic Intermediate Water (AAIW), Antarctic Bottom Water (AABW), and North Atlantic Deep Water (NADW) are marked. Compare this zonal-average section with the hydrographic section along 25° W shown in Fig. 9.9.

oxygen content is slowly used up by biological activity. Hence, oxygen content gives us a sort of clock by which we can get a feel for the "age" of the water (i.e., the time since it left the surface); the lower the content, the "older" the water. Oxygen content (expressed as a percentage of saturation) for the Atlantic and Pacific Oceans is shown in Figs. 11.14, top and bottom, respectively.

In the Atlantic, water in the deep ocean shows a progressive aging from north to south, implying the dominant source is in the far north. However, the water is generally "young" (oxygen saturation > 60%) everywhere except at depths shallower than 1 km in low latitudes, where "old" water is (we infer) slowly upwelling from below (cf. the isopycnals in Fig. 9.7). That

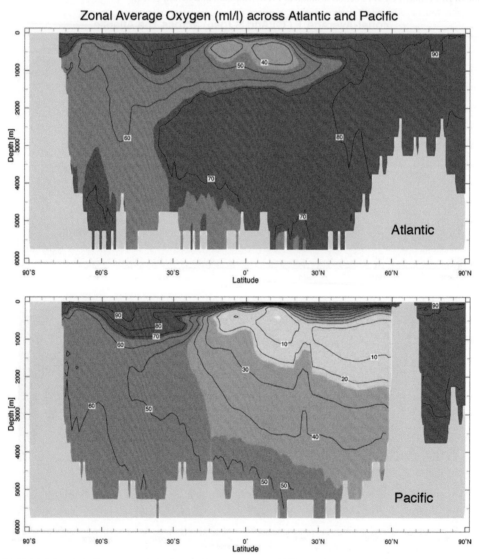

FIGURE 11.14. Zonally averaged oxygen saturation (in mll^{-1}) in the Atlantic (0°–60° W) and Pacific (150–190° W) oceans.

water is relatively young near the Antarctic coast, around 40°–50°S and, especially, in high northern latitudes, is evidence that surface waters are being mixed down in these regions of the Atlantic. In contrast, the Pacific Ocean cross-section, Fig. 11.14 (bottom), shows young water only near the Antarctic. Deep water in high northern latitudes has very low oxygen content, from which we infer that there is no sinking of surface waters in the North Pacific, except in the Arctic basin.

11.2.2. Time scales and intensity of thermohaline circulation

Water whose properties are set (oceanographers use the term "formed") at the source regions marked in Fig. 11.9 must spread out before slowly upwelling back to the surface to complete the circuit of mass flow. Estimates of the strength of the major source, NADW, are about 14 Sv (see Section 11.4). Using this source rate we can make several estimates of parameters indicative of the strength of the circulation. The area of the Atlantic Ocean is about $10^{14}\,\mathrm{m}^2$. The depth of the ocean ventilated by the surface sources is perhaps 3 km (see Fig. 11.13). So one estimate of the time scale of the overturning circulation is $\tau = \frac{\text{ocean volume}}{\text{volume flux}} = \frac{10^{14}\,\mathrm{m}^2 \times 3 \times 10^3\,\mathrm{m}}{1.4 \times 10^7\,\mathrm{m}^3\,\mathrm{s}^{-1}} \simeq 700\,\mathrm{y}$. The net horizontal flow velocity in the deep ocean must be about $v = \frac{\text{volume flux}}{\text{depth} \times \text{width}} = \frac{1.4 \times 10^7\,\mathrm{m}^3\,\mathrm{s}^{-1}}{3 \times 10^3\,\mathrm{m} \times 5 \times 10^6\,\mathrm{m}} \simeq 10^{-3}\,\mathrm{m\,s}^{-1}$. If compensating upwelling occupies almost all of the ocean basin, the upwelling velocity must be about $w = \frac{\text{volume flux}}{\text{area of ocean}} = \frac{1.4 \times 10^7\,\mathrm{m}^3\,\mathrm{s}^{-1}}{10^{14}\,\mathrm{m}^2} \simeq 4\,\mathrm{m\,y}^{-1}(!!)$, ten times smaller than typical Ekman pumping rates driven by the wind (cf. Fig. 10.11).

Thus the interior abyssal circulation is very, very weak, so weak that it is all but impossible to observe directly. Indeed progress in deducing the likely pattern of large-scale abyssal circulation has stemmed as much from the application of theory as direct observation, as we now go on to discuss.

11.3. DYNAMICAL MODELS OF THE THERMOHALINE CIRCULATION

11.3.1. Abyssal circulation schematic deduced from Taylor-Proudman on the sphere

Because of the paucity of direct observations of abyssal flow, theory has been an invaluable guide in deducing likely circulation patterns. The starting point for a theoretical deduction are two important inferences from the observations discussed above:

1. Dense water is formed at the surface in small, highly localized regions of the ocean in polar seas. Thus the abyssal circulation seems to be induced by *local* sources, marked in Fig. 11.9. But for every particle of water that sinks, one must return to the surface. Property distributions reviewed in Section 11.2.1 suggest that the return branch does not occur in one, or a few, geographical locations. It seems reasonable to suppose, therefore, that there is widespread compensating upwelling on the scale of the basin, as sketched in our schematic diagram used to introduce this chapter, Fig. 11.1.

2. The deep flow is sluggish with very long timescales. It will therefore be in geostrophic, hydrostatic, and thermal wind balance. Moreover Eq. 10-12 will be appropriate, telling us that if columns of fluid are stretched or squashed then meridional motion results.

In the region of upwelling (which, we can surmise, is almost all the ocean) fluid columns in the homogeneous abyss must behave like Taylor columns and the geostrophic flow on the planetary scale will

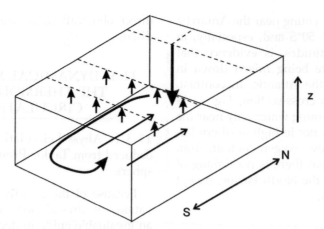

FIGURE 11.15.   Water is imagined to sink in a localized region in polar latitudes (the single long arrow pointing downward) with compensating upwelling distributed over the basin (the many short arrows pointing upward). Taylor columns in the interior are therefore stretched and move toward the pole, satisfying Eq. 11-8. The poleward motion over the interior is balanced by equatorward flow in a western boundary current.

be divergent. Thus integrating Eq. 10-12 from the bottom up to mid-depth in the water column, we find:

$$\int_{bottom}^{mid\text{-}depth} v\,dz = \frac{f}{\beta} w_{mid\text{-}depth} > 0 \qquad (11\text{-}8)$$

where we have set $w_{bottom} = 0$ (assuming a flat bottom) and supposed $w_{mid\text{-}depth} > 0$, directed upwards, as sketched in Figs. 11.1 and 11.15. Thus fluid columns must move poleward, *towards* the deep water mass source!

We can interpret this result in terms of our discussion of T-P on the sphere in Chapter 10. There we found that Taylor columns in the thermocline of the subtropical gyres move equatorward to accommodate Ekman pumping driven from above by the curl of the wind stress. In the present context we deduce that abyssal columns in the interior of the ocean must move *poleward* in response to upwelling at mid-depth, compensating for polar sinking. Referring back to Fig. 10.15b and simply reversing the sign of the arrows, we see that by moving poleward, Taylor columns will stretch in a direction parallel to Ω relative to the spherical, horizontal surfaces, producing the required upwelling.

How can the interior poleward circulation be closed? The answer is that, once again, we have to rely on western boundary currents to close the circuit, as sketched in Fig. 11.15 and discussed in Section 11.3.3. Thus, for example, the poleward interior flow in the North Atlantic is returned to the south, along with source waters, in a *western* boundary current. Such considerations led Henry Stommel[3] to his classic schematic of deep thermohaline circulation

[3] Henry Melson Stommel, 1920–1992, was the most influential dynamical oceanographer of all time. A theorist who made observations at sea and also worked in the laboratory, Stommel shaped the field and contributed many of the seminal ideas that led to the development of our understanding of the wind-driven (Chapter 10) and thermohaline circulation of the ocean (Chapter 11). Stommel joined the Woods Hole Oceanographic Institution (WHOI) in 1944 and left in 1959 to become a professor, first briefly at Harvard, then at MIT before returning to WHOI (his spiritual home) in 1979.

FIGURE 11.16. Schematic of the abyssal circulation in the ocean deduced by Stommel (1958) based on consideration of Eq. 10-12 and its vertical integral, Eq. 11-8. He imagined that the abyssal ocean was driven by sinking in two regions, the Labrador-Greenland Sea and the Weddell Sea, represented by the large black dots. Upwelling over the interior of the ocean led to interior poleward motion with return flows in western boundary currents. The shaded areas represent regions of elevated topography (cf. Fig. 9.1).

shown in Fig. 11.16 driven by sinking in the northern North Atlantic and the Weddell Sea in Antarctica. Note that the interior flow is directed poleward everywhere in accord with Eq. 11-8. Fluid from the sinking regions is carried away in western boundary currents. Thus Stommel hypothesized the existence of a boundary current in the North Atlantic flowing southward at depth, beneath the Gulf Stream! As we shall discuss in Section 11.4, this was subsequently confirmed by direct observation.

11.3.2. GFD Lab XIV: The abyssal circulation

A laboratory experiment can be used to vividly illustrate the dynamical ideas that underpin the abyssal circulation schematic, Fig. 11.16, proposed by Stommel. It was first carried out in Woods Hole by Stommel and collaborators. The apparatus is shown in Fig. 11.17 and described in the legend. We use the same tank as in the wind-driven circulation experiment (GFD Lab XIII, described in Section 10.2.4) with a sloping base to represent spherical effects, but

without the rotating disc above. Dyed water is introduced very slowly, through a diffuser in the shallow end of the tank (representing polar latitudes). The surface of the water thus rises, stretching interior Taylor columns, which therefore move toward the shallow (poleward) end of the tank. Boundary currents develop on the poleward and western boundaries supplying fluid to the deep end (equatorward) of the tank, just as sketched in Fig. 11.15.

Development of the flow and the existence of boundary currents can be clearly seen by charting the evolution of the dye, as shown in Fig. 11.18. Rather than moving away in to the interior directly from the source, we observe westward flow along the northern boundary and then southward flow in a western boundary current. The tank thus fills up via boundary currents which ultimately supply the interior. Note also how turbulent the flow patterns are. Even in this controlled laboratory experiment, in which fluid was introduced very slowly, we do not observe steady, laminar flow. Rather, there is much time-dependence and recirculation of fluid between the interior and

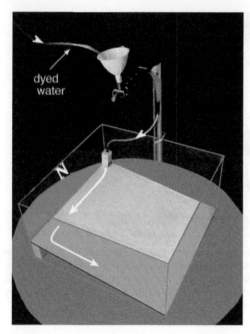

FIGURE 11.17. Apparatus used to illustrate the driving of deep ocean circulation by localized sinking of fluid. A sloping base is used to represent the influence of sphericity on Taylor columns as in GFD XIII. The 50 cm square tank is filled with water and set rotating anticlockwise at a rate of $\Omega = 5$ rpm. (The sense of rotation is thus representative of the northern hemisphere). Dyed water, supplied via a funnel from an overhead bucket, flows slowly into the tank through a diffuser located in the "northeast" corner at a rate of typically $20\,\mathrm{cm}^3\,\mathrm{min}^{-1}$ or so. The circulation of the dyed fluid is viewed from above using a camera.

the boundary currents, a feature of deep boundary currents in the real ocean.

Application of T-P theory to the experiment

We now apply dynamical ideas to the experiment in a way that directly parallels that developed in Section 11.3.1 to infer abyssal circulation patterns in the ocean. If fluid is introduced to the tank of side L at rate S, then the depth of the fluid in the tank, h, (see Fig. 11.19) increases at a rate given by:

$$\frac{dh}{dt} = \frac{S}{L^2}.$$

In the presence of rotation, columns of fluid in steady, slow, frictionless motion must, by the Taylor-Proudman theorem, remain of constant length. Hence, if the free surface rises, an interior column, marked by the thick vertical line in Fig. 11.19, must move toward the shallow end of the tank *conserving its length*. In a time Δt the free surface has risen by Δh and so the vertical velocity is $w = \Delta h / \Delta t = S/L^2$, from above. Given that the displacement of the column must match the geometry of the wedge defined by the upper surface and the sloping bottom, we see that the vertical and horizontal velocities must be in the ratio $w/v = dz/dy = \alpha$, where α is the slope of the bottom and we have used the definitions $w = dz/dt$, $v = dy/dt$. Thus v is given by:

$$v = \frac{w}{\alpha} = \frac{1}{\alpha}\frac{S}{L^2} \qquad (11\text{-}9)$$

exactly analogous to Eq. 11-8.[4] The column moves northwards, towards the shallow end of the tank i.e., "polewards".

Typically we set $S = 20$ $\mathrm{cm}^3\,\mathrm{min}^{-1}$, $\alpha = 0.2$, $L = 50\,\mathrm{cm}$, and so we find that $v = \frac{1}{0.2} \times \frac{20\,\mathrm{cm}^3\,\mathrm{min}^{-1}}{(50\,\mathrm{cm})^2} = 3.3 \times 10^{-4}\,\mathrm{m\,s}^{-1}$, or only 20 cm in 10 min. The boundary currents returning the water to the deep end of the tank are much swifter than this and are clearly evident in Fig. 11.18.

Our experiment confirms the preference for *western* boundary currents. But why are western boundary currents favored over eastern boundary currents? In Chapter 10 we explained the preference of wind-driven ocean gyres for western, as opposed to eastern boundary currents, by invoking an interior Sverdrup balance, Eq. 10-17, and arguing that the sense of the circulation must reflect that implied by the driving wind. But

---

[4]Comparing Eqs. 11-9 and 11-8, we see that α plays the role of $\beta D/f \sim D/a$ (since, see Eq. 10-10, $\beta \sim f/a$) where D is a typical ocean depth and a is the radius of the Earth. It is interesting to note that Ω does not appear in either Eq. 11-9 or 11-8. Neverthless it is important to realize that rotation is a crucial ingredient through imposition of the T-P constraint.

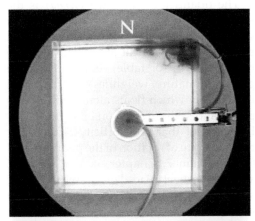

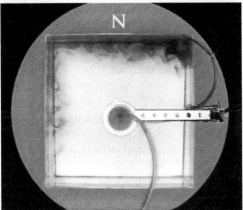

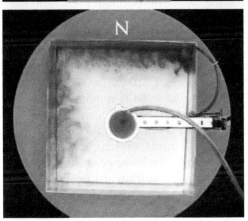

FIGURE 11.18. Three photographs, taken at 10 min intervals, charting the evolution of dye slowly entering a rotating tank of water with a sloping bottom, as sketched in Fig. 11.17. The funnel in the centre carries fluid to the diffuser located at the top right hand corner of the tank. The shallow end of the tank is marked with the 'N' and represents polar latitudes. The dyed fluid enters at the top right, and creates a "northern" and then "western" boundary current.

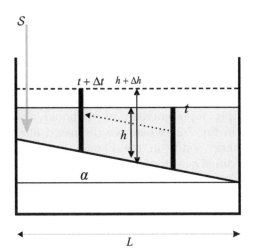

FIGURE 11.19. Fluid entering the tank at rate S leads to a rise in the free surface, h. The Taylor Column marked by the vertical thick line therefore moves toward the shallow end of the tank (dotted arrow), conserving its length. The bottom tilts at an angle α.

in the present context there is no wind-stress to dictate the sense of the circulation. What's going on?

11.3.3. Why western boundary currents?

Let's first review what we can deduce by application of the Taylor-Proudman theorem alone. Taylor-Proudman says (see Section 7.2) that, *if the motion of a rotating, homogeneous fluid is sufficiently slow, steady, and frictionless, then columns parallel to the rotation vector will not tip over or change their length.* Thus if T-P is obeyed, the length of a column is conserved as it moves.

The conditions just mentioned in italics are most likely to be satisfied in the interior of our rotating tank where, in the presence of a rising free surface, interior columns conserve their length by moving poleward (see Fig. 11.19). They are most likely to be violated at the lateral walls where frictional effects begin to play a role and allow, as we shall see, fluid columns to change their length by either shedding or accumulating mass. It turns out that western boundary currents are favored. A jet just cannot make it far if it slides along an eastern boundary, as we now go on to discuss.

Consider the geostrophic relation with f constant, but now let's incorporate a simple representation of frictional drag, as in Eq. 7-28 where we discussed atmospheric winds in the Ekman layer. The modified geostrophic balance, Eq. 7-25, then becomes, written out for convenience here in component form:

$$-fv + \frac{1}{\rho_{ref}}\frac{\partial P}{\partial x} = -k\frac{u}{\delta}$$
$$fu + \frac{1}{\rho_{ref}}\frac{\partial P}{\partial y} = -k\frac{v}{\delta}$$

where k is a drag coefficient representing the rubbing of the Taylor column over the bottom and δ is the Ekman layer depth. Now, just as in the derivation of the T-P theorem in Section 7.2, we cross-differentiate the above to eliminate the pressure p, to obtain:

$$f\left(\frac{\partial u}{\partial x} + \frac{\partial v}{\partial y}\right) = -\varepsilon\left(\frac{\partial v}{\partial x} - \frac{\partial u}{\partial y}\right),$$

where for convenience we have introduced $\varepsilon = k/\delta$. Now, since, by continuity, $(\partial u/\partial x + \partial v/\partial y) = -\partial w/\partial z$, the previous equation can be written:

$$f\frac{\partial w}{\partial z} = \varepsilon\left(\frac{\partial v}{\partial x} - \frac{\partial u}{\partial y}\right). \qquad (11\text{-}10)$$

What is Eq. 11-10 saying physically? If the flow is frictionless ($\varepsilon = 0$), the rhs is zero, and we recover a version of Eq. 7-15, the T-P theorem:

$$f\frac{\partial w}{\partial z} = 0.$$

But the theorem is violated by the presence of friction, which allows columns to stretch or squash at a rate that depends on ε and $\partial v/\partial x - \partial u/\partial y$.

The quantity $\zeta = (\partial v/\partial x - \partial u/\partial y)$ is the vertical component of a vector called the vorticity, which measures the spin of a fluid parcel about a vertical axis relative to the rotating Earth (or table). Imagine that we place a miniature, weightless paddle wheel in to a flow which floats along with a fluid parcel and spins around on its vertical axis. It turns out that the vorticity ζ is exactly twice the rate of rotation of the paddle wheel (see Problem 7, Chapter 6). For example, a paddle wheel placed in a swirling flow such as that sketched in Fig. 7.26 (left) will spin cyclonically, $(\partial v/\partial x - \partial u/\partial y) > 0$. Thus Eq. 11-10 then implies $\partial w/\partial z > 0$: so, since $w = 0$ at the bottom, fluid will upwell away from the boundary and the column will stretch. Cross-isobaric flow at the bottom, where the column rubs over the base of the tank, leads to the requisite acquisition of mass. This is exactly the same process studied in GFD Lab X: fluid is driven into a low pressure system at its base where frictional affects are operative (see Fig. 7.23, top).

Now let us return to the problem of boundary currents. Consider the southward flowing western boundary current sketched in Fig. 11.20. A paddle wheel placed in it will turn cyclonically because flow on its inside flank is faster than on its outside flank. Hence $(\partial v/\partial x - \partial u/\partial y) > 0.$[5] Now using Eq. 11-10 we see that $\partial w/\partial z > 0$. But this is just what is required of a southward flowing boundary current (moving to the deeper end of the tank) because, from Eq. 11-10, it must stretch, $\partial w/\partial z > 0$. Thus the signs in Eq. 11-10 are consistent. But what happens in the southward flowing eastern boundary current sketched on the right of Fig. 11.20? There, $(\partial v/\partial x - \partial u/\partial y) < 0$ and so Eq. 11-10 tells us that $\partial w/\partial z < 0$. But this is the wrong sign if the column is to stretch. Thus we conclude that the southward flowing eastern boundary current cannot satisfy Eq. 11-10 and so is disallowed.

---

[5]This is obvious since, by definition, $|v| \gg |u|$ and $\partial/\partial x \gg \partial/\partial y$ in a meridional boundary current.

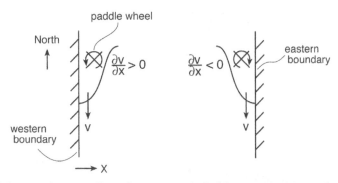

FIGURE 11.20. Schematic diagram of boundary currents in the laboratory experiment shown in Fig. 11.18 and Fig. 11.22. The mass balance of Taylor columns moving in boundary currents on the western margin (whether flowing southwards as sketched here or northward) can be satisfied; the mass balance of Taylor columns moving in boundary currents on the eastern margin (whether running north or south) cannot.

On consideration of the balance of terms in northward flowing boundary currents, we deduce that eastern boundary currents of both signs are prohibited, but western boundary currents of both signs are allowed.

11.3.4. GFD Lab XV: Source sink flow in a rotating basin

The preference for western as opposed to eastern boundary currents can be studied in our rotating tank by setting up a source sink flow using a pump, as sketched in Fig. 11.21. The pump gently draws fluid out of the tank (to create a sink) and pumps it back in through a diffuser (the source). On the way, the fluid is dyed (through an intravenous device) so that when it enters at the source its subsequent path can be followed. Experiments with different arrangements of source and sink are readily carried out. [See Problem 3 at the end of this Chapter.] One example is shown in Fig. 11.22. Fluid is sucked out toward the northern end of the eastern boundary (marked by the black circle) and pumped in on the southern end of the eastern boundary (marked by the white circle). Rather than flow due north along the eastern boundary, fluid tracks west, runs north along the western boundary and then turns eastwards at the "latitude" of the sink. And all because eastern boundary currents are disallowed!

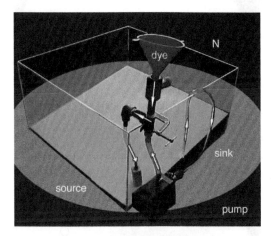

FIGURE 11.21. Source-sink driven flow can be studied with the apparatus shown above. The pump and associated tubing can be seen, together with the intravenous device used to dye the fluid travelling toward the source. Fluid enters the tank through a submerged diffuser. The pumping rate is very slow, only about $20 \, \text{cm}^3 \, \text{min}^{-1}$. An example of the subsequent evolution of the dyed fluid is shown in Fig. 11.22.

11.4. OBSERVATIONS OF ABYSSAL OCEAN CIRCULATION

It is very hard to test whether the circulation schematic, Fig. 11.16, has parallels in the ocean because the predicted mean currents are so very weak and the variability of the ocean so strong. However, one of the key predictions of Stommel's abyssal theory was that there ought to be deep western

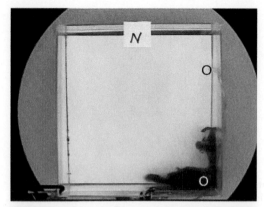

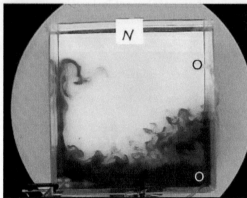

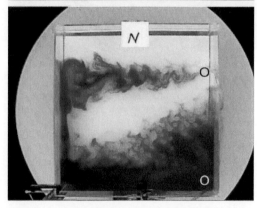

FIGURE 11.22. Three photographs charting the evolution of dye from source (white circle) to sink (black circle) using the apparatus shown in Fig. 11.21. The shallow end of the tank is marked with the 'N' and represents polar latitudes.

boundary currents which are sufficiently strong that they can be directly observed. In a rare case of theory preceding observation, the Deep Western Boundary Current on the Atlantic continental rise south of Cape Cod was indeed subsequently confirmed by direct observation.

There is also definitive evidence of deep western boundary currents in hydrographic sections. For example, Fig. 11.23, a cross section of water properties at 30° S, shows NADW, formed by convective processes in the northern North Atlantic, flowing as a deep, relatively salty, western boundary current in to the South Atlantic. One can also detect Antarctic Bottom Water, flowing northward as a relatively fresh, but very cold body of water, right at the bottom. The entire abyssal ocean is supplied via these western boundary currents, which feed the interior circulation of the open ocean. Other regions of the world ocean are also fed from the Atlantic, but by a more circuitous route which involves the Antarctic Circumpolar Current, as roughly captured by Stommel's remarkable schematic. See also Fig. 11.28.

One very vivid depiction of the abyssal circulation is revealed by the pathways of CFC from the ocean surface in to the interior, shown in Fig. 11.24. Recall, CFCs (see Table 1.2) are man-made substances and only appeared in the 20th Century. The atmospheric concentration of CFC-11 increased dramatically in the middle part of the last century and dissolved into surface waters. They were first observed invading the ocean in the 1960s. At high latitudes (in the Labrador and Greenland Seas for example) CFC-enriched surface waters are mixed by convection to great depth—recall the deep mixed layers in the polar ocean shown in Figs. 9.10 and 11.11—and carried away in to the abyssal ocean by deep western boundary currents. A plume of CFC-11 can be seen in Fig. 11.24 extending down the western margin of the ocean indicating the presence of a strong deep western boundary current which advects the CFC-rich waters away from the source region.

Finally, lest we leave the impression that Stommel's schematic is a detailed representation of the abyssal ocean, the gentle meridional flows sketched in Fig. 11.16

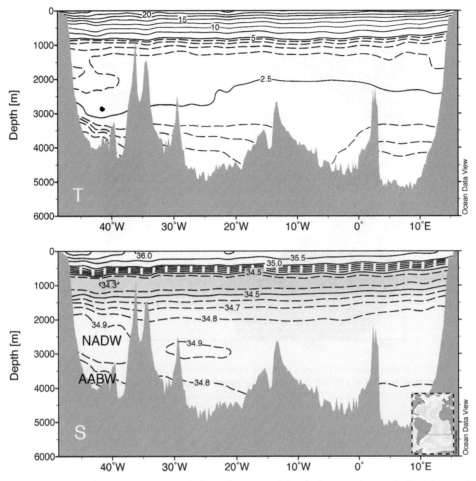

FIGURE 11.23. Zonal section across 30° S from the coast of South America (as marked in the inset). Top: temperature (°C); bottom: salinity (psu). We see North Atlantic Deep Water flowing southwards at a depth of some 2 km, and Antarctic Bottom Water moving northward right at the bottom. Plotted with Ocean Data View.

are not borne out by observations. Instead, as is evident in Fig. 11.25, the mid-depth flow in the ocean appears to show a marked tendency for zonal jets of small meridional scale which connect to the western boundary currents, much as seen in GFD Lab XV, Fig. 11.22. The lateral extent of these jets and their geographical and vertical structure are still largely unknown. However, the predominantly zonal interior is consistent with the idea that there is little mixing and upwelling and so no stretching of Taylor columns, which therefore move zonally to conserve their length if not interrupted by topography.

11.5. THE OCEAN HEAT BUDGET AND TRANSPORT

We now turn to the role of the ocean circulation in meridional heat transport. To maintain an approximately steady climate, the ocean and atmosphere must move excess heat from the tropics to the polar regions. We saw back in Fig. 8.13 that the atmosphere

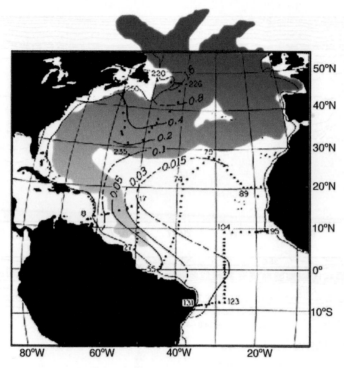

FIGURE 11.24.   Observations of CFCs at a depth of 2 km (contoured). Superimposed in red is a snap-shot for 1983 of the CFC distribution at a depth of 2 km in the North Atlantic, as simulated by a numerical model of ocean circulation and tracer transport. The model results are courtesy of Mick Follows (MIT), the data courtesy of Ray Weiss (Scripps).

transports the larger part of the poleward heat transport with the ocean carrying the remainder. To obtain a quantitative estimate of the role of the ocean in meridional heat transport we must write an equation for the ocean heat budget.

The heat budget for a column of ocean is obtained by integrating Eq. 11-2 vertically over the depth of the ocean:

$$\frac{\partial}{\partial t}(\text{heat content}) = -Q_{\text{net}} - \underbrace{\nabla_h \cdot \vec{\mathcal{H}}_{ocean}}_{\text{by ocean currents}} ,$$

$$(11\text{-}11)$$

where

$$\text{heat content} = \rho_{ref} c_w \int_{bottom}^{top} T dz$$

is the heat stored in the column, Q_{net} is given by Eq. 11-5, $\vec{\mathcal{H}}_{ocean} = \rho_{ref} c_w \int_{bottom}^{top} \mathbf{u} T dz$ is the

(vector) horizontal heat flux by ocean currents integrated over the vertical column, and ∇_h is the horizontal divergence operator. Equation 11-11 says that changes in heat stored in a column of the ocean are induced by fluxes of heat through the sea surface and the horizontal divergence of heat carried by ocean currents. If there is to be a steady state, the global integral of the air-sea flux must vanish (see Section 11.1), because ocean currents can only carry heat from one place to another, redistributing it around the globe.

11.5.1. Meridional heat transport

Let us define $\overline{\mathcal{H}}_{ocean}^{\lambda}$ as the heat flux across a vertical plane extending from the bottom of the ocean to the surface and from the western coast of an ocean basin to the

Float displacements at 2.5 km

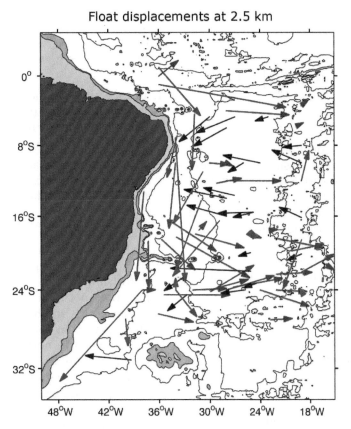

FIGURE 11.25.   Net float displacements over a time of 600–800 d at a depth of 2.5 km, showing the movement of North Atlantic Deep Water in the south Atlantic. Eastward displacements are marked in red, westward in blue, and southward in green. Courtesy of Nelson Hogg, WHOI.

eastern coast (cf. Eq. 8-15 for the analogous expression for the total energy flux, $\overline{\mathcal{H}}_{atmos}^{\lambda}$, in the atmosphere):

$$\overline{\mathcal{H}}_{ocean}^{\lambda} = \rho_{ref} c_w a \cos\varphi \int_{\lambda_{west}}^{\lambda_{east}} \int_{bottom}^{top} vT dz \, d\lambda,$$

where the geometrical factor $a \cos\varphi d\lambda$ is the distance along a latitude circle over an arc $d\lambda$ (see Fig. 6.19). To take account of (slight) compressibility effects, we interpret T in the above expression as the potential temperature.

$\overline{\mathcal{H}}_{ocean}^{\lambda}$ is difficult to measure and not precisely known. However, it can be inferred as follows:

1. as a residual, using atmospheric analyses of velocity and temperature to calculate the heat transport in the atmosphere, which is then subtracted from the total meridional transport calculated from the top-of-the-atmosphere heat flux (incoming solar minus outgoing long-wave radiation) observed directly by satellite.

2. by integrating estimates of air-sea heat fluxes, such as those shown in Fig. 11.4, to obtain the zonal average of the meridional heat flux (an example is shown in Fig. 11.2). If a steady state prevails, the meridional integral of the zonal average of Q_{net} must be balanced by heat transport in the ocean (see

Eq. 11-11) and so yields an estimate of meridional ocean heat transport at latitude φ thus:

$$\overline{\mathcal{H}}^{\lambda}_{ocean}(\varphi) = -a^2 \cos\varphi \int_{\varphi_1}^{\varphi} \int_{\lambda_{west}}^{\lambda_{east}} Q_{net}\,d\lambda\,d\varphi,$$

(11-12)

where the latitude φ_1 is chosen so that $\overline{\mathcal{H}}^{\lambda}_{ocean}(\varphi_1) = 0$. Equation 11-12 simply says that in the steady state the heat flux through the sea surface integrated over an area bounded by two latitude circles and meridional coasts, must be balanced by a horizontal heat flux into (or out of) the region, as depicted in Fig. 11.26. Essentially this is how the oceanographic contribution to Fig. 8.13 was computed. Note that if $Q_{net} < 0$, then $\overline{\mathcal{H}}^{\lambda}_{ocean} > 0$, as drawn in Fig. 11.26.

3. by attempting to directly measure $\overline{\mathcal{H}}^{\lambda}_{ocean}$ from *in situ* ocean observations making use of hydrographic sections, such as Fig. 11.23.

In Fig. 11.27 we show estimates of $\overline{\mathcal{H}}^{\lambda}_{ocean}$ for the world ocean and by ocean basin, computed as a residual using method 1 (note the error bars can reach 0.5 PW). The total peaks at about ± 2 PW at $\pm 20°$, with the magnitude of the northward flux in the northern hemisphere exceeding, somewhat, the southward flux in the southern hemisphere. In the northern hemisphere, both the Atlantic and Pacific Oceans make large contributions. In the southern hemisphere,

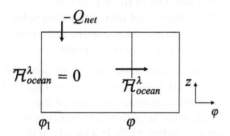

FIGURE 11.26. Schematic of the computation of meridional ocean heat transport from net air-sea heat flux, Eq. 11-12. The sea surface is at the top, the ocean bottom at the bottom. At the latitude φ_1 it is supposed that the meridional heat flux vanishes.

heat is transported poleward by the Pacific and Indian Oceans, but *equatorward* by the Atlantic!

Figure 11.27 suggests that:

1. the overall magnitude of $\overline{\mathcal{H}}^{\lambda}_{ocean}$ is a significant fraction (perhaps 1/4 to 1/3) of the pole-equator heat transport (see Section 11.5.2 below), but

2. the three major ocean basins—the Atlantic, the Pacific, and the Indian Ocean—differ fundamentally in their contribution to $\overline{\mathcal{H}}^{\lambda}_{ocean}$.

In the Pacific Ocean, the heat flux is symmetric about the equator and directed poleward in both hemispheres. In the Indian Ocean (which does not exist north of 25° N) the heat transport is southwards on both sides of the equator. In the Atlantic, however, the heat transport is northward everywhere, implying that in the south Atlantic heat is transported equatorward, up the large-scale temperature gradient. Remarkably, in the Atlantic there is a cross-equatorial heat transport of about 0.5 PW (cf. Fig. 8.13) and convergence of heat transport in the North Atlantic. Indeed poleward of 40° N the Atlantic Ocean is much warmer (by as much as 3°C) than the Pacific (see the map of sea-surface temperature in Fig. 9.3).

Why are heat transports and SST so different in the Atlantic and Pacific sectors? It is thought to be a consequence of the presence of a vigorous overturning circulation in the Atlantic which is largely absent in the Pacific. To see this consider Fig. 11.28 (top) which presents a quantitative estimate of the circulation of the global ocean separated into 3 layers: shallow (top km), deep (2 to 4 km), and bottom (>4 km). The arrows represent the horizontal volume transport (in Sv) in each of the layers across the sections marked. The circles indicate (⊙ for upwelling, ⊗ for downwelling) the vertical transport out of the layer in question marked in Sv. Thus, for example, 15 Sv of fluid sinks out of the shallow layer in the northern North Atlantic: 23 Sv of fluid travel south

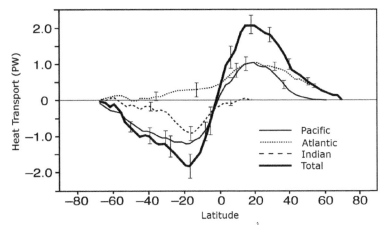

FIGURE 11.27. Northward heat transport in the world ocean, $\overline{\mathcal{H}}^{\lambda}_{ocean}$, and by ocean basin calculated by the residual method using atmospheric heat transport from ECMWF and top of the atmosphere heat fluxes from the Earth Radiation Budget Experiment satellite. The vertical bars are estimates of uncertainty. From Houghton et al. (1996), using data from Trenberth and Solomon (1994).

in the deep layer across 25S in the Atlantic Ocean. Around Antarctica 8 Sv of deep water upwells to the shallow layer, while 21 Sv sinks in to the bottom layer. We see that the Antarctic Circumpolar Current carries a zonal transport of around 140 Sv, evenly distributed between the layers. Note the much weaker overturning circulation in the Pacific with no sinking of fluid in the northern North Pacific.

Figure 11.28 (bottom) is a gross but useful cartoon of the upper branch (shallow to deep) overturning circulation highlighting the asymmetry between the Atlantic and the Pacific and between the northern and southern hemispheres. We see that a major pathway accomplishing northward heat transport in the Atlantic is the advection of warm, near-surface water northward across the equator (see the northward flowing western boundary current crossing the equator in Fig. 10.13, bottom), cooling and sinking in the northern North Atlantic, and thence southwards flow of colder fluid as North Atlantic Deep Water. In contrast, the Pacific Ocean does not support a significant overturning cell induced by polar sinking. This is the key to understanding the different nature of heat transport in the basins evident in Fig. 11.27.

11.5.2. Mechanisms of ocean heat transport and the partition of heat transport between the atmosphere and ocean

Heat is transported poleward by the ocean if, on the average, waters moving polewards are compensated by equatorward flow at colder temperatures. It is useful to imagine integrating the complex three-dimensional ocean circulation horizontally, from one coast to the other, thus mapping it out in a meridional plane, as sketched in Fig. 11.29. If surface waters moving poleward are warmer than the equatorial return flow beneath, then a poleward flux of heat will be achieved.

To make our discussion more quantitative we write down the meridional advection of heat thus (integration across the ocean is implied):

$$\mathcal{H}_{ocean} = c_w \int_{bottom}^{top} \rho_{ref} v \vartheta \, dz \qquad (11\text{-}13)$$

$$= -c_w \int_{bottom}^{top} \frac{\partial \Psi_O}{\partial z} \vartheta \, dz = c_w \int_{bottom}^{top} \Psi_O \frac{\partial \vartheta}{\partial z} \, dz$$

$$= c_w \int_{\vartheta_{bottom}}^{\vartheta_{top}} \Psi_O \, d\vartheta.$$

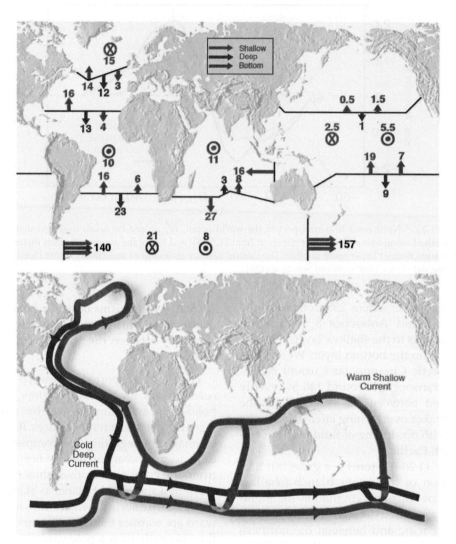

FIGURE 11.28. Top: Estimate of global ocean circulation patterns based on Ganachaud and Wunsch (2000) modified from Alley et al. (2002). The circulation is separated into 3 layers: shallow (red, < 2 km), deep (blue, 2–4 km), and bottom (green, > 4 km). Horizontal arrows across the marked sections represent the volume transport in Sv. Circles (⊙ for upwelling, ⊗ for downwelling) represent the vertical transport out of the layer in question in Sv. Bottom: A cartoon, based on the quantitative estimates shown above, of the ocean's 'shallow to deep' overturning circulation illustrating the asymmetry between the Atlantic and Pacific basins and between northern and southern hemispheres. Blue represents deep flow (2–4 km, red shallow flow (< 2 km). Transitions between shallow and deep are also indicated. This global overturning pattern has become known as the 'conveyer belt'. It is a schematic representation of a highly complex, turbulent flow.

In deriving the above we have written $\rho_{ref}v = -\partial\Psi_O/\partial z$ in which Ψ_O is the streamfunction for the *mass* transport in the meridional plane and made use of the fact that (i) $\Psi_O = 0$ at the top and bottom of the ocean and (ii) the mass transport can be expressed as the product of the density of water multiplied by the *volume* transport of the meridional overturning circulation (MOC), ψ_{MOC} thus:

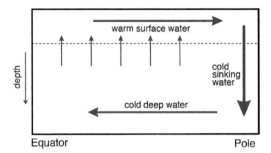

FIGURE 11.29. A schematic diagram of the ocean's meridional overturning circulation, in which warm waters flow poleward at the surface, are cooled by loss of heat to the atmosphere, sink to depth and return equatorward. Such a circulation achieves a poleward transport of heat.

$$\Psi_O = \rho_{ref}\psi_{MOC}. \qquad (11\text{-}14)$$

The relation $\mathcal{H}_{ocean} = c_w \int_{\vartheta_{bottom}}^{\vartheta_{top}} \Psi_O d\vartheta$ tells us that the heat transport can be expressed in terms of the mass transport in temperature layers, or, more generally, noting the multiplication by c_w, the mass transport in energy layers. It is then useful to write down an approximate form thus:

$$\mathcal{H}_{ocean} \simeq c_w \Delta\vartheta \Psi_{O_{max}}, \qquad (11\text{-}15)$$

where $\Delta\vartheta$ is the difference in potential temperature between the poleward and equatorward branches, and $\Psi_{O_{max}}$ is the strength of the overturning mass transport.

Figure 11.30 shows ψ_{MOC} for the global ocean plotted in the (λ, z) and (λ, θ) plane[6]: we see that it has a magnitude of order 20 Sv, implying a meridional mass transport of $\Psi_{O_{max}} = 20 \times 10^9 \, \text{kg s}^{-1}$. If the temperature difference across the MOC is $\Delta\vartheta = 15 \, \text{K}$, typical of the temperature drop across the main thermocline (see Fig. 9.5), then Eq. 11-15 yields a heat transport of 1.2 PW, of the order presented in Fig. 11.27. Thus Eq. 11-15 is a useful vantage point from which to discuss mechanisms of ocean heat transport and—see below—the partition of heat transport between the atmosphere and the ocean.

Both wind-driven and thermohaline circulations play an important role in setting the magnitude and pattern of Ψ_O. Fluid which is pumped down by the wind in middle latitudes is compensated in part by poleward transport of warm surface waters from the tropics in surface Ekman layers. The subsurface equatorial return flow occurs at a colder temperature, resulting in poleward heat transport. Indeed the wind is responsible for the two shallow 'wheels' of overturning circulation symmetrically disposed about the equator in Fig. 11.30 (right), reaching up to ±40°. This is very evident in the Pacific with a heat transport which is symmetric about the equator. In polar latitudes we see cells driven by convective processes feeding the abyssal ocean associated with NADW in the north and AABW in the south. As already mentioned, Atlantic heat transport is northward at all latitudes, consistent with idea that a giant interhemispheric meridional overturning cell associated with polar sinking is a dominant heat transport mechanism, as sketched in Fig. 11.28 (bottom). As we shall discuss in more detail in Section 12.3.5, variability in the MOC in the Atlantic Ocean is often invoked as a player of climate change because of the likely sensitivity of arctic processes to meridional heat transport mediated by the MOC and vice versa.

The framework provided by Eq. 11-15 can also be used to come to some understanding of the processes that set the partition of heat transport between the atmosphere and ocean. We can express the atmospheric heat transport in the same form:

$$\mathcal{H}_{atmos} \simeq c_p \Delta\vartheta_A \Psi_{A_{max}}, \qquad (11\text{-}16)$$

where now $\Psi_{A_{max}}$ is the magnitude of the mass transport overturning circulation of

---

[6]The overturning circulation shown in Fig. 11.30 is derived from a model constrained by obervations, rather than inferred directly from observations, because it is all but impossible to observe ψ_{MOC} directly.

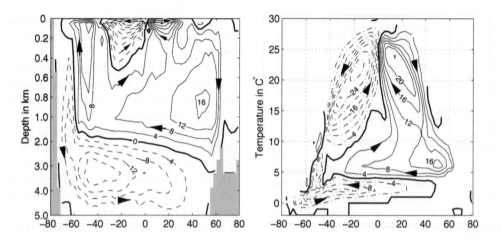

FIGURE 11.30.   The meridional overturning circulation, ψ_{MOC}, in a model of the global ocean plotted in the (λ, z) plane on the left and the (λ, θ) plane on the right. Note that on the left the scale over the top km of the ocean is greatly expanded.

the atmosphere in "energy" layers, appropriately chosen based on moist static energy, such that

$$c_p \Delta \vartheta_A = c_p \Delta T + g \Delta z + L \Delta q$$

is the vertical change in the moist static energy (see Eq. 8-16 and the discussion of meridional transport of heat in Section 8.4) across the overturning circulation.

In Fig. 11.31 we plot annual-mean mass transport stream functions for the atmosphere and ocean, Ψ_A and Ψ_O, in energy layers, $c_p \Delta \vartheta_A$ for the atmosphere and $c_w \Delta \vartheta_O$ for the ocean. The horizontal axis is latitude, the vertical axis is an energy coordinate with units of $c \Delta \vartheta$, i.e., $J \, kg^{-1}$. To facilitate comparison of the strength of the overturning circulation in the two fluids, we have (temporarily) redefined the Sverdrup as an equivalent unit of mass transport: in Fig. 11.31, 1 Sv is the mass transport associated with a volume flux of $10^6 \, m^3 \, s^{-1}$ of water, which is $10^6 \, m^3 \, s^{-1} \times 10^3 \, kg \, m^{-3} = 10^9 \, kg \, s^{-1}$. The first striking feature of Fig. 11.31 is that, in contrast to Fig. 5.21, the atmospheric overturning

circulation comprises one giant cell from equator to pole. This is because, unlike the overturning circulation plotted in Fig. 5.21, Fig. 11.31 includes a large eddy contribution. In middle-to-high latitudes, mass (like heat) is transferred by weather systems rather than by the mean flow. Secondly, note that the intensity of the oceanic cell is much weaker than its atmospheric counterpart. Even at 20°, where Ψ_O reaches its maximum, the atmospheric mass transport is roughly four times that of the ocean. It is only within the deep tropics that the two transports are comparable. The third important feature is that the 'thickness' of the overturning cells in the two fluids are comparable in energy space. In midlatitudes, $c_w \Delta \vartheta_O / c_p \Delta \vartheta_A$ is of order unity, the differences in heat capacity ($c_w / c_p \sim 4$) being compensated by a larger temperature difference across the atmospheric cell ($\Delta \vartheta_A \sim 40$ K compared to $\Delta \vartheta_O \sim 10$ K). The dominance of atmospheric over ocean heat transport in middle-to-high latitudes can thus be rationalized as being a consequence of Ψ_A greatly exceeding Ψ_O (see Fig. 11.31).[7] Finally, note that Ψ_O is

---

[7]It is remarkable that despite the density of air being typically one thousand times less than that of water, the meridional mass transport in the atmosphere ($\sim$100 Sv $= 10^{11} \, kg \, s^{-1}$) exceeds that of the ocean by a factor of 4. This is because meridional wind speeds greatly exceed those in the ocean.

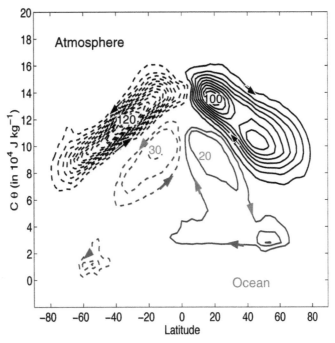

FIGURE 11.31. Annual mean atmospheric (black) and oceanic (green) mass streamfunction within constant energy layers. The contour interval is 10 Sv where 1 Sv = 10^9 kg s^{-1}, dashed when circulating anti-clockwise. The horizontal axis is latitude; the vertical axis is an energy coordinate ($c\Delta\vartheta$) in units of 10^4 J kg^{-1}. From Czaja and Marshall (2006).

is dominated by two large symmetrically disposed cells around the equator. As mentioned previously, this, in the main, is the overturning cell associated with wind-driven subtropical gyres, carrying heat up to 40° N, S or so. The weaker cells at lower temperatures in polar latitudes are the signature of overturning cells directly associated with convectively-induced, polar thermohaline circulations.

11.6. FRESHWATER TRANSPORT BY THE OCEAN

The ocean and atmosphere must move freshwater from regions of excess rainfall to regions with excess evaporation (see Fig. 11.6). Knowledge of water fluxes and transports in the ocean is important for understanding the global hydrological cycle

and climate. For example, variability in fresh water fluxes may have played an important role in the ice ages, as will be discussed in Chapter 12. The plot of evaporation minus precipitation in Fig. 11.6 shows that evaporation exceeds precipitation by more than a meter per year in the trade wind regimes in the eastern parts of the oceans. Here dry air subsides along the poleward edges of the Hadley Cell. The ITCZ is a region of vigorous updrafts and here precipitation exceeds evaporation. As discussed in Chapter 9, these broad patterns of E and P are reflected in the surface salinity distribution of the oceans (see Fig. 9.4).

The transport of fresh water by the ocean can be calculated using the same methods as for heat transport—again there is considerable uncertainty in these estimates. Fig. 11.32 shows an estimate of the meridional transport of freshwater by the Atlantic Ocean. We see that freshwater transport is southwards,

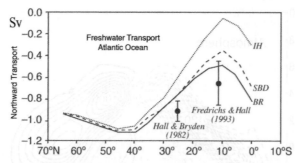

FIGURE 11.32. Meridional transport of fresh water by the Atlantic from three surface-flux calculations: BR-Baumgartner and Reichel (1975); SBD-Schmitt et al. (1989); and IH-Isemer and Hasse's (1987) evaporation estimates combined with Dorman and Bourke's (1981) precipitation values. Also shown are direct measurements at 24° N by Hall and Bryden (1982), and 11° N by Friedrichs and Hall (1993). All are summed relative to an estimated Arctic southward export due to the Bering Strait through flow and the water budget of the Arctic itself. From Schmitt (1994) and Stewart (1995).

corresponding to a transport of saltwater northwards into the North Atlantic. This salinization is thought to play an important role in preconditioning the surface waters of the Atlantic to convection when exposed to cooling in polar latitudes.

11.7. FURTHER READING

A more comprehensive discussion of air-sea interaction can be found in Csanady (2001) and Stewart (2005). Dynamical theories of thermohaline circulation are less well developed than those of the wind-driven ocean circulation. Good accounts of Stommel's theory of abyssal ocean circulation can be found in Pedlosky (1996) and Vallis (2006).

11.8. PROBLEMS

1. It is observed that water sinks in to the deep ocean in polar regions of the Atlantic basin at a rate of 15 Sv.

    (a) How long would it take to fill up the Atlantic basin?

    (b) Supposing that the local sinking is balanced by large-scale upwelling, estimate the strength of this upwelling. Express your answer in $m\,y^{-1}$.

    (c) Assuming that $\beta v = f\,\partial w/\partial z$, infer the sense and deduce the magnitude of the meridional currents in the interior of the abyssal ocean where columns of fluid are being stretched.

    (d) Estimate the strength of the western boundary current.

2. Review Section 11.3.3, but now suppose that boundary currents flow northwards in Fig. 11.20. By considering the role of boundary current friction in inducing Taylor columns to stretch/compress (Eq. 11-10), deduce that northward flowing eastern (western) boundary currents are disallowed (allowed).

3. Consider the laboratory experiment GFD XV: source sink flow in a rotating basin. Use the Taylor-Proudman theorem and that eastern boundary currents are disallowed, to sketch the pattern of flow taking fluid from source to sink for the scenarios given

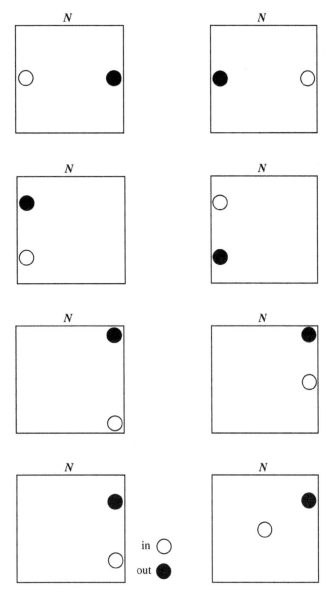

FIGURE 11.33. Possible placement of the source and sink in GFD LabXV: source sink flow in a rotating basin. Note that Fig. 11.22 corresponds to the case at the bottom of the column on the left.

in Fig. 11.33. Note that one of the solutions is given in Fig. 11.22!

4. From Fig. 11.6 one sees that evaporation exceeds precipitation by order $1\,\mathrm{m\,y^{-1}}$ in the subtropics ($\pm\,30°$), but the reverse is true at higher latitudes ($\pm\,60°$).

(a) Estimate the meridional freshwater transport of the ocean required to maintain hydrological balance and compare with Fig. 11.32.

(b) Latent heat is taken from the ocean to evaporate water, which subsequently falls as rain at (predominantly) higher latitudes, as sketched in Fig. 11.5. Given that the latent heat of evaporation of water is $2.25 \times 10^6 \, \text{J kg}^{-1}$, estimate the implied meridional flux of energy in the atmosphere and compare with Fig. 8.13. Comment.

5. In the present climate the volume of freshwater trapped in ice sheets over land is $\sim 33 \times 10^6 \, \text{km}^3$. If all this ice melted and ran into the ocean, by making use of the data in Table 9.1, estimate by how much the sea level would rise. What would happen to the sea level if all the sea-ice melted?

Climate and climate variability

12.1. The ocean as a buffer of temperature change
    12.1.1.   Nonseasonal changes in SST
12.2. El Niño and the Southern Oscillation
    12.2.1.   Interannual variability
    12.2.2.   "Normal" conditions—equatorial upwelling and the Walker circulation
    12.2.3.   ENSO
    12.2.4.   Other modes of variability
12.3. Paleoclimate
    12.3.1.   Climate over Earth history
    12.3.2.   Paleotemperatures over the past 70 million years: The δ^{18}O record
    12.3.3.   Greenhouse climates
    12.3.4.   Cold climates
    12.3.5.   Glacial-interglacial cycles
    12.3.6.   Global warming
12.4. Further reading
12.5. Problems

Climate is frequently defined as the average weather, with an averaging period long enough to smooth out the variability of synoptic systems. Our emphasis in this book has been on understanding the climatological state of the atmosphere and ocean in which the averaging period is over many years. Climate "norms," such as those studied in Chapters 5 and 9, are typically based on instrumental data averaged over several decades. However, these norms themselves change, and we know from the paleorecord that climate fluctuates on all timescales. These timescales cannot all be associated with one component of the climate system, but rather must reflect the interaction of its component parts, the atmosphere, ocean, land-surface, indeed all the elements set out in Fig. 12.1.

There have been ice ages during which the temperature in middle latitudes has dropped by 5°C or more, the ice caps have tripled in volume and even more so in surface area, to cover large tracts of North America and Europe. Ice ages have returned roughly every 100 k y for the last 800 k y or so.[1] Indeed during 90% of the last 800 k y, Earth has been in a glacial climate and only 10% of the time in interglacial conditions similar to those of today.

---

[1] We denote 1000 years by 1 k y, 1 million years by 1 M y, and 1 billion years by 1 B y.

Timescale

Processes	h/d	w	m	y	10 y	10^2y	10^3y	10^4y	10^5y	10^6y	10^8y
Weather	■	■									
Land surface	■	■	■								
Ocean mixed layer	■	■	■								
Sea ice		■	■	■							
Volcanos		■	■	■							
Vegetation	■	■	■	■	■	■	■	■	■	■	
Thermocline					■	■	■				
Mountain glaciers						■	■				
Deep ocean						■	■	■			
Ice sheets						■	■	■	■		
Orbital forcing								■	■		
Tectonics										■	■
Weathering									■	■	■
Solar "constant"				■	■		■	■	■	■	■

The column groups are labelled: **days** (h/d, w, m), **years** (y, 10 y), **thousands of years** (10^2y, 10^3y, 10^4y), **millions of years** (10^5y, 10^6y, 10^8y).

FIGURE 12.1. The instrumental and paleorecord shows that "weather" and "climate" vary on all timescales, from hours and days to millions of years. Here we tabulate the mechanisms operating at different timescales. Greenhouse gases might also be added to the table: natural CO_2 cycles occur on timescales up to 1k–10k y and longer and the human (100 y) timescale for CO_2 is now important; methane changes can occur at 10 y timescales out to 10k y and beyond. It should also be noted that nonlinearities make the true separation of timescales impossible.

During the last glacial period, 15k–60k y ago, dramatic discharges of large quantities of ice from the land ice sheets (Heinrich events) have occurred every 10k y or so. During the same period, abrupt warming events have occurred over Greenland and the northern North Atlantic every 1.5k y (Dansgaard-Oeschger oscillations), as will be discussed in Section 12.3.5. Each of these events caused a warming of some 10°C that occurred abruptly, within 20–50 y, lasted a few hundreds of years, and terminated abruptly again. Regional fluctuations on shorter timescales of order 1–2°C have also occurred, such as the *Little Ice Age* centered over Europe during the seventeenth century. A more recent example of climate fluctuations on timescales of decades and shorter is the dust bowl of the Great Plains in the 1930s. Looking back much further, 100M y ago, ice was in all likelihood totally absent from the planet, and deep ocean temperatures were perhaps more than 10°C warmer than today. Scientists even speculate about whether, during periods in the distant past, Earth was totally frozen over in a "snowball." What is clear is that Earth's climate has always changed, and continues to do so, with the added complication that human activities are now also a contributing factor.

Clearly we must try and understand the nature of climate fluctuations on all timescales. This is a vast undertaking. There is no accepted general theory of climate, but many factors are implicated in the control of climate. The most important processes, and the timescales on which they act, are listed in Fig. 12.1. One important lesson from the paleoclimate record is that the climate can change more rapidly than a known forcing. For example, climate is capable of large changes over a short time, as in the massive reglaciation event known as the *Younger Dryas*, around 12k y ago (see Section 12.3.5). This lasted perhaps 1.5k y or so, but began and ended abruptly. The climate perhaps has preferred states between which it can flip in a discontinuous manner. The mechanisms behind such abrupt climate change are unclear, and are the subject of much current research. Finally, the interpretation of a particular proxy record in terms of climate variables is often uncertain, as is the extent to which it is indicative of regional or global change, issues that are crucial when one is trying to identify mechanisms. In this concluding chapter, then, we begin to explore some of the issues, making use of what we have learned about the observed state of the atmosphere and ocean and the underlying mechanisms. We place an emphasis on the role of the ocean in climate variability.

Heat, water, momentum, radiatively important gases (such as CO_2), and many other substances cross the sea surface, making the ocean a central component of the climate system, particularly on the long timescales associated with ocean circulation. Much of the solar radiation reaching the Earth is absorbed by the ocean and land, where it is stored primarily near the surface. The ocean releases heat and water vapor to the atmosphere and its currents transport heat and salt around the globe. As discussed in Chapter 11, meridional heat transport by the ocean makes an important contribution to the maintenance of the pole-equator temperature gradient, and freshwater transport by the ocean is a central component of the global hydrological cycle. The oceanic heat and salt transports are not steady, and fluctuations in them are thought to be important players in climate variability on interannual to decadal to centennial timescales and upwards. For example, variability in ocean heat and salt transport may have played a role in ice-age dynamics and abrupt climate change, as will be discussed later in Section 12.3.

We discuss in Section 12.1 the buffering of atmospheric temperature changes by the ocean's enormous heat capacity. In Section 12.2 we discuss atmosphere-ocean coupling in the El Niño–Southern Oscillation phenomenon of the tropical Pacific. Finally, in Section 12.3, we briefly review what we know about the evolution of climate over Earth history: warm climates, cold climates, and glacial-interglacial cycles.

12.1. THE OCEAN AS A BUFFER OF TEMPERATURE CHANGE

The oceans have a much greater capacity to store heat than the atmosphere. This can be readily seen as follows. The heat capacity of a "slab" of ocean of depth h is $\gamma_O = \rho_{ref}c_w h$ (i.e., density × specific heat × depth, with units of $J\,K^{-1}\,m^{-2}$). Let us compare this with the heat capacity of the atmosphere, which we may approximate by $\gamma_A = \rho_s c_p H$, where ρ_s is the mean density of air at the surface and H is vertical scale height of the atmosphere (7–8 km). Inserting typical numbers—the ocean is one thousand times more dense than air and its specific heat is about 4 times that of air—and allowing for the fact that the ocean covers about 70% of the Earth's surface area, we find that $\gamma_O/\gamma_A \simeq 40$ if $h = 100$ m, a typical ocean mixed layer depth, and $\gamma_O/\gamma_A \simeq 2000$ if the whole 5 km depth of the ocean is taken in to account.

The thermal adjustment times of the two fluids are also consequently very different. Radiative calculations show that the timescale for thermal adjustment of the atmosphere alone is about a month. Thermal adjustment timescales in the ocean are very much longer. Observations suggest that sea surface temperatures are typically damped at a rate of $\lambda = 15$ W $m^{-2}\,K^{-1}$ according to the equation:

$$\gamma_O \frac{dT}{dt} = -\lambda T + Q_{net}, \qquad (12\text{-}1)$$

where T is the temperature anomaly of the slab of ocean (assumed well mixed) in contact with the atmosphere. Here, as in Chapter 11, Q_{net} is the net air-sea heat flux.

Setting $Q_{net} = 0$ for a moment, a solution to the above equation is $T = T_{init} \exp\left(\frac{-\lambda}{\gamma_O}t\right)$, where T_{init} is an initial temperature anomaly that decays exponentially with an e-folding timescale of γ_O/λ. Inserting typical numbers (see Table 9.3), we obtain a decay timescale of 300 d $\simeq$ 10 months if h is a typical mixed layer depth of 100 m. Setting h equal to the full depth of the ocean of 5 km, this timescale increases to about 40 y. In fact, as described in Section 11.2.2, the time scale for adjustment of the deep ocean is more like 1 k y, because the slow circulation of the abyssal ocean limits the rate at which its heat can be brought to the surface. These long timescales buffer atmospheric temperature changes. Thus even if the ocean did not move and therefore did not transport heat and salt around the globe, the ocean would play a very significant role in climate, reducing the amplitude of seasonal extremes of temperature and buffering atmospheric climate changes.

The ocean also has a very much greater capacity to absorb and store energy than the adjacent continents. The continents warm faster and cool faster than the ocean during the seasonal cycle. In winter the continents are colder than the surrounding oceans at the same latitude, and in summer they are warmer (see Fig. 12.2). This can readily be understood as a consequence of the differing properties of water and land (soil and rock) in thermal contact with the atmosphere. Although the density of rock is three times that of water, only perhaps the upper 1 m or so of land is in thermal contact with the atmosphere over the seasonal cycle, compared to, typically, 100 m of ocean (see Problem 1 at the end of this chapter, in which we study a simple model of the penetration of temperature fluctuations into the underlying soil). Moreover the specific heat of land is typically only one quarter that of water. The net result is that the heat that the ocean exchanges with the atmosphere over the seasonal cycle exceeds that exchanged between the land and the atmosphere by a factor of more than 100. Consequently we observe that the seasonal range in air temperature on land increases with distance from the ocean and can exceed 40°C over Siberia (cf. the surface air–temperature difference field, January minus July, plotted in Fig. 12.2). The amplitude of the seasonal cycle is very much larger over the continents than over the oceans, a clear indication of the ocean "buffer." This buffering is also evident in the very much weaker seasonal cycle in surface air temperature in the southern hemisphere, where there is much more ocean compared to the north. Note also the eastward displacement of air temperature

differences over land in Fig. 12.2 (bottom), evidence of the role of zonal advection by winds.

12.1.1. Nonseasonal changes in SST

Climatological analyses such as those presented in Chapter 5 define the normal state of the atmosphere; such figures represent the average, for a particular month or season, over many years. It is common experience, however, that any individual season will differ from the climatological picture to a greater or lesser extent. One winter, for example, may be colder or warmer, or wetter or drier, than average. Such variability—in both the atmosphere and the ocean—occurs on a wide range of timescales, from a few years to decades and longer. Understanding and (if possible) forecasting this variability is currently one of the primary goals of meteorology and oceanography. Given the large number of processes involved in controlling SST variability (see the discussion in Section 11.1.2), this is a very difficult task. The challenge is made even greater by the relatively short instrumental record (only 50–100 y) and the poor spatial coverage of ocean observations.

In middle to high latitudes, the direct forcing of the ocean by weather systems—cyclones and anticyclones—induce SST changes through their modulation of surface winds, air temperature, and humidity, and hence air-sea fluxes Q in Eqs. 11-6 and 11-7. These short-period atmospheric systems can yield long-lasting SST anomalies, because the large heat capacity of the ocean endows it with a long-term memory. Such SST anomalies do not have a regular seasonal cycle; rather they reflect the integration of "noise" (from weather systems) by the ocean. One can construct a simple model of the process as follows. In Eq. 12-1 we set $Q_{net} = \mathrm{Re}\hat{Q}_\omega e^{i\omega t}$, where $\hat{Q}_\omega$ is the amplitude of the stochastic component of the air-sea flux at frequency ω associated with atmospheric eddies, and the real part of the expression has physical meaning.

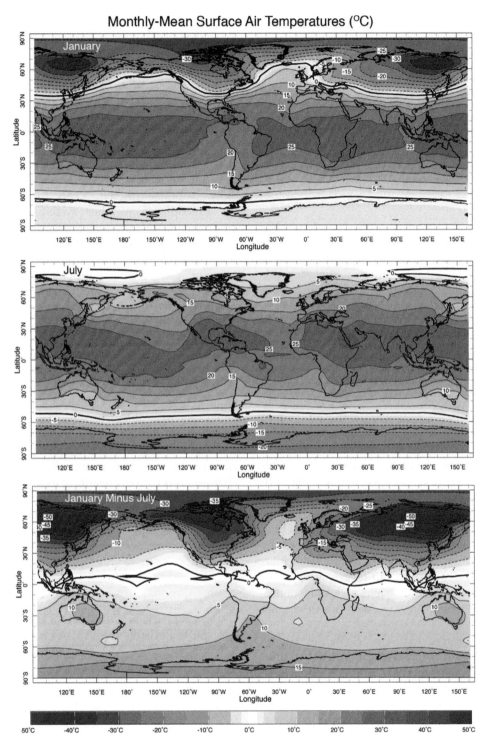

FIGURE 12.2. Monthly mean surface air temperature in January, July, and January minus July: in winter the continents are colder than the surrounding oceans at the same latitude, and in summer they are warmer.

Let us suppose that $\hat{Q}_\omega$ is a constant, so representing a "white-noise" process in which all frequencies have the same amplitude.[2] The response of the SST anomalies that evolve, we assume, according to Eq. 12-1, is given by $T = \text{Re}\, \hat{T}_\omega e^{i\omega t}$, where (see Problem 2 at end of chapter)

$$\hat{T}_\omega^2 = \left(\frac{\hat{Q}_\omega}{\gamma_O} \right)^2 \frac{1}{\omega^2 + \left(\frac{\lambda}{\gamma_O} \right)^2}. \qquad (12\text{-}2)$$

The T spectrum is plotted in Fig. 12.3a and has a very different character from that of the forcing. We see that $\omega_c = \lambda/\gamma_O$ defines a critical frequency that depends on the heat capacity of the ocean in contact with the atmosphere (set by h) and the strength of the air-sea coupling (set by λ). At frequencies $\omega > \omega_c$ the temperature response to white noise forcing decreases rapidly with frequency—the sloping straight line on the log-log plot. At frequencies $\omega < \omega_c$ the response levels out and becomes independent of ω as evident from the grey curve in Fig. 12.3a. For the parameters chosen above, $\omega_c = \frac{1}{300\,\text{d}}$, and so one might expect SST variations with timescales much shorter than 300 d to be damped out, leaving variability only at timescales longer than this. This simple model (first studied by Hasselman, 1976), is the canonical example of how inertia introduced by slow elements of the climate system (in this case thermal inertia of the ocean's mixed layer) smooths out high frequencies to yield a slow response, a "reddening" of the spectrum of climate variability.

In comparison, Fig. 12.3b shows the observed temperature spectra of the atmosphere and of SST. Although the atmospheric spectrum is rather flat, the SST spectrum is much redder, somewhat consistent with the ω^{-2} dependence predicted by Eq. 12-2 above. Simple models of the type described here, explored further in Problem 2 at the end of the chapter, can be used to rationalize such observations.

One aspect of observed air-sea interaction in midlatitudes represented in the above model is that atmospheric changes tend to precede oceanic changes, strongly supporting the hypothesis that midlatitude year-to-year (and even decade-to-decade) variability primarily reflects the slow response of the ocean to forcing by atmospheric weather systems occurring on much shorter timescales. One might usefully call this kind of variability "passive"—it involves modulation by "slower" components of the climate system (in this case the ocean) of random variability of the "faster" component (the atmosphere).

In tropical latitudes, however, changes in SST and tropical air temperatures and winds are more in phase with one another, reflecting the sensitivity of the tropical atmosphere to (moist) convection triggered from below. This sensitivity of the atmosphere to tropical SST can lead to "active" variability—coupled interactions between the atmosphere and the ocean in which changes in one system mutually reinforce changes in the other, resulting in an amplification. In the next section we discuss how such active coupling in the tropical Pacific manifests itself in a phenomenon of major climatic importance known as El Niño.

12.2. EL NIÑO AND THE SOUTHERN OSCILLATION

12.2.1. Interannual variability

Interannual variability of the atmosphere is particularly pronounced in the tropics, especially in the region of the Indian

---

[2]A spectrum, or part of a spectrum, which has less power at higher frequencies, is often called "red," and one that has less power at lower frequencies, "blue." A spectrum that has the same power at all frequencies is called "white." These terms are widely used but imprecisely defined.

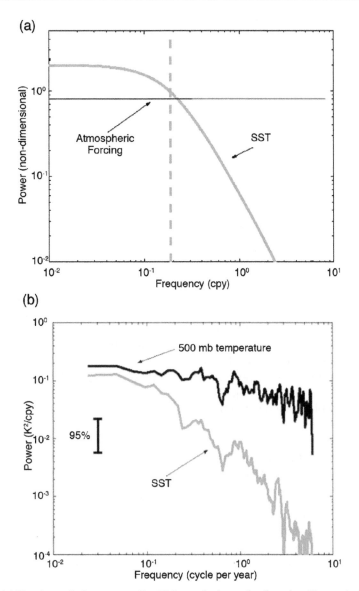

FIGURE 12.3. (a) The theoretical spectrum, Eq. 12-2, graphed on a log-log plot. The vertical grey dotted line indicates the frequency $\omega_c/2\pi$ where $\omega_c = \lambda/\gamma_O$ and is measured in cycles per year (cpy). For the parameters chosen in the text, $\omega_c = \frac{1}{300\,\mathrm{d}}$, and so the grey line is drawn at a frequency of $\frac{1}{2\pi}\frac{365\,\mathrm{d}}{300\,\mathrm{d}} = 0.19$ cpy. (b) Log-log plot of the power spectrum of atmospheric temperature at 500 mbar (black) and SST (grey) associated with the North Atlantic Oscillation, the leading mode of climate variability in the Atlantic sector. See Czaja et al (2003) for more details. The frequency is again expressed in cpy as in (a), and the power in $K^2/$cpy.

and Pacific Oceans, where it is evident, among other things, in occasional failures of the Indian monsoon, extensive droughts in Indonesia and much of Australia, and in unusual rainfall and wind patterns right across the equatorial Pacific Ocean as far

as South America and extending beyond the tropics. This phenomenon has been known for a long time. For example, Charles Darwin, in *Voyage of the Beagle* (1831–1836), noted the tendency for climatic anomalies to occur simultaneously throughout the tropics. Tropical climate variability was first comprehensively described in the 1920s by the meteorologist Gilbert Walker, who gave it the name *Southern Oscillation*.

Manifestations of interannual variability are not, however, confined to the atmosphere. The El Niño phenomenon has been known for centuries to the inhabitants of the equatorial coast of Peru. Until the middle of the 20th century, knowledge of this behavior was mostly confined to the coastal region, where an El Niño is manifested as unusual warmth of the (usually cold) surface waters in the far eastern equatorial Pacific (see Fig. 9.14), and is accompanied (for reasons to be discussed below) by poor fishing and unusual rains.[3] With the benefit of modern data coverage, it is clear that the oceanic El Niño and the atmospheric Southern Oscillation are manifestations of the same phenomenon, which is now widely known by the concatenated acronym ENSO. However, this was not always the case. It was not until the 1960s that Jacob Bjerknes[4] argued that the two phenomena are linked.

12.2.2. "Normal" conditions—equatorial upwelling and the Walker circulation

As discussed in Chapter 7, the lower tropical atmosphere is characterized by easterly trade winds, thus subjecting the tropical ocean to a westward wind stress (see, for example, Figs. 7.28b and 10.2). Let us begin

by considering a hypothetical ocean on an Earth with no continents, and with a purely zonal, steady, wind stress $\tau < 0$ that is independent of longitude, acting on a two-layer ocean with a quiescent deep layer of density ρ_1, capped by a mixed layer of depth h and density $\rho_2 = \rho_1 - \Delta\rho$, as sketched in Fig. 12.4.

The dynamics of Ekman-driven upwelling and downwelling was discussed in Section 10.1. Here, near the equator, we must modify it slightly. As the equator is approached, the Coriolis parameter diminishes to zero and cannot be assumed to be constant. Instead we write $f = 2\Omega \sin\varphi \simeq \beta y$, where $y = a\varphi$ is the distance north of the equator, and $\beta = 2\Omega/a = 2.28 \times 10^{-11} \mathrm{m}^{-1}\,\mathrm{s}^{-1}$ is the equatorial value of the gradient of the Coriolis parameter. The steady zonal

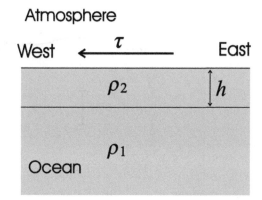

FIGURE 12.4. Schematic of a two-layer ocean model: the upper layer of depth h has a density ρ_2, which is less than the density of the lower layer, ρ_1. A wind stress τ blows over the upper layer.

---

[3]The name "El Niño"—the child (and, by implication, the Christ child)—stems from the observation of the annual onset of a warm current off the Peruvian coast around Christmas. El Niño originally referred to this seasonal warm current that appeared every year, but the term is now reserved for the large-scale warming which happens every few years. The opposite phase—unusually cold SSTs in the eastern equatorial Pacific—is often now referred to as *La Niña*.

[4]Jacob Bjerknes (1897–1975), Norwegian-American meteorologist and Professor at UCLA; son of Vilhelm Bjerknes, the Norwegian pioneer of modern meteorology. Jacob was the first to realize that the interaction between the ocean and atmosphere could have a major impact on the circulation of the atmosphere. He described the phenomenon that we now know as El Niño.

equation of motion is, from Eq. 10-3, anticipating a weak circulation and so neglecting nonlinear advective terms and replacing f by βy:

$$-\beta y v = \frac{1}{\rho_{ref}} \left(-\frac{\partial p}{\partial x} + \frac{\partial \tau_x}{\partial z} \right). \qquad (12\text{-}3)$$

Now, since the wind stress is assumed independent of x, we look for solutions such that all variables are also independent of x, in which case the pressure gradient term in Eq. 12-3 vanishes, leaving

$$-\beta y v = \frac{1}{\rho_{ref}} \frac{\partial \tau_x}{\partial z} .$$

If the deep ocean is quiescent, τ must vanish below the mixed layer; so the above equation can be integrated across the mixed layer to give

$$-\beta y \int_{-h}^{0} v \, dz = \frac{\tau_{wind_x}}{\rho_{ref}} , \qquad (12\text{-}4)$$

which simply states that the Coriolis force acting on the depth-integrated mixed layer flow is balanced by the zonal component of the wind stress τ_{wind}.

In response to the westward wind stress ($\tau < 0$) associated with the easterly tropical trade winds, the flow above the thermocline will be driven northward north of the equator ($y > 0$), and southward for $y < 0$; there is therefore divergence of the flow and consequent upwelling near the equator, as shown in Fig. 12.5. In fact, most of this upwelling is confined to within a few degrees of the equator. In the extratropics, we saw that the adjustment between the mass field and the velocity field in a rotating fluid sets a natural length scale, the deformation radius, $L = \sqrt{g'h/f}$, where

Atmosphere

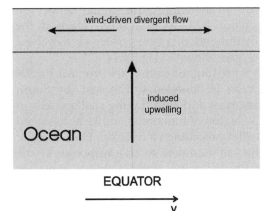

FIGURE 12.5. Schematic meridional cross section of the near-equatorial upwelling induced by a westward wind stress near the equator. Since the upper-layer flow is divergent, mass continuity demands upwelling through the thermocline.

we have replaced 2Ω by f in Eq. 7-23 of Section 7.3.4 and $g' = g\,(\rho_1 - \rho_2)\,/\rho_1$ is the reduced gravity. Here, near the equator, f is no longer approximately constant; for motions of length scale L centered on the equator, $f \sim \beta L$ and so $L^2 \simeq g'h/\beta^2 L^2$, thus defining the *equatorial deformation radius*

$$L_e = \left(\frac{g'h}{\beta^2} \right)^{1/4} .$$

For typical tropical ocean values $h = 100$ m, $g' = 2 \times 10^{-2}$ m s^{-2}, the above gives $L_e \sim 250$ km, or about $2.25°$ of latitude. This is the natural length scale (in the north-south direction) for oceanic motions near the equator.[5]

This upwelling brings cold water up from depth, and thereby cools the upper layers of the near-equatorial ocean. We have in fact

---

[5]Eq. 12-4 is not valid within a distance of order $Y \sim L_e$ of the equator, but this does not affect our deductions about upwelling, since the continuity equation can simply be integrated in y across this region to give, at the base of the mixed layer, $\int_{-Y}^{Y} w \, dy = h \, [v(Y) - v(-Y)]$, yielding the same integrated result without needing to know the detailed variations of $v(y)$ between $\pm Y$.

already seen evidence for this upwelling in the thermal structure: the meridional cross-sections of T in Figs. 9.5 and 9.9 of Chapter 9 show clear upward displacement of the temperature contours near the equator. It shows even more clearly in the distribution of tracers, such as dissolved oxygen, shown in Fig. 11.14.

In reality, of course, the tropical Pacific Ocean is bounded to the east (by South America) and west (by the shallow seas in the region of Indonesia), unlike as assumed in the preceding calculation. One effect of this, in addition to the response to the westward wind stress deduced above, is to make the thermocline deeper in the west, and shallower in the east, as depicted in Fig. 12.6. In consequence, the cold deep water upwells close to the surface in the east, thus cooling the sea surface temperature (SST) there. In the west, by contrast, cold water from depth does not reach the surface, which consequently becomes very warm. This distribution is evident in the equatorial Pacific and (to a lesser degree) Atlantic Oceans (Fig. 9.3). A closeup view of Pacific SST distributions during a warm year (1998) and a cold year (1989) are shown in Fig. 12.7. Note the "cold tongue" in the cold year 1989 extending along the equator from the South American coast.

Its narrowness in the north-south direction is a reflection of the size of the equatorial deformation radius. Aside from being cold, deep ocean water is also rich in inorganic nutrients; thus, the upwelling in the east enriches the surface waters, sustaining the food chain and a usually productive fishery off the South American coast.

The east-west gradient of SST produced by the wind stress in turn influences the atmosphere. The free atmosphere cannot sustain significant horizontal gradients of temperature, because for a finite vertical wind shear, thermal wind balance, Eq. 7-24, demands that $\nabla T \to 0$ as $f \to 0$ at the equator. The regions most unstable to convection are those with the warmest surface temperature. So convection and hence rainfall occurs mostly over the warmest water, which is over the "warm pool" in the western equatorial Pacific, as depicted in Fig. 12.6. The associated latent heating of the air supports net upwelling over this region, which is closed by westerly flow aloft, descent over the cooler water to the east, and a low-level easterly return flow. This east-west overturning circulation is known as the "Walker circulation". While here we have considered this circulation in isolation, in reality it coexists with the north-south Hadley circulation (discussed in Chapter 8). In association

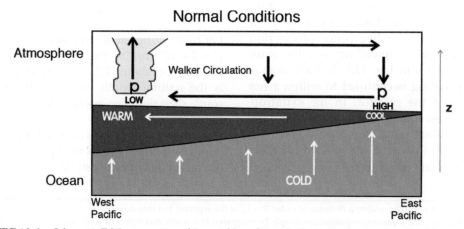

FIGURE 12.6.   Schematic E-W cross section of "normal" conditions in the atmosphere and ocean of the equatorial Pacific basin. The east-west overturning circulation in the atmosphere is called the Walker Circulation.

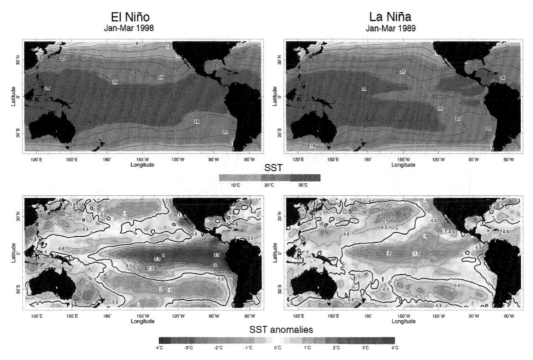

FIGURE 12.7. Sea surface temperatures (SST) (top) and SST anomalies (departure from long-term Jan-Mar mean; bottom) during an El Niño (warm event; left) and La Niña (cold event; right). Green represents warm anomalies, brown cold. Courtesy of NOAA.

with the circulation, there is an east-west equatorial gradient of surface pressure, with low pressure in the west (where the convection occurs) and higher pressure in the east (see Fig. 7.27).

Note that the low-level easterly flow present in the Walker circulation reinforces the trade winds over the equatorial Pacific. In fact, the causal links that were discussed previously can be summarized as in Fig. 12.8. There is thus the potential for positive feedback in the Pacific Ocean-atmosphere system, and it is this kind of feedback that underlies the variability of the tropical Pacific.

12.2.3. ENSO

Atmospheric variability: The Southern Oscillation

As Walker noted, the Southern Oscillation shows up very clearly as a "see-saw" in

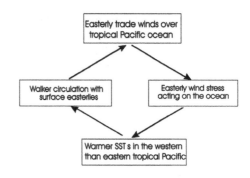

FIGURE 12.8. Schematic of the feedback inherent in the Pacific Ocean-atmosphere interaction. This has become known as the *Bjerknes feedback.*

sea level pressure (SLP) across the tropical Pacific basin. When SLP is higher than normal in the western tropical Pacific region, it tends to be low in the east, and vice versa. This can be made evident by showing how SLP at one location is related to that elsewhere.

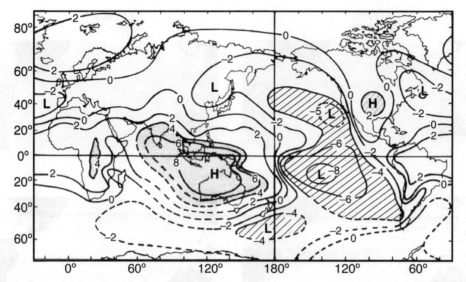

FIGURE 12.9.   Correlations (×10) of the annual-mean sea level pressure with that of Darwin (north of Australia). The magnitude of the correlation exceeds 0.4 in the shaded regions. After Trenberth and Shea (1987).

Fig. 12.9 shows the spatial structure of the temporal correlation[6] of annual-mean SLP with that of Darwin (northern Australia). The correlation reveals a trans-Pacific dipole, with structure roughly similar to that of the Walker cell. Because of the location of the cell that anticorrelates with Darwin SLP, Tahiti SLP is frequently taken to be representative of this cell. It has become conventional to define a "Southern Oscillation Index" (SOI) as

$$SOI = 10 \times \frac{SLP_{Tahiti} - SLP_{Darwin}}{\sigma},$$

where σ is the standard deviation of the pressure difference. The time series of this index for the period 1951–2000 is shown by the solid curve in Fig. 12.10.

The index shows persistent but irregular fluctuations on periods of 2–7 years, with a few outstanding events, such as 1982–1983 and 1997–1998. Since, by definition, $SOI = 0$ under normal conditions—if by "normal"

we mean the long-term average—Fig. 12.10 makes it clear that the tropical atmosphere is rarely in such a state, but rather fluctuates around it.

Such SLP variability is indicative of variability in the meteorology of the region as a whole as well as (to a lesser degree) of higher latitudes. Note, for example, the hint of an impact outside the tropics (such as the L-H-L pattern across N. America) in Fig. 12.9.

Oceanic variability: El Niño and La Niña

Figure 12.10 also shows (dashed curve) a time series of SST anomalies (departures from normal for the time of year) in the far eastern equatorial Pacific Ocean. Like the SOI, the SST shows clear interannual fluctuations on a typical time scale of a few years, and shows a dramatic anticorrelation with the SOI. The correlation coefficient between the two time series, over the period shown, is $C = -0.66$. The negative sign of the correlation, together with the spatial pattern of

---

[6]If two time series are perfectly correlated, then the correlation $C = +1$; if perfectly anticorrelated $C = -1$. If they are uncorrelated, $C = 0$.

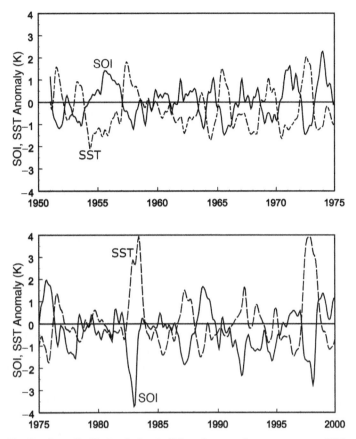

FIGURE 12.10. The Southern Oscillation index (solid) and sea surface temperature (SST) anomaly (K) in the equatorial east Pacific Ocean (dashed), for the period 1951–2000. The SST anomaly refers to a small near-equatorial region off the coast of South America. The two time series have been filtered to remove fluctuations of less than about 3 months.

Fig. 12.9, tells us that warm SST in the east equatorial Pacific approximately coincides with anomalously high pressure in the west and low in the east.

The spatial structure of SST variations is revealed by comparison of the two cases shown in Fig. 12.7. As noted previously, during "normal" or cold conditions, when the SSTs in the eastern equatorial Pacific are at their coldest (illustrated by the case of 1989), the coldest tropical water is concentrated in a narrow tongue extending outward from the South American coast, while the warmest water is found in an extensive warm pool west of the International Date Line. During a warm El Niño

event (illustrated by the case of 1998), the warm water extends much further eastward, and the cold tongue is anomalously weak (in a strong event it may disappear). At such times, the eastern ocean, though still no warmer than the western equatorial Pacific waters, is very much warmer than normal for that time of year. While most of the equatorial Pacific Ocean is anomalously warm, the SST *anomalies* are greatest in the east, where they can be as large as 5 °C. Note from Fig. 12.7 that, for the most part, significant SST variability is concentrated within a few degrees latitude of the equator, consistent with our earlier estimate of the equatorial deformation radius.

Theory of ENSO

The "big picture" of what happens during a warm ENSO event is illustrated schematically in Fig. 12.11. As discussed above, in cold conditions, there is a strong east-west tilt of the thermocline and a corresponding east-west gradient of SST, with cold upwelled water to the east and warm water to the west. Atmospheric convection over the warm water drives the Walker circulation, reinforcing the easterly trade winds over the equatorial ocean. During a warm El Niño event, the warm pool spreads eastward, associated with a relaxation of the tilt of the thermocline. Atmospheric convection also shifts east, moving the atmospheric circulation pattern with it. Pressure increases in the west and decreases in midocean. This adjustment of the Walker circulation, which corresponds to a negative SOI, leads to a weakening or, in a strong event, a collapse of the easterly trade winds, at least in the western part of the ocean. We can summarize the mutual interaction as follows.

First, *the atmosphere responds to the ocean: the atmospheric fluctuations manifested as the Southern Oscillation are mostly an atmospheric response to the changed lower boundary conditions associated with El Niño SST fluctuations.* We should expect (on the basis of our previous discussion) that the Walker circulation and its associated east-west pressure gradient, would be reduced, and the Pacific trade winds weakened, if the east-west contrast in SST is reduced as it is during El Niño. There have been many studies using sophisticated atmospheric general circulation models (GCMs) that have quite successfully reproduced the Southern Oscillation, given the SST evolution as input.

Second, *the ocean responds to the atmosphere: the oceanic fluctuations manifested as El Niño seem to be an oceanic response to the changed wind stress distribution associated with the Southern Oscillation.* This was first argued by Bjerknes, who suggested that the collapse of the trade winds in the west Pacific in the early stages of El Niño (see the lower frame of Fig. 12.11) drives the ocean surface waters

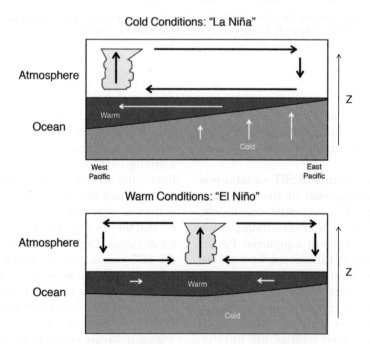

FIGURE 12.11. Schematic of the Pacific Ocean-Atmosphere system during (top) cold La Niña and (bottom) warm El Niño conditions.

eastward (through eastward propagation of a wave of depression on the thermocline); this deepens the thermocline in the east Pacific some two months later. This in turn raises the SST in the east. The basic postulate—that the ocean responds to the atmosphere—has been confirmed in sophisticated ocean models forced by "observed" wind stresses during an El Niño event.

Third, *the El Niño-Southern Oscillation phenomenon arises spontaneously as an oscillation of the coupled ocean-atmosphere system.* Bjerknes first suggested that what we now call ENSO is a single phenomenon and a manifestation of ocean-atmosphere coupling. The results noted previously appear to confirm that the phenomenon depends crucially on feedback between ocean and atmosphere. This is demonstrated in coupled ocean-atmosphere models of varying degrees of complexity, in which ENSO-like fluctuations may arise spontaneously. It appears that stochastic forcing of the system by middle latitude weather systems, which can reach down into the tropics to induce "westerly wind bursts," can also play a role in triggering ENSO events.

Once the El Niño event is fully developed, negative feedbacks begin to dominate the Bjerknes positive feedback, lowering the SST and bringing the event to its end after several months. The details of these negative feedbacks involve some very interesting ocean dynamics. In essence, when the easterlies above the central Pacific start weakening at the beginning of the event, it leads to the formation of an off-equatorial shallower-than-normal thermocline signal, which propagates westward, reflects off the western boundary of the Pacific, and then travels eastwards. After a few months delay the thermocline undulation arrives at the eastern boundary, causing the thermocline to shoal there, so terminating the warm event.

12.2.4. Other modes of variability

The ENSO phenomenon discussed previously is a direct manifestation of strong coupling between the tropical atmosphere and tropical ocean and it gives rise to coherent variability in the coupled climate. There are other modes of variability that arise internally to the atmosphere (i.e., would be present even in the absence of coupling to the ocean below). Perhaps the most important of these is the *annular mode*, a meridional wobble of the subtropical jet stream. The climatological position of the zonal-average, zonal wind, $\bar{u}$, is plotted in Fig. 5.20. But in fact the position and strength of the jet stream maximum varies on all timescales; when it is poleward of its climatological position, $\bar{u}$ is a few $\mathrm{m\,s^{-1}}$ stronger than when it is equatorward. These variations in $\bar{u}$ extend through the depth of the troposphere and indeed right up into the stratosphere. Importantly for the ocean below, the surface winds and air-sea fluxes also vary in synchrony with the annular mode, driving variations in SST and circulation. The manifestation of the annular mode in the northern hemisphere, is known as the *North Atlantic Oscillation*, or NAO for short; the annular mode in the southern hemisphere is known as SAM, for *southern annular mode*. Both introduce stochastic noise into the climate system that can be reddened by interaction with the ocean as discussed in Section 12.1.1.

12.3. PALEOCLIMATE

Here we briefly review something of what is known about the evolution of climate over Earth history. Fig. 12.12 lists standard terminology for key periods of geologic time. Study of paleoclimate is an extremely exciting area of research, a fascinating detective story in which scientists study evidence of past climates recorded in ocean and lake sediments, glaciers and icesheets, and continental deposits. Proxies of past climates are myriad, and to the uninitiated at least, can be bizarre (packrat middens, midges...), including such measurements as the isotopic ratios of shells buried in ocean sediments, thickness and

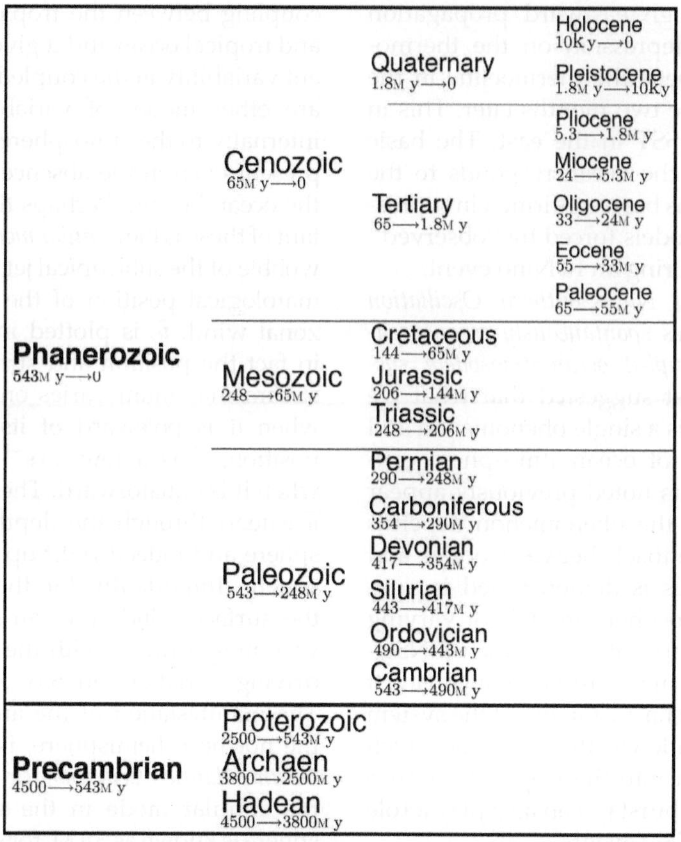

		Holocene 10k y⟶0
	Quaternary 1.8M y⟶0	Pleistocene 1.8M y⟶10Ky
		Pliocene 5.3⟶1.8M y
Cenozoic 65M y⟶0		Miocene 24⟶5.3M y
	Tertiary 65⟶1.8M y	Oligocene 33⟶24M y
		Eocene 55⟶33M y
		Paleocene 65⟶55M y

Phanerozoic 543M y⟶0

Mesozoic 248⟶65M y	Cretaceous 144⟶65M y
	Jurassic 206⟶144M y
	Triassic 248⟶206M y

Paleozoic 543⟶248M y	Permian 290⟶248M y
	Carboniferous 354⟶290M y
	Devonian 417⟶354M y
	Silurian 443⟶417M y
	Ordovician 490⟶443M y
	Cambrian 543⟶490M y

Precambrian 4500⟶543M y

Proterozoic 2500⟶543M y
Archaen 3800⟶2500M y
Hadean 4500⟶3800M y

FIGURE 12.12.   The names and dates of the key periods of geologic time. The unit of time is millions of years before present (My), except during the Holocene, the last 10,000 years (10k y).

density of tree rings, chemical composition of ice, and the radioactivity of corals. Moreover new proxies continue to be developed. What is undoubtedly clear is that climate has been in continual change over Earth history and has often been in states that are quite different from that of today. However, it is important to remember that inferences about paleoclimate are often based on sparse evidence,[7] and detailed descriptions of past climates will never be available. Musing about paleoclimate is nevertheless intellectually stimulating (and great fun), because we can let our imaginations wander, speculating about ancient worlds and what they might tell us about how Earth might evolve in the future. Moreover, the historical record challenges and tests our understanding of the underlying mechanisms of climate and climate change. With such a short instrumental record, paleoclimate observations are essential for evaluating climate variations on timescales of decades and longer. One can be sure that the laws of physics and chemistry (if not biology!) have not changed over time, and so they place strong constraints on what may or may not have happened.

---

[7]One should qualify this statement by recognizing the key role of *observation* in paleoclimate. There are some "hard" paleo observations. For example the glacial terminus (moraines) of North America are incontrovertible evidence of the extent of past glaciations. As our colleague Prof. Ed Boyle reminds us, "Ice Ages would be polite tea-party chit-chat were it not for geologists climbing mountains in muddy boots."

Theory and modeling of paleoclimate and climate change is still rudimentary. This is in part because we must deal not only with the physical aspects of the climate system (difficult enough in themselves) but also with biogeochemical transformations, and on very long timescales, geology and geophysics (cf. Fig. 12.1). Understanding biogeochemistry is particularly important because it is often required to appreciate and quantify the proxy climate record itself. Moreover, because greenhouse gases, such as H_2O, CO_2, and CH_4, are involved in life, we are presented with a much more challenging problem than the mere application of Newton's laws of mechanics and the laws of thermodynamics to the Earth. There are many ideas on mechanisms driving climate change on paleoclimate timescales, only a few on which there is consensus, and even when a consensus forms, there is often little supporting evidence.

Here we have chosen to focus on those aspects of the paleoclimate record for which, it seems to us, there is a broad consensus and are less likely to be challenged as new evidence comes to the fore. In Section 12.3.1 we review what is known about the evolution of climate on the billion year timescale, and then in Section 12.3.2, focus in on the last 70 M y or so. Warm and cold climates are discussed in Sections 12.3.3 and 12.3.4, respectively. We finish by briefly reviewing the evidence for glacial-interglacial cycles and abrupt climate change (Section 12.3.5) and, very briefly, *global warming* (Section 12.3.6).

12.3.1. Climate over Earth history

Earth has supported life of one form or another for billions of years, suggesting that its climate, although constantly changing, has remained within somewhat narrow limits over that time. For example, ancient rocks show markings that are clear evidence of erosion due to running water, and primitive

life forms may go back at least 3.5 B y. One might suppose that there is a natural "thermostat" that ensures that the Earth never gets too warm or too cold. One might also infer that life finds a way to eke out an existence.

It is clear that some kind of thermostat must be in operation because astrophysicists have concluded that 4 B y ago the Sun was burning perhaps 25–30% less strongly than today. Simple one-dimensional climate models of the kind discussed in Chapter 2 suggest that if greenhouse gas concentrations in the distant past were at the same level as today, the Earth would have frozen over for the first two thirds or so of its existence.[8] This is known as the *faint early Sun paradox* (see Problem 4 at the end of the chapter). A solution to the conundrum demands the operation of a thermostat, warming the Earth in the distant past and compensating for the increasing strength of the Sun over time. If the thermostat involved carbon, an assumption that perhaps needs to be critically challenged but is commonly supposed, then we must explain how CO_2 levels in the atmosphere might have diminished over time.

On very long timescales one must consider the exchange of atmospheric CO_2 with the underlying solid Earth in chemical weathering. Carbon is transferred from the Earth's interior to the atmosphere as CO_2 gas produced during volcanic eruptions. This is balanced by removal of atmospheric CO_2 in the chemical weathering of continental rocks, which ultimately deposits the carbon in sediments on the sea floor; see the schematic Fig. 12.13. It is remarkable that the rate of input of CO_2 by volcanic activity and the rate of removal by chemical weathering has remained so closely in balance, even though the input and output themselves are each subject to considerable change.

Volcanic activity is unlikely to be part of a thermostat, because it is driven by heat

---

[8]Indeed there are hints in the paleoclimate record that Earth may have come close to freezing over during several periods of its history (most likely between 500M and 800M y ago), to form what has been called the "snowball Earth."

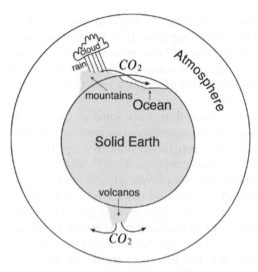

FIGURE 12.13. Carbon from the Earth's interior is injected in to the atmosphere as CO_2 gas in volcanic eruptions. Removal of atmospheric CO_2 on geological timescales is thought to occur in the chemical weathering of continental rocks, being ultimately washed into the ocean and buried in the sediments. The two processes must have been in close, but not exact balance, on geological time scales.

sources deep within the Earth that cannot react to climate change. Chemical weathering of rocks, on the other hand, may be sensitive to climate and atmospheric CO_2 concentrations, because it is mediated by temperature, precipitation, vegetation, and orographic elevation and slope, which are closely tied together (remember the discussion in Section 1.3.2). So, the argument goes, if volcanic activity increased for a period of time, elevating CO_2 levels in the atmosphere, the resulting warmer, moister climate might be expected to enhance chemical weathering, increasing the rate of CO_2 removal and reducing greenhouse warming enough to keep climate roughly constant. Conversely, in a cold climate, arid conditions would reduce weathering rates, leading to a build-up of atmospheric CO_2 and a warming tendency. Scientists vigorously debate whether such a mechanism can regulate atmospheric CO_2; it is currently very difficult to test the idea with observation or models.

Whatever the regulatory mechanism, when the fragmentary paleoclimate record of atmospheric CO_2 levels and temperatures is pieced together over geologic time, a connection emerges. Figure 12.14 shows a synthesis of evidence for continental glaciation plotted along with estimates of atmospheric CO_2 (inferred from the geological record and geochemical models) over the past 600My. Such reconstructions are highly problematical and subject to great uncertainty. We see that CO_2 levels in the atmosphere were thought to have been generally much greater in the distant past than at present, perhaps as much as 10 to 20 times present levels 400–500My ago. Moreover, glaciation appears to occur during periods of low CO_2 and warm periods in Earth history seem to be associated with elevated levels of CO_2. That the temperature and CO_2 concentrations appear to co-vary, however, should not be taken as implying cause or effect.

Factors other than variations in the solar constant and greenhouse gas forcing must surely also have been at work in driving the changes seen in Fig. 12.14. These include changes in the land-sea distribution and orography (driven by plate tectonics), the albedo of the underlying surface, and global biogeochemical cycles. One fascinating idea—known as the Gaia hypothesis—is that life itself plays a role in regulating the climate of the planet, optimizing the environment for continued evolution. Another idea is that ocean basins have evolved on geological timescales through continental drift, placing changing constraints on ocean circulation and its ability to transport heat meridionally. For example, Fig. 12.15 shows paleogeographic reconstructions from the Jurassic (170My ago), the Cretaceous (100My ago), and the Eocene (50My ago). We have seen in Chapters 10 and 11 how the circulation of the ocean is profoundly affected by the geometry of the land-sea distribution and so we can be sure that the pattern of ocean circulation in the past, and perhaps its role in climate, must have been very

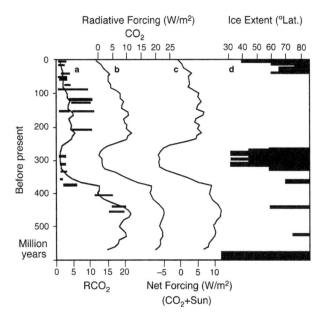

FIGURE 12.14. (a) Comparison of CO_2 concentrations from a geochemical model (continuous line) with a compilation (Berner, 1997) of proxy CO_2 observations (horizontal bars). RCO_2 is the ratio of past atmospheric CO_2 concentrations to present day levels. Thus $RCO_2 = 10$ means that concentrations were thought to be 10 times present levels. (b) CO_2 radiative forcing effects expressed in $W\,m^{-2}$. (c) Combined CO_2 and solar radiance forcing effects in $W\,m^{-2}$. (d) Glaciological evidence for continental-scale glaciation deduced from a compilation of many sources. Modified from Crowley (2000).

different from that of today. It has been hypothesized that the opening and closing of critical oceanic gateways—narrow passages linking major ocean basins—have been drivers of climate variability by regulating the amount of water, heat, and salt exchanged between ocean basins. This, for example, can alter meridional transport of heat by the ocean and hence play a role in glaciation and deglaciation. There are a number of important gateways.

Drake Passage, separating South America from Antarctica, opened up 25–20 M y ago, leaving Antarctica isolated by what we now call the Antarctic Circumpolar Current (Fig. 9.13). This may have made it more difficult for the ocean to deliver heat to the south pole, helping Antarctica to freeze over. However this hypothesis has timing problems. Ice first appeared on Antarctica 35 M y ago, before the opening of Drake Passage, and the most intense glaciation over Antarctica occurred 13 M y ago, significantly

after it opened. Uplift of Central America over the past 10 M y closed a deep ocean passage between North and South America to form the Isthmus of Panama about 4 million years ago. Before then the Isthmus was open, allowing the trade winds to blow warm and possibly salty water between the Atlantic and the Pacific. Its closing could have supported a Gulf Stream carrying tropical waters polewards, as in today's climate, possibly enhancing the meridional overturning circulation of the Atlantic basin and helping to warm northern latitudes in the Atlantic sector, as discussed in Chapter 11. Finally, it has been suggested that closing of the Indonesian seaway, 3–4 M y ago, was a precursor to East African aridification.

12.3.2. Paleotemperatures over the past 70 million years: the $\delta^{18}O$ record

Let us zoom into the last 70 M y period of Fig. 12.14. The paleorecord suggests that

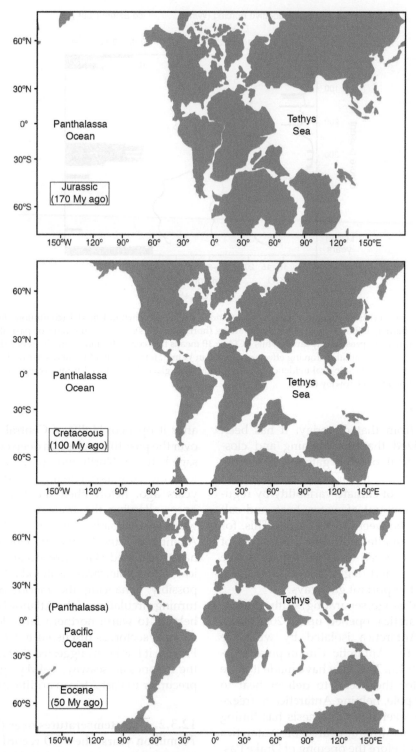

FIGURE 12.15. Paleogeographic reconstructions for (top) the Jurassic (170My ago), (middle) the Cretaceous (100My ago), and (bottom) the Eocene (50My ago). Panthalassa was the huge ocean that in the paleo world dominated one hemisphere. Pangea was the supercontinent in the other hemisphere. The Tethys Sea was the body of water enclosed on three sides (and at times, almost four sides) by the generally "C-shaped" Pangea.

over the last 55 M y there has been a broad progression from generally warmer to generally colder conditions, with significant shorter-term oscillations superimposed. How can one figure this out? Some key supporting evidence is shown in Fig. 12.16 based on isotopic measurements of oxygen. Sediments at the bottom of the ocean provide a proxy record of climate conditions in the water column. One key proxy is $\delta^{18}O$—a measure of the ratio of two isotopes of oxygen, ^{18}O and ^{16}O—which is recorded in sea-bed sediments by the fossilized calcite shells of foraminifera (organisms that live near the surface or the bottom of the ocean). It turns out that the $\delta^{18}O$ in the shells is a function of the $\delta^{18}O$ of the ocean and the temperature of the ocean (see Appendix A.3 for a more detailed discussion of $\delta^{18}O$). The record of $\delta^{18}O$ over the last 55 M y indicates a cooling of the deep ocean by a massive 14°C. In other words, deep ocean temperatures were perhaps close to 16°C (!!) compared to 2°C as observed today (cf. Fig. 9.5). If over

this period of time the abyssal ocean were ventilated by convection from the poles as in today's climate (note how temperature surfaces in the deep ocean thread back to the pole in Fig. 9.5), then one can conclude that surface conditions at the poles must also have been very much warmer. Indeed, this is consistent with other sources of evidence, such as the presence of fossilized remains of palm trees and the ancestors of modern crocodiles north of the Arctic Circle 60 M y ago.

To explain such a large cooling trend, sustained over many millions of years, one needs to invoke a mechanism that persists over this enormous span of time. Following on from the discussion in Section 12.3.1, at least two important ideas have been put forward as a possible cause. Firstly, it has been suggested that the balance in Fig. 12.13 might have changed to reduce CO_2 forcing of atmospheric temperatures over this period due to (a) decreased input of CO_2 from the Earth's interior to the

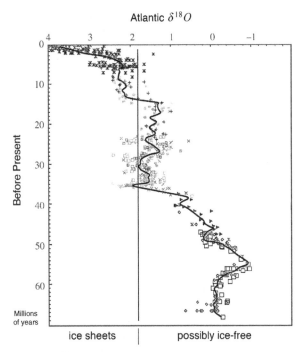

FIGURE 12.16.   A compilation of $\delta^{18}O$ measurements from the fossilized shells of benthic foraminifera analyzed from many sediment cores in the North Atlantic over 70 M y. Modified from Miller et al (1987).

climate system, as the rate of sea-floor spreading decreases over time, reducing volcanic activity, and (b) increased removal of CO_2 from the atmosphere due to enhanced physical and chemical weathering of unusually high-elevation terrain driven by tectonic uplift. Secondly, it has been suggested that poleward ocean heat transport progressively decreased because of changes in the distribution of land and sea, and the opening up and closing of gateways, as briefly discussed in Section 12.3.1.

Whatever the mechanisms at work, as shown in Fig. 12.14, the paleorecord suggests that, over the past 100My or so, the Earth has experienced great warmth and periods of great cold. We now briefly review what "warm" climates and "cold" climates might have been like.

12.3.3. Greenhouse climates

In the Cretaceous period Earth was a "greenhouse world." There were no ice caps and sea level was up to 100–200 m higher than present, largely due to the melting of all ice caps and the thermal expansion of the oceans that were much warmer than today. The great super continent of Pangea had begun to break apart, and by 100My ago one can already recognize present-day continents (Fig. 12.15). High sea level meant that much of the continental areas were flooded and there were many inland lakes and seas. Indeed, the meaning of the word Cretaceous is "abundance of chalk," reflecting the widespread occurrence of limestone from creatures living in the many inland seas and lakes of the period. Broad-leaved plants, dinosaurs, turtles, and crocodiles all existed north of the Arctic Circle.

It is thought that the Cretaceous was a period of elevated CO_2 levels—perhaps as much as five times preindustrial concentrations (see Fig. 12.14)—accounting in part for its great warmth. CO_2 forcing alone is unlikely to account for such warm poles where temperatures were perhaps 25°C warmer than today. One proposed

explanation is that the oceans carried much more heat poleward than today, rendering the poles warmer and the tropics colder. There is speculation that the deep ocean may have been much warmer and saltier than at present, possibly due to convection in the tropics and/or subtropics triggered by high values of salinity, much as observed today in the Eastern Mediterranean. Indeed, the configuration of the continents may have been conducive to such a process; the presence of the Tethys Sea (see Fig. 12.15 middle) and a large tropical seaway extending up in to subtropical latitudes, underneath the sinking branch of the Hadley cell bringing dry air down to the surface, could have increased evaporation and hence salinity to the extent that ocean convection was triggered, mixing warm, salty water to depth. But this is just speculation. It is very difficult to plausibly quantify and model this process and efforts to do so often meet with failure.

Another major challenge in understanding the paleorecord in the Cretaceous is the evidence that palm trees and reptiles were present in the interior of the continents. Crocodiles and (young) palm trees are not frost resistant, indicating that temperatures did not go below freezing even during the peak of winter at some latitudes north of 60° N and in the middle of continents, away from the moderating effects of the ocean. Models, however, simulate freezing conditions in the continental interiors in winter even when CO_2 is increased very dramatically. Note the large seasonal change in temperature in the interior of continents observed in the present climate (Fig. 12.2). Perhaps lakes and small inland seas helped to keep the interior warm.

12.3.4. Cold climates

Most of the time in the last 1My, Earth has been much colder than at present, and ice has encroached much further equatorward (Fig. 12.14). The glacial climate that we know the most about is that at the height of the most recent glacial cycle—the *last glacial maximum* (LGM) between 18 and 23 k y ago, during which ice sheets reached their greatest extent 21 k y ago. Reconstruction of climate at the LGM was carried out in the CLIMAP project[9] employing, in the main, proxy data from ocean sediments. Thick ice covered Canada, the northern United States (as far south as the Great Lakes), northern Europe (including all of Scandinavia, the northern half of the British Isles, and Wales) and parts of Eurasia. The effect on surface elevation is shown in Fig. 12.17, which should be compared to modern conditions shown in Fig. 9.1. Where Chicago, Glasgow, and Stockholm now stand, ice was over 1 km thick. It is thought that the Laurentide Ice sheet covering N. America had roughly the volume of ice locked up in present-day Antarctica. Sea level was about 120–130 m lower than today. Note that the coastline of the LGM shown in Fig. 12.17 reveals that, for example, the British Isles were connected to Europe, and many islands that exist today were joined to Asia and Australia. Most of the population lived in these fertile lowlands, many of which are now under water. Ice sheets on Antarctica and Greenland extended across land exposed by the fall of sea level. Moreover, sea ice was also considerably more extensive, covering much of the Greenland and Norwegian Seas, and persisted through the summer. In the southern hemisphere, Argentina, Chile, and New Zealand were under ice, as were parts of Australia and South America.

Figure 12.17b shows the difference between average August SST centered on the LGM and August SST for the modern era. Many details of this reconstruction have been challenged, but the broad features are probably correct. The average SST was 4°C colder than present and North Atlantic SSTs were perhaps colder by more than 8°C. It appears that low latitude temperatures

were perhaps 2°C lower than today. Winds at the LGM were drier, stronger, and dustier than in the present climate. Ice sheets, by grinding away the underlying bedrock, are very efficient producers of debris of all sizes, which gets pushed out to the ice margin. At the LGM, windy, cold, arid conditions existed equatorward of the ice. Winds scooped up the finer-grained debris, resulting in great dust storms blowing across the Earth's surface with more exposed shelf areas. Indeed, glacial layers in ice cores drilled in both Greenland and Antarctica carry more dust than interglacial layers. Forests shrank and deserts expanded. Today the N. African and Arabian deserts are key sources of dust; at the LGM deserts expanded into Asia. One very significant feature of glacial climates evident in the paleorecord is that they exhibited considerably more variability than warm climates. For example, in an event known as the Younger Dryas, which occurred about 12 k y ago, the climate warmed only to suddenly return to close to LGM conditions for several hundred years; see Fig. 12.23 and the discussion in Section 12.3.5.

Key factors that may explain the dramatically different climate of the LGM are the presence of the ice sheets themselves, with their high albedo reflecting solar radiation back out to space, and (see below) lower levels of greenhouse gases. It is thought that the pronounced climate variability of glacial periods suggested by the paleorecord may have been associated with melting ice producing large inland lakes that were perhaps cut off from the oceans for hundreds of years, but which then intermittently and perhaps suddenly discharged into the oceans. It has been argued that such sudden discharges of buoyant fluid over the surface of the northern N. Atlantic could have had a significant impact on

---

[9]CLIMAP (Climate: Long-range Investigation, Mapping and Prediction), was a major research project of the 1970s and 1980s, which resulted in a map of climate conditions during the last glacial maximum based on proxy data from ocean sediments.

CLIMAP LGM Elevation

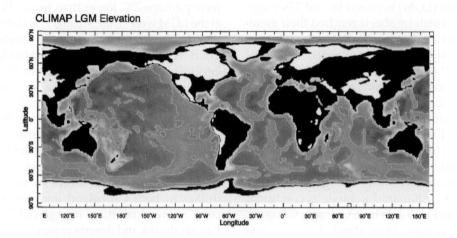

LGM minus Modern August SST

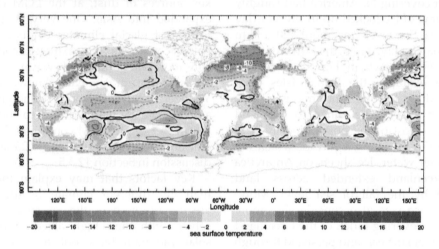

FIGURE 12.17.   (a) CLIMAP reconstruction of elevation at the Last Glacial Maximum (LGM). The white (black) areas represent terrain with a height in excess of (less than) 1.5 km and are indicative of ice-covered areas. The depth of the ocean is represented with a grey scale (dark is deep). The white contour marks the 4 km deep isobath. This figure should be compared with Fig. 9.1. Note the modification of the coast line relative to the modern, due to the 120 m or so drop in sea level. (b) August SST at LGM (from CLIMAP) minus August SST for the modern climate (°C). The brown areas represent negative values, the green areas positive values.

the strength of the ocean's meridional overturning circulation and its ability to transport heat polewards.

12.3.5. Glacial-interglacial cycles

The left frame of Fig. 12.18 shows the $\delta^{18}O$ record over the past 2.5 million years, recorded in the calcite of foraminifera in

sediments of the subpolar North Atlantic. Before about 800 k y ago, one observes remarkable oscillations spanning 2 M y or so, with a period of about 40 k y. After 800 k y ago the nature of the record changes and fluctuations with longer periods are superimposed. These are the signals of great glacial-interglacial shifts on a roughly 100 k y timescale. There have been about 7 such

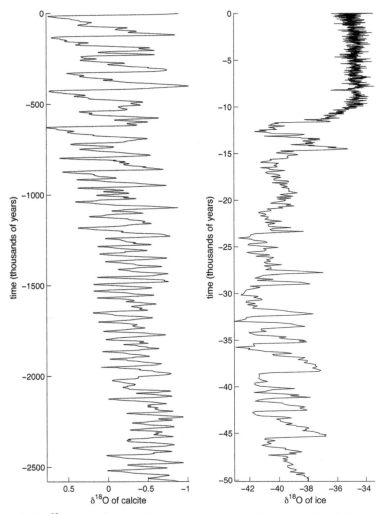

FIGURE 12.18.   Left: δ^{18} O over the last 2.5 million years recorded in the calcite shells of bottom dwelling foraminifera in the subpolar North Atlantic. Shown is the average of tens of δ^{18} O records sampled from various marine sediment cores (Huybers, 2006). Values are reported as the anomaly from the average δ^{18} O over the past million years. More negative values (rightward) indicate warmer temperatures and less ice volume. Right: δ^{18} O of ice over the last 50 k y measured in the GISP2 ice-core (Grootes and Stuiver, 1997). In contrast to the δ^{18} O of marine shells, less negative values in the δ^{18} O of ice indicate warmer atmospheric temperatures, in this case in the vicinity of Greenland.

cycles, during which temperate forests in Europe and North America have repeatedly given way to tundra and ice. Ice has periodically accumulated in the North American and Scandinavian areas until it covered hills and mountains to heights of 2–3 km, as was last observed at the LGM (see Fig. 12.17) and today only in Greenland and Antarctica.

Such glacial-interglacial signals are not limited to the North Atlantic sector. Qualitatively similar signals are evident in different kinds of paleorecords taken from around the world, including deep sea sediments, continental deposits of plants, and ice cores. These reveal a marked range of climate on Earth, cycling between glacial and interglacial conditions.

In particular, ice cores taken from glaciers yield local air temperature,[10] precipitation rate, dust, and direct records of past trace gas concentrations of CO_2 and CH_4. The deepest core yet drilled ($\gtrsim 3$ km), from Antarctica, records a remarkable 700k y history of climate variability shown in Fig. 12.19. The core reveals oscillations of Antarctic air temperature, greenhouse gas concentrations which more-or-less covary with a period of about 100k y. Note, however, that the oscillations do not have exactly the same period. Of the six or seven cycles seen in the Antarctic record, the two most recent have a somewhat longer period than the previous cycles.

The 100k y signals evident in Figs. 12.18 and 12.19 are thought to be representative of climate variability over broad geographical regions. Scientists vigorously debate whether, for example, changes over Antarctica led or lagged those over Greenland, or whether CO_2 changes led or lagged temperature changes. This is very difficult to tie down because of uncertainty in the precise setting of the "clock" within and between records. Here we simply state that at zero order the low frequency signals seem to covary over broad areas of the globe, strongly suggestive of global-scale change.

The oscillations seen in Fig. 12.19 have a characteristic "saw-tooth" pattern, typical of many records spanning glacial-interglacial cycles, with a long period of cooling into the glacial state followed by rapid warming to the following interglacial. Abrupt increases in CO_2 occur during the period of rapid ice melting. Superimposed on the sawtooth are irregular higher frequency oscillations (to be discussed below). Typically, the coolest part of each glacial period and the lowest CO_2 concentrations occur just before the glacial termination. Temperature fluctuations (representative of surface conditions) have a magnitude of about 12°C and CO_2 levels fluctuate between 180 and 300 ppm. The Antarctic dust record also confirms continental aridity. Dust transport was more prevalent during glacial than interglacial times, as mentioned in Section 12.3.4. Finally, it is worthy of note that present levels of CO_2 (around 370 ppm in the year 2000; cf. Fig. 1.3) are unprecedented during the past 700k ys. By the end of this century levels will almost certainly have reached 600 ppm.

Milankovitch cycles

It seems that climate on timescales of 10k y–100k ys is strongly influenced by variations in Earth's position and orientation relative to the Sun. Indeed, as we shall see, some of the expected periods are visible in the paleorecord, but direct association (phasing and amplitude) is much more problematic. Variation in the Earth's orbit over time—known as Milankovitch cycles[11]—cause changes in the amount and distribution of solar radiation

---

[10]Note that $^{18}O/^{16}O$ ratios in ice cores have the opposite relationship to temperature than that of $^{18}O/^{16}O$ ratios in $CaCO_3$ shells (see Appendix A.3). Snow produced in colder air tends to have a lower $\delta^{18}O$ value than snow produced in warmer air. Consequently, the $\delta^{18}O$ value of glacial ice can be used as a proxy for air temperature, with low values indicating colder temperatures than higher values (see Fig. 12.18).

[11]Milutin Milankovitch (1879–1958), the Serbian mathematician, dedicated his career to formulating a mathematical theory of climate based on the seasonal and latitudinal variations of solar radiation received by the Earth. In the 1920s he developed improved methods of calculating variations in Earth's eccentricity, precession, and tilt through time.

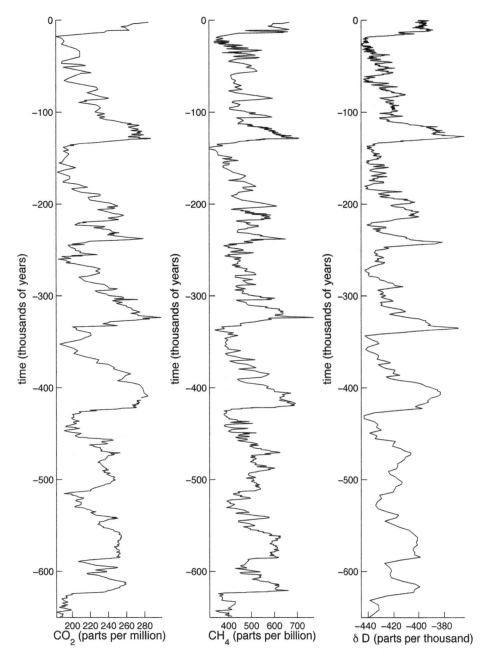

FIGURE 12.19. Ice-core records of atmospheric carbon dioxide (left) and methane (middle) concentrations obtained from bubbles trapped in Antarctic ice. Values to 400 k y ago are from Vostok (Petit et al, 1999), whereas earlier values are from EPICA Dome C (Siegenthaler et al, 2005; Spahni et al, 2005). (right) δ D concentrations from EPICA Dome C (EPICA community members, 2004) measured in the ice, as opposed to the bubbles, are indicative of local air temperature variations, similar to δ^{18} O of ice measurements. A rightward shift corresponds to warming.

reaching the Earth on orbital timescales. Before discussing variations of the Earth's orbit over time, let us return to ideas introduced in Chapter 5 and review some simple facts about Earth's orbit around the Sun and the cause of the seasons.

Imagine for a moment that the Earth travelled around the Sun in a circular orbit, as in Fig. 12.20a (left). If the Earth's spin axis were perpendicular to the orbital plane (i.e., did not tilt), we would experience no seasons and the length of daytime and nighttime would never change throughout the year and be equal to one another. But

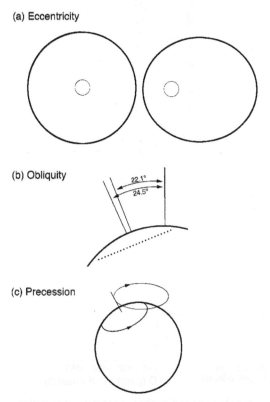

(a) Eccentricity

(b) Obliquity

(c) Precession

FIGURE 12.20. (a) The eccentricity of the Earth's orbit varies on 100k y & 400k y timescales from (almost) zero, a circle, to 0.07, a very slight ellipse. The ellipse shown on the right has an eccentricity of 0.5, vastly greater than that of Earth's path around the Sun. (b) The change in the tilt of the Earth's spin axis—the obliquity—varies between 22.1°and 24.5° on a timescale of 41k y. The tilt of the Earth is currently 23.5°. (c) The direction of the Earth's spin vector precesses with a period of 23k y.

now suppose that the spin axis is tilted as a constant angle, as sketched in Fig. 5.3, and, moreover, that the direction of tilt in space is constant relative to the fixed stars. Now, as discussed in Section 5.1.1, we would experience seasons and the length of daytime would vary throughout the year. When the northern hemisphere (NH) is tilted toward the Sun, the Sun rises high in the sky, daytime is long, and the NH receives intense radiation and experiences summer conditions. When the NH tilts away from the Sun, the Sun stays low in the sky, the daytime is short and the NH receives diminished levels of radiation and experiences winter. These seasonal differences culminate at the summer and winter solstices. In modern times, the longest day of the year occurs on June 21st (the summer solstice) and shortest day of the year on December 21st (the winter solstice) (see Fig. 5.4). The length of the day and night become equal at the equinoxes. Thus we see that seasonality and length of day variations are fundamentally controlled by the tilt of the Earth's axis away from the orbital plane. This tilt of the Earth's axis away from the orbital plane is known as the *obliquity* (see Fig. 12.20b). It varies between 21.1° and 24.5° on about 41k y timescales; at the present time it is 23.5°. Obliquity affects the annual insolation in both hemispheres simultaneously. When the tilt is large, seasonality at high latitudes becomes more extreme but with little effect at the equator.

The Earth's orbit is not exactly circular, however. As shown in Figs. 5.4 and 12.20, Earth moves around the Sun following an elliptical path; the distance from the Sun varies between 153 million km at perihelion (closest distance of the Earth to the Sun) and 158 million km at aphelion (farthest distance between the Earth and Sun). As can be seen in Fig. 5.4, in modern times the Earth is slightly closer to the Sun at the NH winter solstice. Winter radiation is slightly higher than it would be if the Earth followed a perfectly circular orbit. Conversely, at the NH summer solstice the Earth is slightly farther away from the Sun, and so NH summer radiation is slightly lower than it would be if the Earth followed a perfectly circular orbit. This is a rather small effect, however, because the Earth-Sun distance only varies by 3% of the mean. Nevertheless the *eccentricity* of the Earth's orbit around the Sun (see Fig. 12.20a), a measure of its degree of circularity, enhances or reduces the seasonal variation of the intensity of radiation received by the Earth. The eccentricity varies with periods of about 100ky and 400ky. It modulates seasonal differences and precession, the third important orbital parameter.

Precession measures the direction of the Earth's axis of rotation, which affects the magnitude of the seasonal cycle and is of opposite phase in the two hemispheres. Earth's spin axis precesses at a period of 27ky with respect to the fixed stars. However, this is not the climatically relevant period because the direction of the major axis of Earth's eccentric orbit also moves. Thus climatologists define the *climatic precession* as the direction of Earth's spin axis with respect to Earth's eccentric orbit. This has a period of about 23ky. Today the rotation axis points toward the North Star, so setting the dates during the year at which the Earth reaches aphelion and perihelion on its orbit around the Sun (see Fig. 5.4). At the present time, perihelion falls on January 3rd, only a few weeks after the winter solstice, and so the northern hemisphere winter and southern hemisphere summer are slightly warmer than the corresponding seasons in the opposite hemispheres.

We discussed in Chapters 5 and 8 those factors that control the annual-mean temperature as a function of latitude and in particular the importance of the latitudinal dependence of incoming solar radiation. This latitudinal dependence is critically modulated by orbital parameters. Because of their different periodicity (see Fig. 12.21), the composite variations in solar radiation are very complex. They are functions of both latitude and season, as well as time.

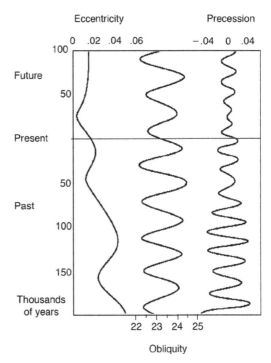

FIGURE 12.21. Variations in eccentricity, precession, and obliquity over 300ky, starting 200ky in the past, through the present day and 100ky in to the future. From Berger and Loutre, (1992).

Variations in summer insolation in middle to high latitudes are thought to play a particularly important role in the growth and retreat of ice sheets: melting occurs only during a short time during the summer and ice surface temperature is largely determined by insolation. Thus cool summers in the northern hemisphere, where today most of the Earth's land mass is located, allow snow and ice to persist through to the next winter. In this way large ice sheets can develop over hundreds to thousands of years. Conversely, warmer summers shrink ice sheets by melting more ice than can accumulate during the winter.

Figure 12.22 shows insolation variations as a function of latitude and seasons during various phases of Earth's orbit. These can be calculated very accurately, as was first systematically carried out by Milankovitch. Note that fluctuations of order 30 $W m^{-2}$ occur in middle to high

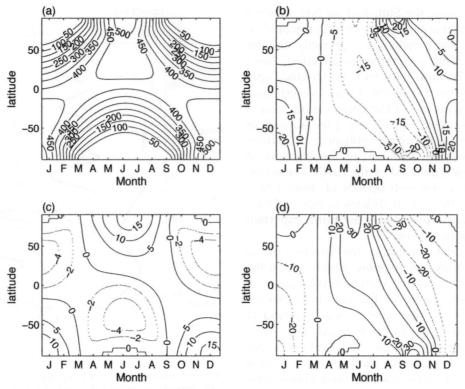

FIGURE 12.22.   Insolation at the top of the atmosphere computed using the orbital solution of Berger and Loutre (1992). (a) Daily average intensity in W m^{-2} contoured against latitude and month, indicating average conditions over the last two million years. (b) Modern insolation plotted as an anomaly from average conditions. (c) Insolation averaged during each maximum of obliquity over the last two million years and shown as an anomaly from average conditions. (d) Similar to (c) but for when Earth is closest to the Sun during northern hemisphere summer solstice.

latitudes, a significant signal comparable, for example, to the radiative forcing due to clouds.

Astronomical forcing is an immensely appealing mechanism, offering a seemingly simple explanation of climate variability on timescales of tens to hundreds of thousands of years. It is widely applied in an attempt to rationalize the paleorecord. One of the most convincing pieces of evidence of astronomical periods showing up in the paleorecord are the fluctuations in $\delta^{18}O$ of calcite found in North Atlantic deep sea cores shown in Fig. 12.18 (left) over the past 2.5My. As discussed previously, an oscillation with a period of about 40ky, that of obliquity, can be seen by eye for the first 2My of the record. However, the 100ky cycles at the end of the record (see also Fig. 12.19), which

are signatures of massive glacial-interglacial cycles, may have little directly to do with orbital forcing, which has very little power at this period. Perhaps the 100Ky cycle is being set by internal dynamics of the ice sheets, almost independently of orbital forcing. Whatever the extent of orbital forcing, it must be significantly amplified by positive feedbacks involving some or all of the following: water vapor, ice-albedo interactions, clouds, ocean circulation, internal ice-sheet dynamics, among many other processes.

In summary, many theories have been put forward to account for the shape and period of oscillations of the kind seen in, for example, Figs. 12.18 and 12.19, but none can account for the observed record and none is generally accepted.

Abrupt Climate Change

As we have seen, over Earth history the climate of the planet has been in markedly different states, ranging from a "greenhouse" to an "icehouse." Moreover the paleorecord suggests that there have been very rapid oscillations between glacial and interglacial conditions. For example, the 50 k y record of $\delta^{18}O$ shown in Fig. 12.18 (right), taken from an ice core in Greenland, reveal many large rightward spikes on millennial timescales (indicating frequent abrupt transitions to warmer followed by a return to colder conditions). These are called Dansgaard-Oeschger events (or D-O for short, after the geochemists Willi Dansgaard and Hans Oeschger who first noted them) and correspond to abrupt warmings of Greenland by 5–10°C, followed by gradual cooling and then an abrupt drop to cold conditions again. They were probably confined to the N. Atlantic and are less extreme than the difference between glacial and interglacial states. Scientists have also found evidence of millennial timescale fluctuations in the extent of ice-rafted debris deposited in sediments in the North Atlantic—known as Heinrich events (after the marine geologist Hartmut Heinrich). They are thought to be the signature of intermittent advance and retreat of the sea-ice edge. D-O and Heinrich events are examples of what are called "abrupt" climate changes, because they occur on timescales very much shorter (10, 100, 1000 y) than that of external climate forcings, such as Milankovitch cycles, but long compared to the seasons. It is important to realize then that the LGM was not just much colder than today, but that it repeatedly and intermittently swung between frigid and milder climates in just a few decades. Indeed such erratic behavior is a feature of the last 100 k y of climate history.

The general shift from colder, dustier conditions to warmer, less dusty conditions over the last 10 k y or so seen in Fig. 12.23, is generally interpreted as the result of orbital-scale changes in obliquity and precession. Obliquity reached a maximum

10 k y ago (see Fig. 12.21), enhancing the seasonal cycle and producing a maximum of summer insolation at all latitudes in the northern hemisphere (Fig. 12.22c), so making it less likely that ice survives the summer. Atmospheric CO_2 concentrations may also have played a role (although it is not known to what extent they are a cause or an effect), increasing from 190 ppm to 280 ppm (Fig. 12.19). The combination of increased summer insolation and increased CO_2 concentrations probably triggered melting of the massive northern ice sheets, with ice-albedo feedbacks helping to amplify the shifts. It is thought that huge inland lakes were formed, many times the volume of the present Great Lakes, which may have intermittently and suddenly discharged into the Arctic/Atlantic Ocean. As can be seen in Fig. 12.23, the warming trend after the last ice age was not monotonic but involved large, short-timescale excursions. Evidence from the deposits of pollen of the plant *Dryas octopetala*, which thrives today in cold tundra in Scandinavia, tells us that 12 k y ago or so, warming after the LGM was punctuated by a spell of bitter cold, a period now known as the Younger Dryas. Further evidence for this cold period, together with numerous other fluctuations, come from Greenland ice cores such as that shown in Fig. 12.18 (right). Along with the longer-term trends, one observes (Fig. 12.23) spectacular, shorter-term shifts, of which the Younger Dryas is but one.

After the last ice age came to an end, the climate warmed up dramatically to reach present day conditions around 10 k y ago. Since then climate has settled in to a relatively quiescent mode up until the present day. This period—the last 10 k y—is known as the Holocene. There was a warm climatic optimum between 9k and 5k y ago, during which, for example, El Niño appears to have been largely absent. The relatively benign climate of the Holocene is perhaps the central reason for the explosion in the development of human social and

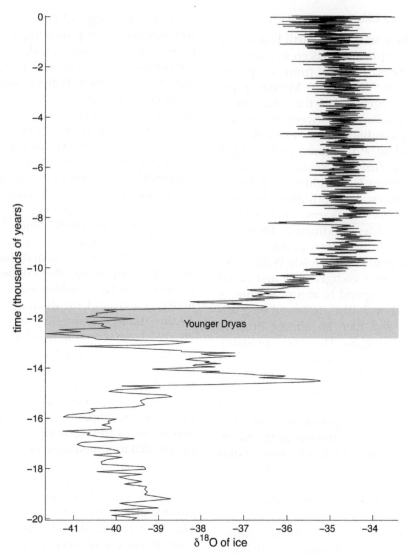

FIGURE 12.23.   The transition from the Last Glacial Maximum to the relatively ice-free conditions of the Holocene took roughly ten thousand years. In certain regions this transition was punctuated by rapid climate variations having timescales of decades to millennia. Shown is the GISP2 ice-core (Grootes and Stuiver, 1997) with shading indicating the return to glacial-like conditions, a period known as the Younger Dryas. The Younger Dryas is a prominent feature of many North Atlantic and European climate records and its presence can be detected in climate records across much of the Northern Hemisphere.

economic structures, farming, and agriculture. Before the Holocene, agriculture was perhaps impossible in much of Northern Europe, because the variance in climate was so great.

A commonly held view is that an important mechanism behind rapid climate shifts is fluctuation of the ocean's thermohaline circulation discussed in Chapter 11.

The thermohaline circulation may have been sensitive to freshwater discharge from inland lakes formed from melting ice. The discharge of fresh water may have occurred intermittently and perhaps involved large volumes of fresh water sufficient to alter the surface salinity and hence buoyancy of the surface ocean. Let us return to Fig. 11.28 (bottom), which shows the ocean's meridional overturning circulation (MOC), with a deep-sinking branch in the northern North Atlantic. Warm, salty water is converted into colder, fresher water by heat loss to the atmosphere and fresh water supply from precipitation and ice flow from the Arctic. As discussed in Chapter 11, in the present climate the MOC carries heat poleward, helping to keep the North Atlantic ice-free. But what might have happened if, for some reason, fresh water supply to polar convection sites was increased, as was likely in the melt after the LGM, reducing salinity and so making it more difficult for the ocean to overturn? One might expect the MOC to decrease in strength,[12] with a concomitant reduction in the supply of heat to northern latitudes by ocean circulation, perhaps inducing cooling and accounting for the abrupt temperature fluctuations observed in the record.

Theories that invoke changes in the ocean's MOC as an explanation of abrupt climate change signals, although appealing, are not fully worked out. Sea ice, with its very strong albedo and insulating feedbacks that dramatically affect atmospheric temperature, is a potential amplifier of climate change. Moreover sea ice can also grow and (or) melt rapidly because of these positive feedbacks and so is likely to be an important factor in abrupt climate change. A wind field change could also account for the observed correlation between reconstructed Greenland temperatures and deep sea cores, with changes in ocean circulation being driven directly by the wind. Moreover, the wind field is likely to be very sensitive to the presence or absence of ice, because of its elevation, roughness, and albedo properties.

12.3.6. Global warming

Since the 1950s, scientists have been concerned about the increasing atmospheric concentrations of CO_2 brought about by human activities (cf. Fig. 1.3). The problem, of course, is that the carbon locked up in the oil and coal fields, the result of burial of tropical forests over tens of millions of years, are very likely to be returned to the atmosphere in a few centuries. As already mentioned, by the end of this century atmospheric CO_2 concentrations are likely to reach 600 ppm, not present on the Earth for perhaps 10 M y (Fig. 12.14). There is concern that global warming will result and indeed warming induced by human activity appears to be already underway. Figure 12.24 shows temperature reconstructions of northern hemisphere surface air temperature during the last 1100 y together with the instrumental record over the past 150 y or so. The spread between the reconstructions indicates a lower bound on the uncertainty in these estimates. Even after taking due note of uncertainty and that the temperature scale is in tenths °C, the rapid rise in the late twentieth century is alarming and, should it continue, cause for concern.

Global warming could occur gradually, over the course of a few centuries. However, some scientists speculate that the climate might be pushed into a more erratic state that could trigger abrupt change. If the atmosphere were to warm, so the argument goes, it would contain more water vapor, resulting in an enhancement of meridional water vapor transport, enhanced precipitation over the pole, a suppression of ocean convection, a reduction in the intensity of the MOC, thence a reduction in the meridional ocean heat transport, and so an abrupt cooling of the high-latitude climate.

Even though the possibility of imminent abrupt climate change is small and, as far as we know, even less likely to occur in warm periods such as our own, it must be taken very seriously because the impacts on the environment and humanity would be so large, the more so if the transition were to be very abrupt. As we have seen,

---

[12]There is a commonly held misconception that a weakening of the Atlantic MOC is synonymous with a weakening of the Gulf Stream. As discussed in detail in Chapter 10, the Gulf Stream is a wind-driven phenomenon whose strength depends on the wind and is not directly related to buoyancy supply.

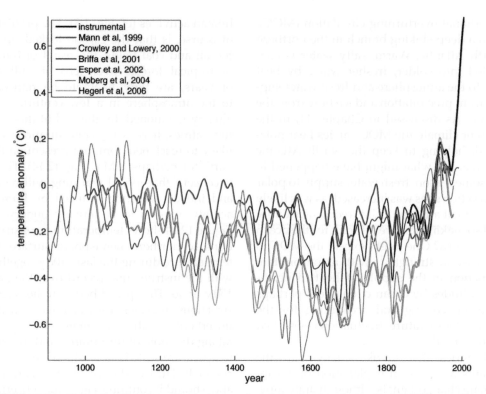

FIGURE 12.24.   Estimates of Northern Hemisphere surface air temperature during the last 1100 years. Temperatures obtained from instruments (Jones and Moberg, 2003) are shown in black. Colored curves indicate different
proxy reconstructions of temperature. Proxies, such as tree rings, ice cores, and corals, are necessary for estimating
temperature before widespread instrumental coverage, before about 1850. The spread between the reconstructions
indicates a lower-bound on the uncertainty in these estimates. All records have been smoothed using a 20-year
running average and adjusted to have zero-mean between 1900 and 1960.

the paleorecord suggests that such events
have happened very rapidly in the past (on
timescales as short as a decade). Moreover,
climate models support the idea that the
ocean's MOC, with its coupling to ice and
the hydrological cycle, is a sensitive component of the climate system. We simply do
not know the likelihood of an abrupt climate
shift occuring in the future or, should it do
so, the extent to which human activities may
have played a role.

12.4. FURTHER READING

A good, basic discussion of the physics
of El Niño can be found in Philander (1990).
A comprehensive introductory account of

climate over Earth history from the perspective of the paleoclimate record is given in
Ruddiman (2001). Burroughs (2005) brings
a fascinating human perspective to his
account of climate change in prehistory.

12.5. PROBLEMS

1. Consider a homogeneous slab of
   material with a vertical diffusivity, k_v,
   subject to a flux of heat through its
   upper surface, which oscillates at
   frequency ω given by $Q_{net} = \mathrm{Re}$
   $\hat{Q}_\omega e^{i\omega t}$, where $\hat{Q}_\omega$ sets the amplitude of
   the net heat flux at the surface. Solve
   the following diffusion equation for
   temperature variations within the slab,

$$\frac{dT}{dt} = k_v \frac{d^2T}{dz^2}$$

assuming that $k_v\, dT/dz = Q_{net}/\rho c$ at the surface ($z = 0$, where ρ is the density of the material and c is its specific heat) and that $T \longrightarrow 0$ at great depth ($z = -\infty$).

(a) Use your solution to show that temperature fluctuations at the surface have a magnitude of $\widehat{Q}_\omega/\rho c\gamma\omega$ where $\gamma = \sqrt{k_v/\omega}$ is the e-folding decay scale of the anomaly with depth.

(b) Show that the phase of the temperature oscillations at depth lag those at the surface. On what does the lag depend?

(c) For common rock material, $k_v = 10^{-6}$ m^2 s^{-1}, $\rho = 3000$ kg m^{-3} and $c = 1000$ J kg^{-1} K^{-1}. Use your answers in (a) to estimate the vertical scale over which temperature fluctuations decay with depth driven by (i) diurnal and (ii) seasonal variations in $\widehat{Q}_\omega$. If $\widehat{Q}_\omega = 100$ Wm^{-2}, estimate the magnitude of the temperature fluctuations at the surface over the diurnal and seasonal cycles. Comment on your results in view of the fact that the freezing depth—the depth to which soil normally freezes each winter—is about 1 m in the NE of the U.S. In areas of central Russia, with extreme winters, the freezing depth can be as much as 3 m, compared to, for example, San Francisco where it is only a few centimeters. In the Arctic and Antarctic the freezing depth is so deep that it becomes year-round permafrost. Instead, there is a thaw line during the summer.

2. Imagine that the temperature of the ocean mixed layer of depth h,

governed by Eq. 12-1, is forced by air-sea fluxes due to weather systems represented by a white-noise process $Q_{net} = \widehat{Q}_\omega e^{i\omega t}$, where $\widehat{Q}_\omega$ is the amplitude of the forcing at frequency ω. Solve Eq. 12-1 for the temperature response $T = \text{Re}\widehat{T}_\omega e^{i\omega t}$, and show that:

$$\widehat{T}_\omega = \frac{\widehat{Q}_\omega}{\gamma_O \left(\frac{\lambda}{\gamma_O} + i\omega \right)}.$$

Hence show that it has a spectrum, $\widehat{T}_\omega \widehat{T}_\omega^*$, where $\widehat{T}_\omega^*$ is the complex conjugate, given by Eq. 12-2. Graph the spectrum using a log-log plot and hence convince yourself that fluctuations with a frequency greater than λ/γ_O are damped.

3. For the one-layer "leaky greenhouse" model considered in Fig. 2.8 of Chapter 2, suppose that, all else being fixed, the atmospheric absorption depends linearly on atmospheric CO_2 concentration as

$$\epsilon = \epsilon_0 + [CO_2]\,\epsilon_1\,,$$

where $[CO_2]$ is CO_2 concentration (in ppm), $\epsilon_0 = 0.734$, and $\epsilon_1 = 1.0 \times 10^{-4}$ (ppm)$^{-1}$. Calculate, for this model, the surface temperature:

(a) for the present atmosphere, with $[CO_2] = 380$ ppm (see Table 1.2);

(b) in pre-industrial times, with $[CO_2] = 280$ ppm; and

(c) in a future atmosphere with $[CO_2]$ doubled from its present value.

4. Faint early Sun paradox

The emission temperature of the Earth at the present time in its history is 255 K. Way back in the early history of the solar system, the radiative output of the Sun was thought to be 25% less than it is now. Assuming all else (Earth-Sun distance, Earth albedo, atmospheric concentration of greenhouse gases, etc.) has remained fixed, use the one-layer "leaky greenhouse" model explored in Problem 3 to:

(a) determine the emission temperature of the Earth at that time if greenhouse forcing then was the same as it is now. Hence deduce that the Earth must have been completely frozen over.

(b) if the early Earth were not frozen over because of the presence of elevated levels of CO_2, use your answer to Problem 3 to estimate how much CO_2 would have had to have been present. Comment on your answer in view of Fig. 12.14.

5. Bolide impact

There is strong evidence that a large meteorite or comet hit the Earth about 65 My ago near the Yucatan Peninsula, extinguishing perhaps 75% of all life on Earth—the K-T extinction marking the end of the Cretaceous (K) (see Fig. 12.12). It is speculated that the smoke and fine dust generated by the resulting fires would have resulted in intense radiative heating of the midtroposphere with substantial surface cooling (by as much as 20°C) which could interrupt plant photosynthesis and thus destroy much of the Earth's vegetation and animal life.

A slight generalization of the one-dimensional problems considered in Chapter 2 provide insights in to the problem.

By assuming that a fraction 'f' of the incoming solar radiation in Fig. 2.8 is absorbed by a dust layer and that, as before, a fraction 'ϵ' of terrestrial wavelengths emitted from the ground is absorbed in the layer, show that:

$$T_s = \left(\frac{2-f}{2-\epsilon}\right) T_e,$$

where T_e is the given by Eq. 2-4. [Hint: write down expressions for the radiative equilibrium of the dust layer and the ground.]

Investigate the extreme case where the dust layer is so black that it has zero albedo (no radiation reflected, $\alpha_p = 0$) and is completely absorbing ($f = 1$) at solar wavelengths.

6. Assuming that the land ice over the North American continent at the Last Glacial Maximum shown in Fig. 12.17 had an average thickness of 2 km, estimate the freshwater flux into the adjacent oceans (in Sv) that would have occurred if it had completely melted in 10 y, 100 y, 1000 y. Compare your estimates to the observed freshwater meridional flux in the ocean, Fig. 11.32. Another useful comparative measure is the flux of the Amazon river, 0.2 Sv.

Appendices

A.1. Derivations
    A.1.1. The Planck function
    A.1.2. Computation of available potential energy
    A.1.3. Internal energy for a compressible atmosphere
A.2. Mathematical definitions and notation
    A.2.1. Taylor expansion
    A.2.2. Vector identities
    A.2.3. Polar and spherical coordinates
A.3. Use of foraminifera shells in paleo climate
A.4. Laboratory experiments
    A.4.1. Rotating tables
    A.4.2. List of laboratory experiments
A.5. Figures and access to data over the web

A.1. DERIVATIONS

A.1.1. The Planck function

A blackbody is a theoretical construct that absorbs 100% of the radiation that hits it. Therefore it reflects no radiation and appears perfectly black. It is also a perfect emitter of radiation. Planck showed that the power per unit area, per unit solid angle, per unit wavelength, emitted by a black body is given by:

$$B_\lambda(T) = \frac{2hc^2}{\lambda^5 \left(\exp\left[\frac{hc}{\lambda kT} \right] - 1 \right)}, \qquad \text{(A-1)}$$

where h is Planck's constant, c is the speed of light, k is Boltzmann's constant and λ is the wavelength of the radiation. Figures 2.2 and 2.3 are plots of $B_\lambda(T)$ against wavelength for various T's.

If B_λ is integrated over all wavelengths, one obtains the blackbody radiance:

$$\int_0^\infty B_\lambda(T)d\lambda = \frac{\sigma}{\pi} T^4$$

where $\sigma = 2\pi^5 k^4 / 15h^3 c^2$ is the Stefan-Boltzmann constant. The above can be written as an integral over $\ln \lambda$ thus:

$$T^{-4} \int_0^\infty \lambda B_\lambda(T) d\ln \lambda = \frac{\sigma}{\pi}. \qquad \text{(A-2)}$$

So if $T^{-4}\lambda B_\lambda$ is plotted against $\ln \lambda$ then the area under the curve is independent of T. This is the form plotted in Fig. 2.6. For more details see Andrews (2000).

A.1.2. Computation of available potential energy

Consider the two-layer fluid shown in Fig. 8.9 in which the interface is given by:

$$h = \frac{1}{2}H + \gamma y.$$

The potential energy of the system is (noting that $\gamma < H/2L$, i.e., the interface does not intersect the upper or lower boundaries)

$$P = \int_0^H \int_{-L}^L g\rho z \, dy \, dz$$

$$= g \int_{-L}^L dy \left[\int_0^{h(y)} \rho_2 z \, dz + \int_{h(y)}^H \rho_1 z \, dz \right]$$

$$= g\rho_1 \int_{-L}^L dy \int_0^H z \, dz + g \, \Delta\rho \int_{-L}^L dy \int_0^{h(y)} z \, dz$$

$$= gH^2 L \rho_1 + g\frac{\Delta\rho}{2} \int_{-L}^L h^2(y) \, dy.$$

Substituting for h we have

$$\int_{-L}^L h^2(y) \, dy = \int_{-L}^L \left[\frac{1}{4}H^2 + \lambda y + \lambda^2 y^2 \right] dy$$

$$= \frac{1}{2}H^2 L + \frac{2}{3}\lambda^2 L^3,$$

and so

$$P = gH^2 L \left(\rho_1 + \frac{\Delta\rho}{4} \right) + g\frac{\Delta\rho}{3}\gamma^2 L^3,$$

which, when expressed in terms of reduced gravity, $g' = g(\rho_1 - \rho_2)/\rho_1$, leads to Eq. 8-9.

A.1.3. Internal energy for a compressible atmosphere

Internal energy for a perfect gas is defined as in Eq. 8-12 of Section 8.3.4:

$$IE = c_v \int \rho T \, dV = \frac{c_v}{R} \int p \, dV$$

$$= \frac{c_v}{R} \int dA \int_{z_s}^\infty p \, dz,$$

where the ideal gas law has been used, dA is an area element such that $dV = dA \, dz$, and

where z_s is the height of the Earth's surface. If we neglect surface topography, so that $z_s = 0$, then, integrating by parts,

$$\int_0^\infty p \, dz = [zp]_0^\infty - \int_0^\infty \frac{\partial p}{\partial z} z \, dz.$$

Since we saw in Chapter 3 that pressure decays approximately exponentially with height, $(zp) \to 0$ as $z \to \infty$. Therefore, using hydrostatic balance, we have

$$\int_0^\infty p \, dz = g \int_0^\infty \rho z \, dz,$$

and so the internal energy is

$$IE = \frac{c_v}{R}g \int z\rho \, dV.$$

This is the form given in Eq. 8-12.

A.2. MATHEMATICAL DEFINITIONS AND NOTATION

A.2.1. Taylor expansion

We assume that the function $r(x)$ can be expressed as an expansion at the point x_A in terms of an infinite series thus:

$$r(x) = c_0 + c_1 (x - x_A) + c_2 (x - x_A)^2$$
$$+ c_3 (x - x_A)^3 + \ldots,$$

where the cs are constants. To find c_0 we set $x = x_A$ to yield $c_0 = r(x_A)$. To find c_1 we differentiate once with respect to x and then set $x = x_A$ to yield $c_1 = (dr/dx)_A$, where the subscript A indicates that dr/dx is evaluated at $x = x_A$. Carrying on we see that we can write $(d^m r/dx^m)_A = m!c_m$ and hence:

$$r(x) = r(x_A) + (x - x_A) \left(\frac{dr}{dx} \right)_A +$$
$$\frac{(x - x_A)^2}{2!} \left(\frac{d^2 r}{dx^2} \right)_A + \ldots$$

So, if $x = x_A + \delta x$, where δx is a small increment in x, then, to first order in δx, we may write:

$$r(x_A + \delta x) \simeq r(x_A) + \delta x \left(\frac{dr}{dx} \right)_A .$$

Taylor expansions were often used in Chapter 4 (in the derivation of vertical stability criteria) and in Chapter 6 (in the derivation of the equations that govern fluid motion).

A.2.2. Vector identities

Cartesian coordinates are best used for getting our ideas straight, but occasionally we also make use of polar (sometimes called cylindrical) and spherical polar coordinates (see Section A.2.3).

The Cartesian coordinate system is defined by three axes at right angles to each other. The horizontal axes are labeled x and y, and the vertical axis is labeled z, with associated unit vectors $\hat{\mathbf{x}}, \hat{\mathbf{y}}, \hat{\mathbf{z}}$ (respectively). To specify a particular point we specify the x coordinate first (abscissa), followed by the y coordinate (ordinate), followed by the z coordinate, to form an ordered triplet (x, y, z).

If ϕ is a scalar field and $\mathbf{a}$, a 3-dimensional vector field thus:

$$\mathbf{a} = a_x \hat{\mathbf{x}} + a_y \hat{\mathbf{y}} + a_z \hat{\mathbf{z}}$$

where a_x, a_y and a_z are magnitudes of the projections of $\mathbf{a}$ along the three axes, then we have the following definitions:

I. $\nabla \phi = \hat{\mathbf{x}} \frac{\partial \phi}{\partial x} + \hat{\mathbf{y}} \frac{\partial \phi}{\partial y} + \hat{\mathbf{z}} \frac{\partial \phi}{\partial z}$ [the "gradient of ϕ" or "grad ϕ," a vector]

II. $\nabla \cdot \mathbf{a} = \frac{\partial a_x}{\partial x} + \frac{\partial a_y}{\partial y} + \frac{\partial a_z}{\partial z}$ [the "divergence of $\mathbf{a}$" or "div $\mathbf{a}$," a scalar]

III. $\nabla \times \mathbf{a} = \hat{\mathbf{x}} \left(\frac{\partial a_z}{\partial y} - \frac{\partial a_y}{\partial z} \right) + \hat{\mathbf{y}} \left(\frac{\partial a_x}{\partial z} - \frac{\partial a_z}{\partial x} \right) + \hat{\mathbf{z}} \left(\frac{\partial a_y}{\partial x} - \frac{\partial a_x}{\partial y} \right)$ [the "curl of $\mathbf{a}$," a vector]

Inspection of I, II, and III shows that we can define the operator ∇ uniquely as:

IV. $\nabla \equiv \hat{\mathbf{x}} \frac{\partial}{\partial x} + \hat{\mathbf{y}} \frac{\partial}{\partial y} + \hat{\mathbf{z}} \frac{\partial}{\partial z}$

The quantity $\nabla \cdot (\nabla \phi)$ is:

$$\nabla \cdot (\nabla \phi) = \frac{\partial^2 \phi}{\partial x^2} + \frac{\partial^2 \phi}{\partial y^2} + \frac{\partial^2 \phi}{\partial z^2} = \nabla^2 \phi$$

V. $\mathbf{a} \cdot \mathbf{b} = a_x b_x + a_y b_y + a_z b_z$
[the scalar product of $\mathbf{a}$ and $\mathbf{b}$, a scalar]

VI. $\mathbf{a} \times \mathbf{b} = \begin{vmatrix} \hat{\mathbf{x}} & \hat{\mathbf{y}} & \hat{\mathbf{z}} \\ a_x & a_y & a_z \\ b_x & b_y & b_z \end{vmatrix} =$
$\hat{\mathbf{x}} \left(a_y b_z - a_z b_y \right) + \hat{\mathbf{y}} \left(a_z b_x - a_x b_z \right) + \hat{\mathbf{z}} \left(a_x b_y - a_y b_x \right)$
[the vector product of $\mathbf{a}$ and $\mathbf{b}$, a vector].

Here $\begin{vmatrix} & \\ & \end{vmatrix}$ is the determinant. Thus, for example, in Eq. 7-17,

$$\hat{\mathbf{z}} \times \nabla \sigma = \begin{vmatrix} \hat{\mathbf{x}} & \hat{\mathbf{y}} & \hat{\mathbf{z}} \\ 0 & 0 & 1 \\ \frac{\partial \sigma}{\partial x} & \frac{\partial \sigma}{\partial y} & \frac{\partial \sigma}{\partial z} \end{vmatrix}$$

$$= \hat{\mathbf{x}} \left(-\frac{\partial \sigma}{\partial y} \right) + \hat{\mathbf{y}} \left(\frac{\partial \sigma}{\partial x} \right),$$

and so, $\partial \mathbf{u}_g / \partial z = \frac{-g}{f \rho_{ref}} \hat{\mathbf{z}} \times \nabla \sigma$ yields the component of the thermal wind equation, Eq. 7-16.

Some useful vector identities are:

1. $\nabla \cdot (\nabla \times \mathbf{a}) = 0$

2. $\nabla \times (\nabla \phi) = 0$

3. $\nabla \cdot (\phi \mathbf{a}) = \phi \nabla \cdot \mathbf{a} + \mathbf{a} \cdot \nabla \phi$

4. $\nabla \times (\phi \mathbf{a}) = \phi \nabla \times \mathbf{a} + \nabla \phi \times \mathbf{a}$

5. $\nabla \cdot (\mathbf{a} \times \mathbf{b}) = \mathbf{b} \cdot (\nabla \times \mathbf{a}) - \mathbf{a} \cdot (\nabla \times \mathbf{b})$

6. $\nabla \times (\mathbf{a} \times \mathbf{b}) = \mathbf{a} (\nabla \cdot \mathbf{b}) - \mathbf{b} (\nabla \cdot \mathbf{a}) + (\mathbf{b} \cdot \nabla) \mathbf{a} - (\mathbf{a} \cdot \nabla) \mathbf{b}$

7. $\nabla (\mathbf{a} \cdot \mathbf{b}) = (\mathbf{a} \cdot \nabla) \mathbf{b} + (\mathbf{b} \cdot \nabla) \mathbf{a} + \mathbf{a} \times (\nabla \times \mathbf{b}) + \mathbf{b} \times (\nabla \times \mathbf{a})$

8. $\nabla \times (\nabla \times \mathbf{a}) = \nabla (\nabla \cdot \mathbf{a}) - \nabla^2 \mathbf{a}$

In (8), $\nabla^2 \mathbf{a} = \hat{\mathbf{x}} \nabla^2 a_x + \hat{\mathbf{y}} \nabla^2 a_y + \hat{\mathbf{z}} \nabla^2 a_z$.

An important special case of (7) arises when $\mathbf{a} = \mathbf{b}$:

$$(\mathbf{a} \cdot \nabla)\mathbf{a} = \nabla \left(\frac{1}{2}\mathbf{a} \cdot \mathbf{a} \right) - \mathbf{a} \times (\nabla \times \mathbf{a}).$$

These relations can be verified by the arduous procedure of applying the definitions of ∇, $\nabla\cdot$, $\nabla\times$ (and $\mathbf{a} \cdot \mathbf{b}$, $\mathbf{a} \times \mathbf{b}$ as necessary) to all the terms involved.

A.2.3. Polar and spherical coordinates

Polar coordinates

Any point is specified by the distance $\mathbf{r}$ (radius vector pointing outwards) from the origin and θ (vectorial angle) measured from a reference line (θ is positive if measured counterclockwise and negative if measured clockwise), as shown in Fig. 6.8.

$$x = r\cos\theta; \; y = r\sin\theta$$

The velocity vector is $\mathbf{u} = \hat{\mathbf{r}}v_r + \hat{\theta}v_\theta = (v_r, v_\theta)$ where $v_r = Dr/Dt$ and $v_\theta = rD\theta/Dt$ are the radial and azimuthal velocities, respectively, and $\hat{\mathbf{r}}$, $\hat{\theta}$ are unit vectors.

In polar coordinates:

$$\nabla\phi = \left(\frac{\partial\phi}{\partial r}, \frac{1}{r}\frac{\partial\phi}{\partial\theta} \right)$$

$$\nabla \cdot \mathbf{u} = \frac{1}{r}\frac{\partial}{\partial r}(ru) + \frac{1}{r}\frac{\partial v}{\partial\theta}$$

$$\nabla^2\phi = \frac{1}{r}\frac{\partial}{\partial r}\left(r\frac{\partial\phi}{\partial r} \right) + \frac{1}{r^2}\frac{\partial^2\phi}{\partial\theta^2}$$

Spherical polar coordinates

In spherical coordinates a point is specified by coordinates (λ, φ, z) where, as shown in Fig. 6.19, z is radial distance, φ is latitude and λ is longitude. Velocity components (u, v, w) are associated with the coordinates (λ, φ, z), so that $u = z\cos\varphi D\lambda/Dt$ is in the direction of increasing λ (eastward), $v = zD\varphi/Dt$ is in the direction of increasing φ (northward) and $w = Dz/Dt$ is in the direction of increasing z (upward, in the direction opposite to gravity).

In spherical coordinates:

$$\nabla\phi = \left(\frac{1}{z\cos\varphi}\frac{\partial\phi}{\partial\lambda}, \frac{1}{z}\frac{\partial\phi}{\partial\varphi}, \frac{\partial\phi}{\partial z} \right)$$

$$\nabla \cdot \mathbf{u} = \frac{1}{z\cos\varphi}\frac{\partial u}{\partial\lambda} + \frac{1}{z\cos\varphi}\frac{\partial v\cos\varphi}{\partial\varphi} + \frac{\partial w}{\partial z}$$

$$\nabla^2\phi = \frac{1}{z^2\cos^2\varphi}\frac{\partial^2\phi}{\partial\lambda^2} + \frac{1}{z^2\cos\varphi}\frac{\partial}{\partial\varphi}\left(\cos\varphi\frac{\partial\phi}{\partial\varphi} \right) + \frac{1}{z^2}\frac{\partial}{\partial z}\left(z^2\frac{\partial\phi}{\partial z} \right)$$

A.3. USE OF FORAMINIFERA SHELLS IN PALEO CLIMATE

A key proxy record of climate is $\delta^{18}O$, the ratio of ^{18}O to ^{16}O in the shells of surface (planktonic) and bottom (benthic) dwelling foraminifera which are made of calcium carbonate, $CaCO_3$. The $\delta^{18}O$ in the shell, measured and reported as

$$\delta^{18}O = \frac{\left(\frac{^{18}O}{^{16}O}\right)_{smpl} - \left(\frac{^{18}O}{^{16}O}\right)_{std}}{\left(\frac{^{18}O}{^{16}O}\right)_{std}} \times 1000^o/_{oo},$$

where $\left(^{18}O/^{16}O\right)_{smpl}$ is the ratio of ^{18}O to ^{16}O in the sample and $\left(^{18}O/^{16}O\right)_{std}$ is the ratio in a standard reference, is controlled by the $\delta^{18}O$ value of the water and the temperature at which the shell formed. It turns out that ^{18}O is increasingly enriched in $CaCO_3$ as the calcification temperature decreases, with a 1°C decrease in temperature resulting in a $0.2^o/_{oo}$ increase in $\delta^{18}O$. However, in order to use oxygen isotope ratios in foraminifera as an indicator of temperature, one must take into account the isotopic signal associated with ice volume stored on the continents and local precipitation-evaporation-runoff conditions, as we now describe.

The vapor pressure of $H_2^{16}O$ is higher than that of $H_2^{18}O$, and so $H_2^{16}O$ becomes concentrated in the vapor phase (atmosphere) during evaporation of seawater. The residual water therefore becomes enriched in ^{18}O relative to water vapor. Similar physics (that of Rayleigh distillation) results in the condensate (rain or snow) from clouds being depleted in ^{18}O relative to ^{16}O. Therefore as an air mass loses water, for example as it is advected to higher latitudes or altitudes, or transported into a continental interior, it becomes progressively depleted in ^{18}O. Precipitation in polar latitudes is commonly 20–40°/$_{oo}$ depleted in ^{18}O relative to average ocean water. During glacial periods, when perhaps 120 m of sea level-equivalent water is stored in polar ice sheets as highly isotopically depleted ice, the entire ocean becomes enriched in ^{18}O by approximately 1°/$_{oo}$. This whole-ocean change in δ^{18}O has been a powerful tool for reconstructing ice age cycles and paleoclimatology in general. The same principal results in regions of the ocean that receive continental runoff having low δ^{18}O values in their surface water, and those regions of the ocean that experience an excess of evaporation over precipitation to have high surface water δ^{18}O values.

A.4. LABORATORY EXPERIMENTS

A.4.1. Rotating tables

The experiments described throughout this text come to life when they are carried out live, either in demonstration mode in front of a class or in a laboratory setting in which the students are actively involved. Indeed a subset of the experiments form the basis of laboratory-based, hands-on teaching at both undergraduate and graduate level at MIT. As mentioned in Section 0.1, the experiments have been chosen not only for their relevance to the concepts under discussion, but also for their transparency and simplicity. They are not difficult to carry out and "work" most, if not every, time. The key piece of equipment required is a turntable (capable of rotating at speeds between 1 and 30 rpm) on which the experiment is placed and viewed from above by a co-rotating camera. We have found it convenient to place the turntable on a mobile cart (see Fig. A.1) equipped with a water tank and assorted materials such as ice buckets, cans, beakers, dyes, and so on. The cart can be used to transport the equipment and as a platform to carry out the experiment.

More details about the equipment required to carry out these experiments,

FIGURE A.1. A turntable on a mobile cart equipped with a water storage tank and pump, power supply, and monitor. A tank filled with water can be seen with an ice bucket in the middle, seated on the turntable, and viewed through a co-rotating overhead camera.

including turntables and fluid carts, can be obtained from Professor John Marshall at MIT.

Measurement of table rotation rates

The rotation rate can be expressed in terms of period, revolutions per minute, or units of 'f,' as described below and set out in Table A.1:

1. The angular velocity of the tank, Ω, in radians per second

2. The Coriolis parameter (f) defined as $f = 2\Omega$

3. The period of one revolution of the tank is $\tau = 2\pi/\Omega$

4. Revolutions per minute, rpm = $60/\tau$

TABLE A.1.   Various measure of rotation rates. If Ω is the rate of rotation of the tank in radians per second, then the period of rotation is $\tau_{\text{tank}} = 2\pi/\Omega$ s. Thus if $\Omega = 1$, $\tau_{\text{tank}} = 2\pi$ s.

Ω (rad/s)	0	0.10	0.25	0.5	1	1.5	2
$f = 2\Omega$ (rad/s)	0	0.21	0.5	1	2	3	4
τ (s)	∞	10.46	25.1	12.6	6.3	4.2	3.14
rpm	0	1	2.4	4.7	9.5	14.3	19.1

A.4.2. List of laboratory experiments

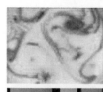

GFD 0: Rigidity imparted to rotating fluids—Section 0.2.2.

GFD I: Cloud formation on adiabatic expansion—Section 1.3.3.

GFD II: Convection—Section 4.2.4.

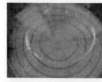

GFD III: Radial inflow—Section 6.6.1.

GFD IV: Parabolic surfaces—Section 6.6.4.

GFD V: Inertial circles—Section 6.6.4.

GFD VI: Perrot's bathtub experiment—Section 6.6.6.

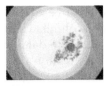

GFD VII: Taylor columns—Section 7.2.1.

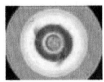

GFD VIII: The Hadley circulation and thermal wind—Section 7.3.1.

GFD IX: Cylinder 'collapse'—Section 7.3.3.

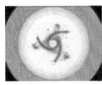

GFD X: Ekman layers—Section 7.4.1.

GFD XI: Baroclinic instability—Section 8.2.2.

GFD XII: Ekman pumping and suction—Section 10.1.2.

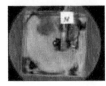

GFD XIII: Wind-driven ocean gyres—Section 10.2.4.

GFD XIV: Abyssal ocean circulation—Section 11.3.2.

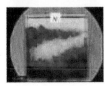

GFD XV: Source sink flow—Section 11.3.4.

A.5. FIGURES AND ACCESS TO DATA OVER THE WEB

The vast majority of the data and figures presented in this book were accessed and plotted using the Climate Data Library of the International Research Institute of the Lamont-Doherty Earth Observatory of Columbia University. The library contains numerous datasets from a variety of Earth science disciplines and climate-related topics, which are accessible free over the web— see http://iridl.ldeo.columbia.edu/. Web-based tools permit access to data sets, analysis software for manipulation of the data and graphical presentation, together with the ability to download the data in numerous formats. An excellent specialized tool for plotting oceanographic hydrographic data is Ocean Data View: http://odv.awi.de.

References

Textbooks and reviews

Andrews, D. G. (2000). An introduction to atmospheric physics. Cambridge University Press, Cambridge.

Burroughs, W. (2005). Climate change in prehistory: the end of the reign of chaos. Cambridge University Press, Cambridge.

Csanady, G. T. (2001). Air-sea interaction. Cambridge University Press, Cambridge.

Cushman-Roisin, B. (1994). Introduction to geophysical fluid dynamics. Prentice Hall, Englewood Cliffs, New Jersey.

Emanuel, K. A. (1994). Atmospheric Convection. Oxford University Press, New York.

Goody, R. M. and Yung, Y. L. (1989). Atmospheric radiation: theoretical basis. 2nd ed. Oxford University Press, Oxford.

Gill, A. E. (1982). Atmosphere-ocean dynamics. International Geophysics, vol. 30. Academic Press, San Diego, CA.

Green, J. S. A. (1999). Atmosphere dynamics. Cambridge University Press, Cambridge.

Hartmann, D. L. (1994). Global physical climatology. International Geophysics, vol. 56. Academic Press, San Diego, CA.

Holton, J. R. (2004). An introduction to dynamic meteorology. International Geophysics, vol. 88. 4th ed. Academic Press, Boston, MA.

Houghton, J. T. (1986). The physics of atmospheres. 3rd ed. Cambridge University Press, Cambridge.

Houghton, J. T. et al. (1996). The science of climate change. Cambridge University Press, Cambridge.

Lamb, H. (1932). Hydrodynamics. 6th ed. Cambridge Mathematical Library Series. Cambridge University Press, Cambridge.

Lorenz, E. (1967). The nature and theory of the general circulation of the atmosphere. World Meteorological Organization.

Pedlosky, J. (1996). Ocean circulation theory. Springer-Verlag, Berlin.

Peixoto, J. P., and Oort, A. H. (1992). Physics of climate. American Institute of Physics Press, New York, NY.

Philander, S. G. H. (1990). El Niño, La Niña and the Southern Oscillation. Academic Press, San Diego, CA.

Pickard, G. L., and Emery, W. J. (1990). Descriptive physical oceanography. 5th ed. Butterworth-Heinemann, Oxford.

Pond, S., and Pickard, G. (1983). Introductory dynamical oceanography. Butterworth-Heinemann, Oxford.

Rhines, P. B. (1993). "Ocean general circulation: wave and advection dynamics." In Modelling oceanic climate interactions, edited by J. Willebrand and D. L. T. Anderson, 67–149. Springer-Verlag, Berlin.

Ruddiman, W. F. (2001). Earth's climate, past and future. W. H. Freeman and Company, New York, NY.

Stewart, R. H. (2005). Introduction to physical oceanography. *http://oceanworld.tamu.edu/resources/ocng_textbook/contents.html*.

Stommel, H. (1965). The Gulf Stream: a physical and dynamical description. 2nd ed. University of California Press, Berkeley and Cambridge University Press, London.

The Open University (1989). Ocean circulation. Pergamon Press, New York, NY.

Vallis, G. K. (2006). Atmospheric and oceanic fluid dynamics: fundamentals and large-scale circulation. Cambridge University Press, Cambridge.

Wallace, J. M., and Hobbs, P. V. (2006). Atmospheric science: an introductory survey. 2nd ed. Elsevier Academic Press, Boston, MA.

Wells, N. (1997). The atmosphere and oceans: a physical introduction. 2nd ed. John Wiley and Sons, New York, NY.

Wunsch, C. (1996). The ocean circulation inverse problem. Cambridge University Press, Cambridge.

Other references

Alley, R. et al. (2002). Abrupt climate change: inevitable surprises. Committee on Abrupt Climate Change, National Research Council, The National Academies Press, Washington D. C.

Baumgartner, A., and Reichel, A. (1975). The world water balance. Elsevier, New York, NY.

Conkright, M. S., Levitus, S., and Boyer, T. (1994). World ocean atlas 1994, vol. 1: nutrients. NOAA Atlas NESDIS 1, U. S. Department of Commerce, Washington D. C.

Czaja, A., Robertson, A., and Huck, T. (2003). The role of coupled processes in producing NAO variability. *Geophys. Monogr.* **134**, 147–172.

Czaja, A., and Marshall, J. (2006). The partitioning of poleward heat transport between the atmosphere and ocean. *J. Atmos. Sci.* **63**, 1498–1511.

Davis, R. E., Sherman, J. T., and Dufour, J. (2001). Profiling ALACEs and other advances in autonomous subsurface floats. *J. Atmos. Ocean. Tech.* **18**, 982–993.

Dorman, C. E., and Bourke, R. H. (1981). Precipitation over the Atlantic Ocean, 30°S to 70°N. *Mon. Wea. Rev.* **109**: 554–563.

daSilva, A., Young, A. C., and Levitus, S. (1994). Atlas of surface marine data, vol. 1: algorithms and procedures, NOAA Atlas NESDIS 6, U. S. Department of Commerce, Washington, D. C.

Durran, D. R., and Domonkos, S. K. (1996). An apparatus for demonstrating the inertial oscillation. *Bull. Amer. Meteo. Soc.* **77**, 557–559.

Eady, E. (1949). Long waves and cyclone waves. *Tellus* **1**(3), 33–52.

Ekman, V. W. (1905). On the influence of the Earth's rotation on ocean currents. *Arch. Math. Astron. Phys.* **2**(11), 1–52.

Friedrichs, M. A. M., and Hall, M. M. (1993). Deep circulation in the tropical North Atlantic. *J. Mar. Res.* **51**(4), 697–736.

Ganachaud, A., and Wunsch, C. (2000). Improved estimates of global ocean circulation, heat transport and mixing from hydrographic data. *Nature* **6811**, 453–456.

Hall, M. M., and Bryden, H. L. (1982). Direct estimates and mechanisms of ocean heat transport. *Deep Sea Res.* **29**, 339–359.

Hasselmann, K. (1976). Stochastic climate models, part I: theory. *Tellus* **28**, 473.

Hide, R., and Titman, C. W. (1967). Detached shear layers in a rotating fluid. *J. Fluid. Mech.* **29**(1), 39–60.

Isemer, H., and Hasse, L. (1987). The Bunker climate atlas of the North Atlantic Ocean, vol. 2: air-sea interactions, Springer-Verlag, Berlin Heidelberg.

Kalnay, E. et al. (1996). The NCEP/NCAR 40-year reanalysis project. *Bull. Amer. Meteor. Soc.*, **77**, 437–471.

Levitus, S., and Boyer, T. (1994). World ocean atlas 1994, vol. 2: oxygen. NOAA Atlas NESDIS 2, U. S. Department of Commerce. Washington, D. C.

Levitus, S., and Boyer, T. (1994). World ocean atlas 1994, vol. 3: nutrients. NOAA Atlas NESDIS 3, U. S. Department of Commerce. Washington, D. C.

Levitus, S., and Boyer, T. (1994). World ocean atlas 1994, vol. 4: temperature. NOAA Atlas NESDIS 4, U. S. Department of Commerce. Washington, D. C.

Marshall, J., and Schott, F. (1999). Open ocean deep convection: observations, models and theory. *Rev. Geophys.* **37**(1), 1–64.

Menemenlis, D., Hill, C., Adcroft, A., Campin, J.-M, et al. (2005). NASA supercomputer improves prospects for ocean climate research. EOS Transactions, *Amer. Geophys. Union* **86**(9), 89–95.

Niiler, P. P. (2001). "The world ocean surface circulation." *In* Ocean circulation and climate, International Geophysics, vol. 77, edited by G. Siedler, J. Church, and J. Gould, 193–204.

Schmitt, R. W., Bogden, P., and Dorman, C. (1989). Evaporation minus precipitation and density fluxes for the North Atlantic. *J. Phys. Oceanogr.* **10**, 1210–1221.

Schmitt, R. W. (1994). The ocean freshwater cycle. JSC Ocean Observing System Development Panel, Texas A&M University, College Station, TX.

Stommel, H., The westward intensification of wind-drive ocean currents. *Trans. Am. Geophys. Union* **29**(2), 202–206.

Stommel, H. (1958). The abyssal circulation. *Deep Sea Res.* **5**, 80.

Sverdrup, H. U. (1947). Wind-driven currents in a baroclinic ocean: with application to the equatorial currents of the eastern Pacific. *Proc. Nat. Acad. Sci.* **33**(11), 318–326.

Taylor, G. I. (1921). Experiments with rotating fluids. *Proc. Roy. Soc. Lond. A* **100**, 114–121.

Trenberth, K. E., and Shea, D. J. (1987). On the evolution of the Southern Oscillation. *Mon. Wea. Rev.* **115**, 3078–3096.

Trenberth, K. E., Olson, J., and Large, W. (1989). A global ocean wind stress climatology based on ECMWF analyses. Tech. Rep. NCAR/TN-338+STR, National Center for Atmospheric Research, Boulder, CO.

Trenberth, K. E., and Solomon, A. (1994). The global heat balance: heat transports in the atmosphere and ocean. *Clim. Dynam.* **10**(3), 107.

Trenberth, K. E., and Caron, J. M. (2001). Estimates of meridional atmosphere and ocean heat transports. *J. Clim.* **14**, 3433–3443.

Whitehead, J. A., and Potter, D. L. (1977). Axisymmetric critical withdrawal of a rotating fluid. *Dynam. Atmos. Oceans* **2**, 1–18.

Wunsch, C. (2002). What is the thermohaline circulation? *Science* **298**(5596), 1179–1180.

References to paleo-data sources (Chapter 12)

Berger, A., and Loutre, M. F. (1992). Astronomical solutions for paleoclimate studies over the last 3 million years. *Earth Planet. Sci. Lett.* **111**, 369–382.

Berner, R. A. (1997). The rise of plants and their effect on weathering and atmospheric CO_2. *Science* **276**, 544–546.

Briffa, K. R., Osborn, T. J., Schweingruber, F. H. et al. (2001). Low-frequency temperature variations from a northern tree-ring density network. *J. Geophys. Res.* **106**, 2929–2941.

Crowley, T. (2000). "Carbon dioxide and Phanerozoic climate." *In* Warm climates in Earth history, edited by B. T. Huber, K. G. MacLeod, and S. L. Wing, 425–444. Cambridge University Press, Cambridge.

Crowley, T., and Lowery, T. (2000). Northern hemisphere temperature reconstruction. *Ambio.* **29**, 51–54.

EPICA community members. (2002). Eight glacial cycles from an Antarctic ice core. *Nature* **429**(6992), 623–628. doi:10.1038/nature02599.

Esper, J., Cook, E., and Schweingruber, F. (2002). Low-frequency signals in long tree-ring chronologies for reconstructing past temperature variability. *Science* **295**(5563), 2250–2253.

Grootes, P. M., and Stuiver, M. (1997). Oxygen 18/16 variability in Greenland snow and ice with 10^3 to 10^5-year time resolution. *J. Geophys. Res.* **102**, 26455–26470.

Hegerl, G., Crowley, T., Hyde, W., and Frame, D. (2006). Climate sensitivity constrained by temperature reconstructions over the past seven centuries. *Nature* **440**, 1029–1032. doi:10.1038/nature04679.

Huybers, P. (2007). Glacial variability over the last 2Ma: an extended depth-derived agemodel, continuous obliquity pacing, and the Pleistocene progression, *Quat. Sci. Rev.* **26**, 37–55.

Jones, P., and Moberg, A. (2003). Hemispheric and large-scale surface air temperature variations: an extensive revision and an update to 2001. *J. Clim.* **14**, 206–223.

Mann, M. E., Bradley, R. S., and Hughes, M. K. (1999). Northern hemisphere temperatures during the past millennium: inferences, uncertainties, and limitations. *Geophys. Res. Lett.* **26**(6), 759–761.

Miller, K. G., Janacek, T. R., Holmgren, K., and Keil, D. J. (1987). Abyssal circulation and benthic foraminiferal changes near the Paleocene/Eocene boundary. *Paleoceanography* **2**, 741–761.

Moberg, A., Sonechkin, D., Holmgren, K., et al. (2005). Highly variable northern hemisphere temperatures reconstructed from low- and high-resolution proxy data. *Nature* **443**, 613–617. doi:10.1038/nature03265.

Siegenthaler, I., Stocker, T. F., Monnin, E., et al. (2005). Stable carbon cycle-climate relationship during the late Pleistocene. *Science* **310**, 1313–1317.

Spahni, R., Chappellaz, J., Stocker, T. F., et al. (2005). Atmospheric methane and nitrous oxide of the late Pleistocene from Antarctic ice cores. *Science* **310**, 1317–1321.

Index

A

Abrupt climate change, 260, 289–291
Absorption of radiation by
  carbon dioxide, 14
  nitrogen, 14
  oxygen, 14
  ozone, 25
  water vapor, 14
Absorptivity, 16
Abyssal circulation of ocean, 234
  observations of, 245–247
  Stommel's schematic of, 241f
Acceleration
  centrifugal, 136
  due to gravity, 7, 100–101
  terms in momentum equation, 84–87, 110
Adiabatic
  compression, 144
  lapse rate, 30, 39–41
  motion, 49–41
Adiabats
  dry, 40
  moist, 49–50
Age
  of water parcels, 238
Ageostrophic flow, 129–135
  definition, 129
  driven by rotating disc, 201–203
  in high and low pressure systems, 130–133
Air
  density of, 4
  dry, 5
  moist, 5
  physical properties of, 4–6
  pressure, 4

Air temperature at surface
  global maps of, 263f
  seasonal range of, 262
Air-sea fluxes, 224–231
  of buoyancy, 231f
  of heat
    balance in tropics, 231–232
    contributions to, 225–228
    global maps, 228f
    zonal-average of analyzed fields, 226
Albedo, 12
  table of values of desert, snow etc, 11t
Altimetry, 184, 189
Amazon river, 215
Angular momentum, 142–143
  absolute, 142
  conservation of
    implied wind distribution, 143
  day length and, 143
  in radial inflow experiment, 93
  of ring of air, 142f
Angular velocity of earth, 1
Annular modes, 273
Annulus experiment, 146–147
Antarctic
  Bottom Water (AABW), 236, 237f
  Circumpolar Current (ACC), 177f, 179, 194, 220
  Intermediate Water (AAIW), 234, 237f
  surface pressure around, 134
Anthropogenic
  forcing of climate, 291–292
Anticyclonic flow, 93, 132, 201, 219
Aphelion, 63, 286
Archimedes, 34

Astronomical forcing of climate, 284–288
Atlantic Ocean
    oxygen distribution in, 238f
    sections of temperature and salinity, 173f,
        188f, 236f, 237f, 247f
    SST, 179f
    surface currents, 179f
    mid-depth circulation, 181f, 249f
Atmosphere
    characteristics of, 1–8
    chemical composition of, 2–4
    circulation of, 139–161
    convection in, 39–42
    density of, 1
    energetics of, 153–154
    energy balance of, 64
    gases in, 2–4
    geometry of, 1–2
    heat transport in, 251–255
    instability of, 40, 49–50
    isothermal, 28
    mean meridional circulation of, 74–77
    measurements of, 5t
    meridional structure of, 61–79
    non-isothermal, 28–29
    pressure of, 4
    stability in, 42–46
    vertical structure of, 23–30
Atmospheric absorption spectrum, 13–14, 13f
Available Potential Energy (APE)
    baroclinic instability and, 139, 146–147, 147f,
        152–153
    computation of, 296
    definition, 149
    in atmosphere, 150
    in ocean 219
        schematic of creation mechanism, 219f
Azimuthal velocity, 91, 144–145

B
Balanced motion, 109–138
    in radial inflow experiment, 116
Ball on curved surface, 33
Baroclinic fluid, xv, 122
Baroclinic instability, 146–149
    APE in, 139, 146–147, 147f, 152–153
    in atmosphere, 147
    in ocean, 218–220
    laboratory experiment, 146
    wedge of, 152–154
Barotropic, xv
    fluid, 117, 122

Bath-tub experiment, 103
Bathymetry, 164f
Bergen school, 77
Beta effect, 206
    representation in laboratory, 211f, 212
Bjerknes
    feedback, 269
        schematic diagram of, 269f
    Jacob, 266
    Vilhelm, 77
Black body, 10, 13
    energy emitted from, 11f, 13f, 295
Bolide impact, 294
Bort (de), Leon, 25
Boundary
    conditions, 89
    current 178f, 179f, 206, 243
        deep western, 245–247, 248f
    layer, xvf, 131
Boyle, Robert, 4
Brine rejection, 224
Brunt-Vaisala frequency, 44
Budget
    energy, 140f, 154–156, 247–255
    mass, 87–89
    momentum, 140f, 156–157
Buoyancy, 34–35, 43, 224
    air-sea fluxes and, 224, 229
    anomaly, 225
    convection and, 197
    in Cu, 52, 54f
    evolution equation for in ocean, 225
    flux at sea surface, 225, 231f
    frequency, 44, 126
        definition in atmosphere, 44
        definition in ocean, 174
    hydrostatic balance and, 34f
    mixed layer and, 233
    in oceans, 164
Buys-Ballot's law, 111

C
Carbon dioxide, 2, 275–276, 276f
    concentrations of, 3f
    through history, 277n, 279–280, 284, 285f, 289
    increases in, 4, 291
    IR and, 9
Centrifugal
    acceleration, 95
    force, 92
CFC's—see Chlorofluorocarbons
Charney, Jule, 153

Chlorofluorocarbons (CFC's), 27
    in atmosphere, 3t
    in ocean, 246
        observations of, 248f
Circulation
    angular momentum and, 156
    of atmosphere, 139–161
    GCMs, 272
    mean meridional, 74–77
    MOC, 252–253, 291
    ocean density and, 216–218
    in oceans, 140, 163–194, 252f
        Atlantic, 177f, 179f, 181f
        global, 177f
        Indian, 180f
        Pacific, 178f
    oxygen and, 236–239
    temperature and, 156
    in tropics, 142–145
    in troposphere, 141
    winds and, 197–222
Circulation and vorticity, 106
Clausius, Rudolf, 39
Clausius-Clapeyron relationship, 6
CLIMAP, 281
Climate, 259–294
    abrupt change, 289–291
    as average weather, 259
    through earth history, 275–277
    ENSO and, 264–273
    feedbacks, 19–20, 288, 291
    forcing as a function of timescale, 260f
    forecasting variability in, 262
    greenhouse, 280
    latitude and, 157–158
    norms, 259
    paleoclimate, 273–292
    role of ocean in, 261–264
    variability
        stochastic model of, 264
Climatic optimum, 289
Climatic precession, 287
Cloud
    albedo of, 11t, 19–20
    cumulonimbus, 51f, 52f, 53f, 53
    cumulus, 51f, 52–53
    formation experiment, 7
    lenticular, 44, 45f
    supercell, 53f
    top, 48
    vertical heat transport by, 51, 53
Cold front, 147, 148f
Cold tongue, 268

Compressible
    atmosphere
        convection in, 39–42
        energetics of, 153–154
        instability of, 40
    flow, 88–89
    fluid, 5
Concorde, 105
Condensation
    level, 48, 51f
    nucleii, 5, 7
Conditional stability, 50
Conservation
    of angular momentum, 93, 142
    of energy, 139, 154, 247–255
    of mass, 87–89
Continental divide, 45
Continuity equation, 88
    in pressure coordinates, 88
Convection, 19, 31–60, 32–33
    in atmosphere, 46–56
    buoyancy and, 197
    Cb and, 53–55
    in compressible atmosphere, 39–42
    Cu and, 52–53
    deep, 51, 233–239
    in deserts, 42
    dry, 39–42
    drying due to, 73f
    energetics of, 36
    heat transport in, 38–39
    humidity and, 47–48
    instability and, 33–34
    laboratory experiment, 36, 57
    moist, 46–50
    in oceans, 164, 234–237, 235f, 236f
    precipitation and, 6
    radiation and, 56–57
    shallow, 32–33, 56
    SST and, 224
    stability and, 35–36
    study of, 37f
    in tropics, 51
    in water, 34–39
    where does it occur in Atmosphere?, 55
    where does it occur in Ocean?, 232, 233f
Conveyor belt
    schematic diagram of, 252
Coordinate system
    Cartesian, 101, 297
    polar, 92f, 298
    spherical, 101, 298

Coriolis
    acceleration, 93, 95
    force
        components on the sphere, 101
    laboratory experiment, 96
    Gustave, 95
    parameter, 100, 102
Counter-current
    equatorial, 214f
Cretaceous, 278f, 280
Crocodiles, north of the Arctic circle, 279
Cryosphere, 165
Cumulonimbus, 51, 52f, 53
Cumulus, 51, 54f
Cyclonic flow, 93
Cyclostrophic balance, 116
Cylinder collapse experiment, 123, 126

D
Dalton's law, 5
Dansgaard-Oeschger (D-O) events, 260, 289
Darwin, Charles, 265
Day, variations in length of, 143
Deacon, George, 179
Defant, Albert, 146
Deflecting force, 98
Delta O18, 277–279
    in benthic foraminifera
        last 2.5 million years, 283f
        last 70 million years, 279f
    in foraminifera shells, 298
    of ice
        last 20 thousand years, 290f
        last 50 thousand years, 283f
Density
    anomaly, 119, 166
    depth of density surfaces in ocean, 170, 173f
    of air, 4
    potential density distribution in ocean, 172f
    surface outcrops in ocean, 173
    vertical variation in atmosphere, 29
Desert, 42
    subtropical, 157
Dew point, 7
Diabatic heating rate, 89
Differentiation following the motion, 82–83
Downdraft, 73f
Drake Passage, 164
    closing and opening of, 277
Dry adiabatic lapse rate, 39–41
Dry convection
    schematic, 43f

Dry static energy, 154
Dryas Octopetala, 289
Dust, 295
Dust bowl, 260
Dye stirring, xviii, 117f
Dynamic height, 187–188
Dynamic method, 187–188

E
Eady
    Eric, 149
    growth rate, 149
Earth
    axis of, 62f, 63, 286, 286f
    blackbody spectrum for, 13
    climate history of, 275–277
    energy from, 9f
    orbit, 62–63, 284
    orbital changes of, 284–288, 286f
    parameters, 1t
    rotating fluids and, 103–104
    as snowball, 260, 275n
    solar constant for, 10t
    spin, 1, 286
        in opposite direction, 221
    temperature of, 14
Eccentricity, 287
Eddies
    atmospheric, 77, 146
        momentum transport by, 156
        time and space scales, 147–148
        banana-shaped, 157f
        ocean, 181, 198
        observations of, 185f, 188
        time and space scales, 219
    large-scale, xv
    stirring by, 147
Effective planetary temperature, 11
Ekman
    layer, 129
        balance of forces in, 199
        depth of, 200
        in atmosphere on planetary scale, 133
        in laboratory experiment, 220
        mass transport of, 200
        theory of, 198
        vertical motion in, 131
    pumping, 198f
        and suction, 201
        driven by wind patterns, 203
        expression for, 204
        plotted from analyzed fields, 205f

response of the interior ocean to, 206
schematic of driving by large-scale wind
field, 204f
spiral, 131, 201f
theory, 130
transport
in tropical ocean, 234
Ekman, Vagn Walfrid, 201
El Niño-Southern Oscillation (ENSO), 269–272,
289
theory of, 272
Emission temperature, 12
Energy
balance
of atmosphere, 9, 64
top of the atmosphere, 64f
budget
of atmosphere, 139
dry static, 154
internal
of atmosphere, 139
kinetic, 154
moist static, 154
potential, 36
transport northward, 64f
Energetics
in compressible atmosphere, 153–154
convection and, 36
of thermal wind equation, 149–154
ENSO, 269–272: see also El Niño-Southern
Oscillation,
Bjerknes positive feedback, 273
theory of, 272
Entropy, 64
Eocene, 278f
Equation of state
of air, 4
of water, 168
Equations of motion, 84–89
in a rotating frame, 90, 94–95, 182
on sphere in local Cartesian system, 102
in Cartesian coordinates, 87
in spherical coordinates, 100–102
non-rotating, 84–89
Equatorial
counter-current, 214f
currents, 178
Pacific, 265
atmospheric convection in, 55
upwelling, 267f
Equilibrium, between radiation and convection,
56–57
Equinox, 63

Equipotential surface
parabola, 96
Equivalent potential temperature, 50
Euler, Leonhard, 83
Eulerian derivative, 83
Evaporation
annual-mean map of, 230
as principal heat loss term, 232
Expansion coefficient
of water, 166
Extra-tropical circulation
of atmosphere, 145
of ocean, 176

F
Faint early sun paradox, 275, 293
Feedback, 18
due to black body radiation, 19, 22
due to clouds, 20
evaporative, 232
ice-albedo, 19, 289
water vapor, 19
Ferrel
Cell, 77
William, 77
Fisheries
off South America, 268
Float
PALACE, 180
displacements, 249f
Fluid—See also Rotating fluids
compressible, 39, 88–89
continuum, 4
convection in, 32–33, 33f
dynamics
geophysical, xvii
friction and, 86–87
gravity and, 85–86
hydrostatic balance and, 87
incompressible, 88
motion of, 81–107
natural, xv
nonrotating, 84–87
Fog, 8
Foraminifera, 279f, 298
Foucault pendulum, 100
Frame of reference, 90
absolute, 91
Freshwater
transport by ocean, 256f, 257
Friction, 86, 129–130
in western boundary currents, 207, 243

Fronts, 147–148
  Cold, 148f
  Warm, 148f

G

Gaia hypothesis, 276
General circulation
  of atmosphere, 139, 141f
  of ocean, 176–182, 177f
Geoid, 183
  of a rotating fluid, 97
  of earth, 101
Geologic time
  table of names and dates, 274f
Geometric height, 101
Geopotential
  height, 67
  surfaces on the sphere, 100
Geostrophic
  adjustment, 125
  balance, 110–116
    at (near) the equator, 267
    in Cartesian coordinates, 113
    in ocean, 182
  current
    divergence of, 206
  flow
    at depth in ocean, 184
    at ocean surface, 183
  motion, 110
  streamlines, 113
  wind, 111, 112
    divergence of, 112
    in pressure coordinates, 113
Geothermal heating, 164
Glacial-interglacial cycles, 282–288
  astronomical forcing of, 284–288
    caveats,
Global overturning circulation in ocean, 252f,
    254f, 254
Global warming, 291–292
  trend from proxy and instrumental records,
    292f
Gravity, 85
  modified by centrifugal forces, 95, 97, 101f
  reduced, 226
Gravity waves
  in atmosphere, 42, 44
  in the ocean, 174–175
  internal, 44
  surface, 44
Greenhouse
  climates, 280

climate feedbacks and, 19–20
  domestic, 15f
  effect, 9, 14–20
    models of, 14–20
  gases, 14
    distribution of, 23, 25
  leaky, 16, 17f
  opacity and, 16–19, 17f
Gulf Stream, 169
  dynamical balance in, 193
  rings, 189
Gyres
  in laboratory experiment, 212f
  ocean, 176
  subpolar, 178
  subtropical, 179

H

Hadley circulation, 75, 76f, 141f, 142, 143f, 156,
    229, 268
  schematic, 73f
Hadley, George, 73
Hailstones, 55
Heat budget
  of atmosphere, 154–156, 253–255
  of ocean, 247–255
Heat capacity
  of slab of ocean, 261
  of soil and rock, 262
Heat engine, xvi
Heat transport
  in atmosphere, 140f, 154–156
    mechanisms of, 156
  by fluid motions, 64
  in ocean, 248
    mechanisms of, 253
    methods of computation, 250
    by ocean gyres, 221
    observations of basin by basin, 250
  meridional
    atmosphere, 156f
    ocean, 156f
  partition between atmosphere and ocean, 255f
Height of free surface
  observations of (ocean), 185f, 186f
Heinrich events, 260, 289
High pressure, 114
  schematic, 111f
Holicism, xix
Holocene, 289–290
  transition from LGM, 290f
Howard, Luke, 51

Humidity
   convection and, 47–48
   relative, 48
   saturation specific, 48
   specific, 47–48
   temperature and, 70–73
Hydrographic section
   25W, 173f
   across 30S in Atlantic, 247f
   across Labrador Sea, 236f
   through Gulf Stream along 38N, 188f
Hydrostatic balance, 26–28, 33, 68, 92
   buoyancy and, 34f
   conditions for validity of, 87
   fluids and, 87
   internal energy and, 153
   inferences from, 182–188

I
Ice
   particles, in clouds, 7, 54
   puck, 97
   sheets, 165
      at Last Glacial Maximum, 281, 282f
Ice ages, 260, 280–291
Ice cores, 282–284, 283f
   record over last 600 thousand years, 285f
Incandescent material, 10
Incompressible flow, 88
Inertial
   circles, 98–100
      laboratory experiment, 97
      observations of, 100f
   frame, 99
   period, 100
Infrared radiation (IR), 2, 9, 14
   carbon dioxide and, 9, 14
   water vapor and, 2, 9, 14
Instability. See also Baroclinic instability
   of atmosphere, 49–50
   of ball on corrugated surface, 33
   of compressible atmosphere, 40
   convection and, 33–34
Instrumental record, 262
Interglacials, 259, 282–284
Internal energy, 153
   for compressible atmosphere, 296
Inversion, 44–46
   trade, 46f
Ionosphere, 24
Isentropic surfaces, 64
Isopycnal slope, 186
Isothermal atmosphere, 28, 68

Isotopes, of oxygen, 279, 279f
Isthmus of Panama, 277

J
Jet Stream, 273
Jupiter, 10t, 20
Jurassic, 278f

K
K-T extinction, 294
Kinetic energy, xv
   ratio to potential energy, 151
Kirchhoff's law, 16
Kuroshio, 169, 177f, 178f

L
La Niña, 269f, 269–271
Laboratory experiment
   abyssal ocean circulation, 241
   baroclinic instability, 146
   cloud formation, 7
   complete list, 300–301
   convection, 36
   cylinder collapse, 123
   dye stir, xvii, 117
   on earth's rotation: Perrot bathtub, 103
   Ekman layers, 130
   Ekman pumping and suction, 201
   Hadley circulation, 120, 144
   parabolic equipotential surfaces, 96
   radial inflow, 90
   rotation stiffens fluid, xvii
   source-sink flow in a rotating system, 245
   Taylor columns, 118
   thermal wind, 219
   visualizing the Coriolis force, 97
   wind-driven gyres, 211
Lagrangian derivative, 82–84
Lake
   discharge in to ocean, 290
   freezing of, 166
   inland, at last glacial maximum, 281
Lapse rate
   dry-adiabatic, 40
   moist, 49
Last glacial maximum, 281, 282f
Latent heat, 154
   content of air parcel, 49
   flux
      bulk formula for, 226
      schematic of mechanism, 228
Laurentide Ice sheet, 281

Laws
  of mechanics, 84
  of thermodynamics, 89
Lee waves, 45
Little Ice Age, 260
Local thermodynamic equilibrium, 4
Longwave radiation
  emitted, 63–64
Low pressure, 114–116
  schematic, 111f

M

Margules
  Max, 125
  relation, 123–125
Mass conservation, 87
Material derivative, 83
Mauna Loa, 4
Maury, Matthew, 74
Meridional overturning circulation
  of ocean (MOC), 253, 254f
  of atmosphere, 74
Mesopause, 24, 65
Mesosphere, 24
Mid-Atlantic ridge, 216
Milankovitch cycles, 63, 284, 288f
Milankovitch, Milutin, 284
Mixed layer
  buoyancy and, 233
  in ocean, 171
    model of, 191
  observed depth of in ocean, 174f
  processes in, 175f
  schematic diagram of, 175f
MOC (meridional overturning circulation), 253
  and climate change, 290–291
Modified gravitational potential, 101
Moist convection, 46–50
Moist potential temperature, 50
  climatological profile, 41f
  observed, 67f
Moist static energy, 49, 254
Moisture, 5–6
  distribution of, 6, 70–73
Momentum
  budget of atmosphere, 140f, 156–157
  equation, 84–87, 102, 182
Mountain waves. See Lee waves
Mountains, clouds and, 44, 45f

N

Nansen, Fridtjof, 201
Nitrogen, 2, 14

Non-isothermal atmosphere, 28–29
North Atlantic Current, 207
North Atlantic Deep Water, 236, 237f, 246, 247f
North Atlantic Oscillation, 265f, 273
North Star, 163, 287
North Pole, 125f

O

Obliquity, 286
Ocean,
  adjustment timescale of, 261
  albedo of, 11t
  APE in, 151n
  baroclinic instability in, 218–220
  basins in, 164–165
  evolution of, 276
  as a buffer of climate change, 262–262
  buoyancy in, 164
  circulation
    at mid-depth, 181f, 249f
    Atlantic, 179f
    global, 177f
    Indian, 180f
    Pacific, 178f, 215f
    schematic, 214f
  climate and, 261
  compared to atmosphere, 163–164
  convection in, 164
  currents in, 177f, 178f, 179f, 180f, 181f
    classification of, 214
  density of, circulation and, 216–218
  Ekman pumping and, 205n
  equator-to-pole temperature gradient and, 261
  evaporation in, 172
  key facts, 165
  freshwater transport by, 255–256
  gateways in, 277
  global overturning circulation in, 252f, 254f, 255f
  heat transport by, 247–255
  measurements of, 165t
  mixed layer in, 171–176
  radiation in, 171–172
  role of heat transport by in paleo climate, 176, 290–291
  salinity of, 168f, 171f
  salt in, 7
  solar radiation and, 261
  surface, 183, 185f
    geostrophic flow and, 183–184
    height of, 184, 185f
      steric effects on, 186–187
    salinity at, 169

stratification of, 216–218, 233f
temperature in, 170f, 261–264
thermohaline circulation of, 223–258
two-layer model of, 266
wind stress on, 164,199f, 215f
Ocean surface temperature. See Sea surface
    temperature
Ocean gyres
    effects of stratification and topography on,
        217–218
    schematic diagram of, 214f
    theory of, 213–216
    volume transport in, 216
Outgoing radiation, 55–56, 63–64
Oxygen, 2
    role in absorption of radiation, 14
    distribution in Atlantic vs Pacific ocean, 238f,
        239
Ozone, 14
    absorption by, 25
    profile of, 25f

P
Pacific Ocean
    oxygen distribution in, 238f
    SST, 178f
    surface currents, 178f
    Sverdrup balance in, 215
Paleoclimate, 273–292
    proxy data, 273
Paleogeographic reconstructions, 278
Paleotemperatures, 277–280
Pangea, 278
Parabolic surface, 92, 96–97
    Coriolis forces and, 96–100
Parcel of fluid, 34
Pascal, Blaise, 27
Perfect gas law, 4
Perihelion, 63, 286
Perrot, 103
    bathtub experiment, 103–104
Photodissociation, 24
Photolysis, 24
Pip flicking, 211
Planck
    function, 295
    spectrum, 11
Planck, Max, 10
Planets
    properties of, 10
Plumb line, 101
Polar
    coordinates, 92

night, 63
oceans
    convection in, 232–234
Pollution, 46–47
Porridge/polenta
    convection of, 33
Potential energy, 149
    available, 150–154
        creation by winds in ocean, 219
        in two-layer fluid, 151
    in atmosphere, 153
    release of, 352
Potential temperature, 41
    atmospheric profile, 41f
    equation for evolution of, 89
    moist, 50
    observed (atmosphere), 66f, 67f
Precession, 286
Precipitation
    convection and, 6
    annual-mean map of, 230
Preconditioning
    of deep convection in ocean, 233–234
        schematic diagram of, 237f
Pressure, 4–6
    at center of planet, 29
    at surface of earth
        climatological map, 133f
        instantaneous map, 132f
    coordinates, 67–68
        volume element in, 88
    gradients of, 73, 85f, 86, 92
    loading of ocean, 182
    partial, 5
    saturated vapor, 6
    surfaces
        500mb, 69f
        tilt of, 70, 113
        unit of, 4
Pressure gradient force, 73, 85–86
Profiling float, 180n
Pseudo-adiabatic lapse rate, 49n

R
Radial force balance, 92–93
Radial inflow experiment, 90
Radiation
    absorption of, 11, 13f
    convection and
    entering the ocean, 171
    IR, 10
    net, 63
    reflection of, 11

by season, 62–63
solar, 10
UV, 10, 14
Radiative
convective equilibrium, 31, 56–57
equilibrium temperature, 9
forcing, temperature and, 62–67
profile, 18f, 26
Rainfall rate, 59, 230f
Reflection
clouds and, 11
deserts and, 11
of radiation, 11
snow and, 11
Relative humidity, 48, 71
by zones, 72f
Rayleigh (Lord), 32
Red spectrum, 264
Reduced gravity, 126, 267
Reference ellipsoid, 101, 183
Relative humidity, 71, 72f
Respiration, 3
Rossby
adjustment problem, 125–126
number, 19, 110, 116, 182, 199
computed from data, 115f
in atmosphere, xix
in Gulf Stream, 182
in ocean, xix
in radial flow experiment, 116
radius of deformation, 126, 148, 219
at the equator, 267
ratio of between atmosphere and ocean, 151
Rossby, Carl–Gustav, 110
Rotating
coordinates, 93–96
transformation rule, 94
turntable, xiii, 81, 299
Rotating fluids, xvii, 81
Earth and, 103–104
equations for, 90–104
parabolic shape of, 96–100
radial inflow and, 90–93, 90n
rotating coordinates, 93–94
on spheres, 100–103
Rotation, gravity and, 123–125

S
Sahara, 59
Salinity
distribution at sea surface, 168f
distribution in ocean, 171f
evolution equation for in ocean, 225

transport by ocean, 255
units of, 165
Saturated vapor pressure, 5–6
Saturated specific humidity, 48
Scale height (atmospheric), 28
Sea floor spreading, 280
Sea level
air pressure at, 134f
in Cretaceous, 280
at LGM, 282f
Sea surface
elevation
altimetric observations of, 185
temperature, 167f
anomalies in the Pacific, 269f, 270–271
damping timescale of, 261
during ENSO, 269f
during glacial periods, 281
factors controlling, 231
instantaneous map in Gulf Stream, 191f
non-seasonal changes, 262
sensitivity of atmosphere to, 264
SOI and, 271f
Sea water
density of, 166f
equation of state, 168
major constituents, 168t
Seasonal cycle
in ocean, 175f, 176
Seasons, 62
Sensible heat flux, 227f
bulk formula for, 226
Shallow atmosphere, 101
Shortwave
flux at sea surface, 225, 226f
Sierra Nevada, 44
Smoke, 7
Snow
albedo of, 11t
reflection and, 11
Snowball earth, 260, 275
SOI. see Southern Oscillation Index
Solar
constant, 9
radiation, 62
absorbed, 63f
distribution of, 62f
oceans and, 261
seasonal cycle, 62
spectrum, 10
Solstice, 63f, 286
Sonic boom, 8f

Southern hemisphere
    geostrophic balance in, 136
Southern ocean, 163
Southern Oscillation, 264–273
    Index, 270
Specific humidity, 70–73
    meridional section of, 71f
    saturated, 72f
    temperature and, 71
Spectroscopic measurements, 30
Spheres
    centrifugal force on, 101f
    Coriolis forces on, 101–103
    rotating fluids on, 100–103
    Taylor-Proudman theorem on, 208–211, 224
Spherical coordinates, 102
Sphericity
    representation of in laboratory experiment,
        211
Spinning top
    as analogue of wind-driven circulation, 211
Squall front, 54f
Stability
    condition
        dry adiabatic, 40
        conditional, 50
    to vertical displacements, 35–36
Standard temperature and pressure, 2
Stefan–Boltzmann
    constant, 295
    law, 11
Steric effect, 186
Stommel, Henry, 240
Stommel's abyssal theory, 245–246
Stratification
    observations of in ocean, 233f
Stratopause, 24f, 63f
Stratosphere, 24, 25, 65, 67f
    winter polar, 137
Streamfunction
    for geostrophic motion 113
    for meridional overturning circulation, 76,
        251–253
    for Sverdrup flow, 214
Subduction
    in ocean circulation, 172
Subgeostrophic flow, 129–135
Subpolar gyres, 178
Substantial derivative, 82–84
Subtropics, evaporation in, 232
Subtropical jet, 74, 144
Sulfate aerosols, 7
Summer pole, 63

Sun, 9
    energy emitted from, 10
Supercell, 53f
Surface air temperatures, 263f
    history of, 292
Surface currents
    global, 177f
    in Atlantic Ocean, 179f
    in high resolution ocean model, 192f
    in Indian Ocean, 180f
    in Pacific Ocean, 178f
    measured by drifters in Atlantic, 208f
Surface drifters, 176
    in Antarctic Circumpolar Current, 194f
    speeds from, 190f
Surface elevation of ocean
    observations of, 185f
Surface pressure, 27
    climatological map of, 133f
    instantaneous map of , 132f
    zonal-average, 134f
Surface temperature
    global average of, 10f
    in greenhouse model, 15
Sverdrup, Harald, 213
Sverdrup
    relation, 213
    theory of ocean circulation, 214–216
    transport, 216
        computed in Pacific, 215
    unit of volume transport, 215, 253
    redefined as a unit of mass transport, 253

T
Taylor Column, xviii, 118–119, 210, 217, 217f
    gyroscopic rigidity of, 209
Taylor, G.I., 118
Taylor expansion, 33, 296
Taylor-Proudman theorem, 117–119
    in a layered ocean, 217–218
    breakdown of, 243
    on sphere, 207–211, 224, 239–241
Tectonic uplift
    role in chemical weathering, 275–276
Temperature. See also Radiative equilibrium
        temperature; Sea surface temperature
    absolute, 4
    circulation and, 156
    and CO2
        in the paleo record, 277, 279f
    of Earth, global average mean, 14
    emission, 12
    evolution equation for, 225

humidity and, 70–73
inversions in, 44–46
meridional structure in atmosphere, 64–66,
    65f, 66f, 67f, 127f, 128f at 500mb, 127f
of mesopause, 24f
of mesosphere, 24f
of ocean surface, 167f
distribution in ocean, 170f, 173f, 188f, 237f,
    247f
paleotemperatures, 277–280
pole-equator difference, 64
of radiative equilibrium, 9
radiative forcing and, 62–67
salinity and, 168–171
saturated vapor pressure and, 6
specific humidity and, 71
of stratosphere, 65–67
of thermosphere, 24f
of troposphere, 64–65
vertical distribution of, 23–26, 24f
winds and, 73–78
Terrestrial radiation
from troposphere, 31
water vapor and, 19
Tethys Sea, 278f
Thermal wind, 119–128
    and pole-equator temperature difference, 127f
    and Taylor-Proudman theorem, 122–123
    energetics of, 149–154
    equation for incompressible fluid, 120
    in ocean, 186
    in pressure coordinates, 126
    physical content of, 122
    summary equations, 135
Thermals, 37
Thermobaric effect, 231
Thermocline, 169, 171–172, 175f, 197
    schematic diagram, 198f
Thermodynamic equation, 89
Thermodynamics, first law of, 39
Thermohaline circulation, 197
    dynamical models of, 239–241
    inference from tracer observations, 234
    role in climate change, 290–291
    schematic, 224
    time scales and intensity of, 239
Thermosphere, 23, 24
    temperature of, 24f
Thermostat, operating over earth history,
    275–276
Thickness, 70, 79
Thunderstorms, 56
Topography, 2

Torque
    on atmosphere, 143
Torricelli, Evangelista, 68
Total derivative, 82–84
Trade
    inversion, 144
    winds, 134f, 143f, 144f, 157
Transport
    computation of in ocean, 187
Triatomic molecules, 14
Tropical
    rain forest, 157
Tropics. *See also* El Niño-Southern Oscillation
    circulation in, 142–145
    convection in, 51
    moist air in, 6
Tropopause, 24f, 25
    gap, 65
Troposphere, 23, 25
    circulation in, 141
    radiative equilibrium temperature profile for,
        26f
    temperature of, 24f, 64–65
    terrestrial radiation from, 31
    weather in, 26
Turbulence
    shear-induced, xv
    within ocean mixed layer, 175f

U
Ultraviolet, 19, 24
Universal gas constant, 4
Updraft, 55, 71

V
Vapor pressure
    saturated, 6
Venus, 10t
Vertical velocity
    in pressure coordinates, 84
Visible spectrum, 10, 14
Volcanic activity, 275–276
    and CO2, 276
Vorticity, 106, 244
Voyage of the Beagle (Darwin), 265

W
Walker
    circulation, 268, 272
    Gilbert, 266
Warm front, 148f
Warm pool, 268
Water
    convection in, 34–39

equation of state, 168
fresh, 166
physical properties of, 165
Water vapor, xvif, xix, 2, 5–6, 26f
  infrared radiation and, 2
  saturated with, 5, 6f
  vertical distribution of, 26f
  vertical transport, 73f
Wavelength, energy flux and, 10
Weather systems
  in atmosphere, 147–149
  in ocean, 188–189
Weathering of rocks
  and atmospheric CO2, 276
Wedge of instability, 152
Westerly wind bursts, 273
Western boundary currents, 176, 243–244
  breakdown of geostrophy in, 244
  laboratory demonstration, 212f, 243f, 246f
  reason for existence of, 243
  schematic diagram of, 207f
  why western?, 206–207
White noise, 262
Wien's displacement law, 13
Wind shear, 120, 122, 128, 139, 152, 183, 198, 268
Wind-driven ocean circulation, 197–222
Wind-stress
  annual-mean distribution of, 199f
  bulk formula for, 198
  zero curl line, 204–205

Winds. *See also* Thermal wind; Trade winds
  circulation and, 197–222
  distribution of, 74–78, 158f
  easterlies, 74
  eddies, 77–78, 146
  in Ekman layer, 130–133
  Ekman pumping and, 205n
  global pattern of
    schematic, 158f
  gyres and, 206–207
  mean meridional circulation of, 74–77
  mean zonal, 74
  meridional cross section
    instantaneous, 128f
  on oceans, 164
  temperature and, 73–78
  westerly, 127f, 157, 204f
  zonal average
    observed, 75f, 76f
Work
  done by wind on ocean, 207
Wust, Georg, 187

Y
Younger Dryas, 260, 281, 290f
Yucatan Peninsula, 294

Z
Zonal average, 64

equation of state, 168
fresh, 166
physical properties of, 165
Water vapor, xvii, xix, 2, 5–6, 261
infrared radiation and, 2
saturated with, 5, 6f
vertical distribution of, 2d
vertical transport, 73l
Wavelength, energy flux and, 10
Weather systems
in atmosphere, 147–149
in ocean, 168–184
Weathering of rocks
and atmospheric CO2, 276
Wedge of instability, 152
Westerly wind bursts, 223
Western boundary currents, 176, 243–264
breakdown of geostrophy in, 244
laboratory demonstration, 212l, 243l, 244l
reason for existence of, 243
schematic diagram of, 207f
why westerly, 205–207
White noise, 262
Wien's displacement law, 13
Wind speed, 120, 122, 128, 136, 182, 183, 198, 208
Wind-driven ocean circulation, 193–222
Wind-stress
annual-mean distribution of, 198l
bulk formula for, 198
zero-curl line, 204–205

Winds. See also Thermal wind; Trade winds
circulation and, 192–222
distribution of, 74–78, 158
easterlies, 74
eddies, 77–78, 146
in Ekman layer, 130–133
Ekman pumping and, 205n
global pattern of
schematic, 188f
gyres and, 205–207
mean meridional circulation of, 74–77
mean zonal, 74
meridional cross section
instantaneous, 128f
on oceans, 164
temperature and, 73–78
westerly, 127f, 152, 204f
zonal average
observed, 75f, 76f
Work
done by wind on ocean, 207
Wüst, Georg, 187

Y
Younger Dryas, 280, 281, 290f
Yucatan Peninsula, 204

Z
Zonal average, 64

International Geophysics Series

EDITED BY

RENATA DMOWSKA
Division of Applied Science
Harvard University
Cambridge, Massachusetts

DENNIS HARTMANN
Department of Atmospheric Sciences
University of Washington
Seattle, Washington

H. THOMAS ROSSBY
Graduate School of Oceanography
University of Rhode Island
Narragansett, Rhode Island

Volume 1   BENNO GUTENBURG. Physics of the Earth's Interior. 1959*

Volume 2   JOSEPH W. CHAMBERLAIN. Physics of the Aurora and Airglow. 1961*

Volume 3   S. K. RUNCORN (ed.). Continental Drift. 1962*

Volume 4   C. E. JUNGE. Air Chemistry and Radioactivity. 1963*

Volume 5   ROBERT G. FLEAGLE AND JOOST A. BUSINGER. An Introduction to Atmospheric Physics. 1963*

Volume 6   L. DEFOUR AND R. DEFAY. Thermodynamics of Clouds. 1963*

Volume 7   H. U. ROLL. Physics of the Marine Atmosphere. 1965*

Volume 8   RICHARD A. CRAIG. The Upper Atmosphere: Meteorology and Physics. 1965*

Volume 9   WILLIS L. WEBB. Structure of the Stratosphere and Mesosphere. 1966*

Volume 10   MICHELE CAPUTO. The Gravity Field of the Earth from Classical and Modern Methods. 1967*

Volume 11   S. MATSUSHITA AND WALLACE H. CAMPBELL (eds.). Physics of the Geomagnetic Phenomena (In two volumes). 1967*

Volume 12   K. YA KONDRATYEV. Radiation in the Atmosphere. 1969*

Volume 13   E. PALMÅN AND C. W. NEWTON. Atmospheric Circulation Systems: Their Structure and Physical Interpretation. 1969*

Volume 14   HENRY RISHBETH AND OWEN K. GARRIOTT. Introduction to Ionospheric Physics. 1969*

Volume 15   C. S. RAMAGE. Monsoon Meteorology. 1971*

Volume 16   JAMES R. HOLTON. An Introduction to Dynamic Meteorology. 1972*

Volume 17   K. C. YEH AND OWEN K. GARRIOTT. Theory of Ionospheric Waves. 1972*

Volume 18   M. I. BUDYKO. Climate and Life. 1974*

Volume 19   MELVIN E. STERN. Ocean Circulation Physics. 1975

Volume 20   J. A. JACOBS. The Earth's Core. 1975*

Volume 21   DAVID H. MILLER. Water at the Surface of the Earth: An Introduction to Ecosystem Hydrodynamics. 1977

Volume 22   JOSEPH W. CHAMBERLAIN. Theory of Planetary Atmospheres: An Introduction to their Physics and Chemistry 1978*

Volume 23   JAMES R. HOLTON. An Introduction to Dynamic Meteorology, Second Edition. 1979*

Volume 24   ARNETT S. DENNIS. Weather Modification by Cloud Seeding. 1980*

Volume 25   ROBERT G. FLEAGLE AND JOOST A. BUSINGER. An Introduction to Atmospheric Physics, Second Edition. 1980*

Volume 26   KUO-NAN LIOU. An Introduction to Atmospheric Radiation. 1980*

Volume 27   DAVID H. MILLER. Energy at the Surface of the Earth: An Introduction to the Energetics of Ecosystems. 1981

Volume 28   HELMUT G. LANDSBERG. The Urban Climate. 1991

Volume 29   M. I. BUDKYO. The Earth's Climate: Past and Future. 1982*

Volume 30   ADRIAN E. GILL. Atmosphere-Ocean Dynamics. 1982

Volume 31   PAOLO LANZANO. Deformations of an Elastic Earth. 1982*

Volume 32   RONALD T. MERRILL AND MICHAEL W. McELHINNY. The Earth's Magnetic Field: Its History, Origin, and Planetary Perspective. 1983*

Volume 33   JOHN S. LEWIS AND RONALD G. PRINN. Planets and Their Atmospheres: Origin and Evolution. 1983

Volume 34   ROLF MEISSNER. The Continental Crust: A Geophysical Approach. 1986

Volume 35   M. U. SAGITOV, B. BODKI, V. S. NAZARENKO, AND KH. G. TADZHIDINOV. Lunar Gravimetry. 1986

Volume 36   JOSEPH W. CHAMBERLAIN AND DONALD M. HUNTEN. Theory of Planetary Atmospheres, 2nd Edition. 1987

Volume 37   J. A. JACOBS. The Earth's Core, 2nd edition. 1987*

Volume 38   J. R. APEL. Principles of Ocean Physics. 1987

Volume 39   MARTIN A. UMAN. The Lightning Discharge. 1987*

Volume 40   DAVID G. ANDREWS, JAMES R. HOLTON AND CONWAY B. LEOVY. Middle Atmosphere Dynamics. 1987

Volume 41   PETER WARNECK. Chemistry of the National Atmosphere. 1988*

Volume 42   S. PAL ARYA. Introduction to Micrometeorology. 1988*

Volume 43   MICHAEL C. KELLEY. The Earth's Ionosphere. 1989*

Volume 44   WILLIAM R. COTTON AND RIHCARD A. ANTHES. Storm and Cloud Dynamics. 1989

Volume 45   WILLIAM MENKE. Geophysical Data Analysis: Discrete Inverse Theory, Revised Edition. 1989

Volume 46   S. GEORGE PHILANDER. El Niño, La Niña, and the Southern Oscillation. 1990

Volume 47   ROBERT A. BROWN. Fluid Mechanics of the Atmosphere. 1991

Volume 48   JAMES R. HOLTON. An Introduction to Dynamic Meteorology, Third Edition. 1992

Volume 49   ALEXANDER A. KAUFMAN. Geophysical Field Theory and Method.
Part A: Gravitational, Electric, and Magnetic Fields. 1992*
Part B: Electromagnetic Fields I. 1994*
Part C: Electromagnetic Fields II. 1994*

Volume 50   SAMUEL S. BUTCHER, GORDON H. ORIANS, ROBERT J. CARLSON, AND GORDON V. WOLFE. Global Biogeochemical Cycles. 1992

Volume 51   BRIAN EVANS AND TENG-FONG WONG. Fault Mechanics and Transport Properties of Rocks. 1992

Volume 52   ROBERT E. HUFFMAN. Atmospheric Ultraviolet Remote Sensing. 1992

Volume 53   ROBERT E. HOUZE, JR. Cloud Dynamics. 1993

Volume 54   PETER V. HOBBS. Aerosol-Cloud-Climate Interactions. 1993

Volume 55   S. J. GIBOWICZ AND A. KIJKO. An Introduction to Mining Seismology. 1993

Volume 56   DENNIS L. HARTMANN. Global Physical Climatology. 1994

Volume 57   MICHAEL P. RYAN. Magmatic Systems. 1994

Volume 58   THORNE LAY AND TERRY C. WALLACE. Modern Global Seismology. 1995

Volume 59   DANIEL S. WILKS. Statistical Methods in the Atmospheric Sciences. 1995

Volume 60   FREDERIK NEBEKER. Calculating the Weather. 1995

Volume 61   MURRY L. SALBY. Fundamentals of Atmospheric Physics. 1996

Volume 62   JAMES P. McCALPIN. Paleoseismology. 1996

Volume 63   RONALD T. MERRILL, MICHAEL W. McELHINNY, AND PHILLIP L. McFADDEN. The Magnetic Field of the Earth: Paleomagnetism, the Core, and the Deep Mantle. 1996

Volume 64   NEIL D. OPDYKE AND JAMES E. T. CHANNELL. Magnetic Stratigraphy. 1996

Volume 65   JUDITH A. CURRY AND PETER J. WEBSTER. Thermodynamics of Atmospheres and Oceans. 1988

Volume 66   LAKSHMI H. KANTHA AND CAROL ANNE CLAYSON. Numerical Models of Oceans and Oceanic Processes. 2000

Volume 67   LAKSHMI H. KANTHA AND CAROL ANNE CLAYSON. Small Scale Processes in Geophysical Fluid Flows. 2000

Volume 68   RAYMOND S. BRADLEY. Paleoclimatology, Second Edition. 1999

Volume 69   LEE-LUENG FU AND ANNY CAZANAVE. Satellite Altimetry and Earth Sciences: A Handbook of Techniques and Applications. 2000

Volume 70   DAVID A. RANDALL. General Circulation Model Development: Past, Present, and Future. 2000

Volume 71   PETER WARNECK. Chemistry of the Natural Atmosphere, Second Edition. 2000

Volume 72   MICHAEL C. JACOBSON, ROBERT J. CHARLSON, HENNING RODHE, AND GORDON H. ORIANS. Earth System Science: From Biogeochemical Cycles to Global Change. 2000

Volume 73   MICHAEL W. McELHINNY AND PHILLIP L. McFADDEN. Paleomagnetism: Continents and Oceans. 2000

Volume 74   ANDREW E. DESSLER. The Chemistry and Physics of Stratospheric Ozone. 2000

Volume 75   BRUCE DOUGLAS, MICHAEL KEARNY AND STEPHEN LEATHERMAN. Sea Level Rise: History and Consequences. 2000

Volume 76   ROMAN TEISSEYRE AND EUGENIUSZ MAJEWSKI. Earthquake Thermodynamics and Phase Transformation in the Interior. 2001

Volume 77   GEROLD SIEDLER, JOHN CHURCH, AND JOHN GOULD. Ocean Circulation and Climate: Observing and Modelling the Global Ocean. 2001

Volume 78   ROGER A. PIELKE, SR. Mesoscale Meteorological Modeling, 2nd Edition. 2001

Volume 79   S. PAL ARYA. Introduction to Micrometeorology. 2001

Volume 80   BARRY SALTZMAN. Dynamical Paleoclimatology: Generalized Theory of Global Climate Change. 2002

Volume 81A   WILLIAM H. K. LEE, HIROO KANAMORI, PAUL JENNINGS, AND CARL KISSLINGER. International Handbook of Earthquake and Engineering Seismology, Part A. 2002

Volume 81B   WILLIAM H. K. LEE, HIROO KANAMORI, PAUL JENNINGS, AND CARL KISSLINGER. International Handbook of Earthquake and Engineering Seismology, Part B. 2003

Volume 82   GORDON G. SHEPPERD. Spectral Imaging of the Atmosphere. 2002

Volume 83   ROBERT P. PEARCE. Meteorology at the Millennium. 2001

Volume 84   KUO-NAN LIOU. An Introduction to Atmospheric Radiation, 2nd Ed. 2002

Volume 85   CARMEN J. NAPPO. An Introduction to Atmospheric Gravity Waves. 2002

Volume 86   MICHAEL E. EVANS AND FRIEDRICH HELLER. Environmental Magnetism: Principles and Applications of Enviromagnetics. 2003

Volume 87   JOHN S. LEWIS. Physics and Chemistry of the Solar System, 2nd Ed. 2003

Volume 88   JAMES R. HOLTON. An Introduction to Dynamic Meteorology, 4th Ed. 2004

Volume 89   YVES GUÉGUEN AND MAURICE BOUTÉCA. Mechanics of Fluid Saturated Rocks. 2004

Volume 90   RICHARD C. ASTER, BRIAN BORCHERS, AND CLIFFORD THURBER. Parameter Estimation and Inverse Problems. 2005

Volume 91   DANIEL S. WILKS. Statistical Methods in the Atmospheric Sciences, 2nd Ed. 2006

Volume 92   JOHN M. WALLACE AND PETER V. HOBBS. Atmospheric Science: An Introductory Survey, 2nd Ed. 2006

Volume 93   JOHN MARSHALL AND R. ALAN PLUMB. Atmosphere, Ocean, and Climate Dynamics: An Introductory Text. 2008

*Out of Print

Printed and bound by CPI Group (UK) Ltd, Croydon, CR0 4YY

03/10/2024

01040321-0018